Achieving sustainable cultivation of soybeans

Volume 2: Diseases, pests, food and other uses

It is widely recognised that agriculture is a significant contributor to global warming and climate change. Agriculture needs to reduce its environmental impact and adapt to current climate change whilst still feeding a growing population, i.e. become more 'climate-smart'. Burleigh Dodds Science Publishing is playing its part in achieving this by bringing together key research on making the production of the world's most important crops and livestock products more sustainable.

Based on extensive research, our publications specifically target the challenge of climate-smart agriculture. In this way we are using 'smart publishing' to help achieve climate-smart agriculture.

Burleigh Dodds Science Publishing is an independent and innovative publisher delivering high quality customer-focused agricultural science content in both print and online formats for the academic and research communities. Our aim is to build a foundation of knowledge on which researchers can build to meet the challenge of climate-smart agriculture.

For more information about Burleigh Dodds Science Publishing simply call us on +44 (0) 1223 839365, email info@bdspublishing.com or alternatively please visit our website at www.bdspublishing.com.

Related titles:

Achieving sustainable cultivation of soybeans Volume 1: Breeding and cultivation techniques
Print (ISBN 978-1-78676-112-5); Online (ISBN 978-1-78676-114-9, 978-1-78676-115-6)

Achieving sustainable cultivation of grain legumes Volume 1:
Advances in breeding and cultivation techniques
Print (ISBN 978-1-78676-136-1); Online (ISBN 978-1-78676-139-2, 978-1-78676-138-5)

Achieving sustainable cultivation of grain legumes Volume 2:
Improving cultivation of particular grain legumes
Print (ISBN 978-1-78676-140-8); Online (ISBN 978-1-78676-142-2, 978-1-78676-143-9)

Chapters are available individually from our online bookshop: https://shop.bdspublishing.com

climate•SMART•publishing
IN AGRICULTURAL SCIENCE

BURLEIGH DODDS SERIES IN AGRICULTURAL SCIENCE

NUMBER 30

Achieving sustainable cultivation of soybeans

Volume 2: Diseases, pests, food and other uses

Edited by Professor Henry T. Nguyen,
University of Missouri, USA

burleigh dodds
SCIENCE PUBLISHING

Published by Burleigh Dodds Science Publishing Limited
82 High Street, Sawston, Cambridge CB22 3HJ, UK
www.bdspublishing.com

Burleigh Dodds Science Publishing, 1518 Walnut Street, Suite 900, Philadelphia, PA 19102-3406, USA

First published 2018 by Burleigh Dodds Science Publishing Limited

Library of Congress Control Number: 2017956486

British Library Cataloguing in Publication Data
A catalogue record for this book is available from the British Library

ISBN 978-1-78676-116-3 (print)
ISBN 978-1-78676-119-4 (online)
ISBN 978-1-78676-118-7 (online)
ISSN 2059-6936 (print)
ISSN 2059-6944 (online)

Typeset by Deanta Global Publishing Services, Chennai, India
Printed by Lightning Source

Contents

Series list viii

Acknowledgements xii

Introduction xiii

Part 1 Diseases, pests and weeds

1 Oomycete and fungal pathogens of soybean 3
Anne E. Dorrance, The Ohio State University, USA
1 Introduction 3
2 Case studies on diseases caused by Oomycetes 6
3 Case studies on diseases caused by true fungi: fungal pathogens that affect foliage 10
4 Case studies on diseases caused by true fungi: fungal pathogens that affect roots 13
5 Summary 15
6 Future trends in research 15
7 Where to look for further information 16
8 References 16

2 Bacterial and viral diseases affecting soybean production 27
Glen L. Hartman, USDA-ARS and University of Illinois, USA
1 Introduction 27
2 Bacterial diseases: blight, pustule and tan spot 28
3 Bacterial diseases: wilts and other bacterial diseases 32
4 Viral diseases: mosaic and mottle viruses 34
5 Other viral diseases 37
6 Summary 39
7 References 40

3 Nematode pathogens of soybean 47
T. L. Niblack and H. D. Lopez-Nicora, The Ohio State University, USA
1 Introduction 47
2 General nematode biology 48
3 Soybean cyst nematodes 50
4 Root-knot nematodes, *Meloidogyne* spp. 59
5 Lesion nematodes, *Pratylenchus* spp. 63
6 Reniform nematodes 65
7 Lance nematodes, *Hoplolaimus* spp. 68
8 Other nematodes 69
9 Conclusion 70
10 Where to look for further information 71
11 References 72

4 Key factors limiting sustainable insect pest management in soybeans 87
M. E. O'Neal, Iowa State University, USA; and R. Cox, EarthEmpower Consulting and Investment, Mexico
1 Introduction 87
2 Invasive species 90
3 Insect resistance management (IRM), especially for Bt-soybeans 95
4 Conclusion 97
5 Future trends 98
6 Where to look for further information 99
7 References 99

5 Advances in disease-resistant varieties of soybean 105
David R. Walker, USDA-ARS, USA
1 Introduction 105
2 Resistance to major soybean diseases and nematode pests 108
3 Selection for disease resistance 126
4 Contribution of soybean genomics research to breeding strategies 127
5 Pathogen genomics 128
6 Disease resistance from *G. soja* and perennial *Glycine* species 128
7 Transgene-mediated disease resistance 129
8 Summary and future trends 129
9 Where to look for further information 131
10 References 131

6 Advances in pest-resistant varieties of soybean 145
Shichen Zhang and Dechun Wang, Michigan State University, USA
1 Introduction 145
2 Mechanisms of HPR 147
3 Insect resistance in the soybean germplasm 148
4 Breeding challenges 154
5 *Bacillus thuringiensis* (Bt) soybean 155
6 Case study: developing aphid-resistant varieties 156
7 Conclusion 161
8 Future trends 166
9 Where to look for further information 166
10 References 166

7 Integrated weed management in soybean cultivation 175
Bob Hartzler, Iowa State University, USA
1 Introduction 175
2 Establishing weed-management goals 176
3 Management tactics: crop rotation, competitiveness of soybean and use of cover crops 178
4 Further management tactics 181
5 Interactions between weeds and other pest complexes 184
6 Factors limiting the adoption of diversified weed management 185
7 Conclusion 185
8 Where to look for further information 186
9 References 186

Part 2 Food and other uses

8 Nutritional, nutraceutical and functional properties of soybeans 195
Suzanne Hendrich, Iowa State University, USA
1 Introduction 195
2 Nutritional effects of soybean oils 196
3 Nutritional and nutraceutical functions of soybean proteins in cardiovascular disease 197
4 Health effects of soybean proteins and associated constituents 198
5 Other health or adverse effects of soybean foods and ingredients 202
6 Conclusion and future trends 204
7 References 204

9 Improving the nutritional value of soybean 209
Istvan Rajcan and Jocelyne Letarte, University of Guelph, Canada
1 Introduction 209
2 Soybean protein content and profile 210
3 Soybean oil profile 214
4 Tocopherols 217
5 Isoflavones 220
6 Saponins 222
7 Future trends and conclusion 223
8 References 224

10 Allergens in soybean 235
Eliot M. Herman, University of Arizona, USA
1 Introduction 235
2 Clinical allergies to soybean 236
3 Infant food allergies and intolerance 237
4 Soybean IgE-binding proteins 238
5 Mitigating soybean allergens 241
6 Soybean intolerance in aquaculture 245
7 Food allergy, biotechnology and the risk of extrinsic introduced allergens 246
8 Summary 248
9 Where to look for further information 248
10 Acknowledgements and dedication 248
11 References 248

11 Nutritional considerations for soybean meal use in poultry diets 257
Justin Fowler, University of Georgia, USA
1 Introduction 257
2 Nutritional content of SBM 258
3 Anti-nutritive compounds 260
4 Genetically modified soybeans 263
5 Conclusion and future trends 264
6 Where to look for further information 265
7 References 265

Index 267

Series list

Title	Series number
Achieving sustainable cultivation of maize - Vol 1 From improved varieties to local applications Edited by: Dr Dave Watson, CGIAR Maize Research Program Manager, CIMMYT, Mexico	001
Achieving sustainable cultivation of maize - Vol 2 Cultivation techniques, pest and disease control Edited by: Dr Dave Watson, CGIAR Maize Research Program Manager, CIMMYT, Mexico	002
Achieving sustainable cultivation of rice - Vol 1 Breeding for higher yield and quality Edited by: Prof. Takuji Sasaki, Tokyo University of Agriculture, Japan	003
Achieving sustainable cultivation of rice - Vol 2 Cultivation, pest and disease management Edited by: Prof. Takuji Sasaki, Tokyo University of Agriculture, Japan	004
Achieving sustainable cultivation of wheat - Vol 1 Breeding, quality traits, pests and diseases Edited by: Prof. Peter Langridge, The University of Adelaide, Australia	005
Achieving sustainable cultivation of wheat - Vol 2 Cultivation techniques Edited by: Prof. Peter Langridge, The University of Adelaide, Australia	006
Achieving sustainable cultivation of tomatoes Edited by: Dr Autar Mattoo, USDA-ARS, USA & Prof. Avtar Handa, Purdue University, USA	007
Achieving sustainable production of milk - Vol 1 Milk composition, genetics and breeding Edited by: Dr Nico van Belzen, International Dairy Federation (IDF), Belgium	008
Achieving sustainable production of milk - Vol 2 Safety, quality and sustainability Edited by: Dr Nico van Belzen, International Dairy Federation (IDF), Belgium	009
Achieving sustainable production of milk - Vol 3 Dairy herd management and welfare Edited by: Prof. John Webster, University of Bristol, UK	010
Ensuring safety and quality in the production of beef - Vol 1 Safety Edited by: Prof. Gary Acuff, Texas A&M University, USA & Prof.James Dickson, Iowa State University, USA	011
Ensuring safety and quality in the production of beef - Vol 2 Quality Edited by: Prof. Michael Dikeman, Kansas State University, USA	012
Achieving sustainable production of poultry meat - Vol 1 Safety, quality and sustainability Edited by: Prof. Steven C. Ricke, University of Arkansas, USA	013
Achieving sustainable production of poultry meat - Vol 2 Breeding and nutrition Edited by: Prof. Todd Applegate, University of Georgia, USA	014
Achieving sustainable production of poultry meat - Vol 3 Health and welfare Edited by: Prof. Todd Applegate, University of Georgia, USA	015

Achieving sustainable production of eggs - Vol 1 016
Safety and quality
Edited by: Prof. Julie Roberts, University of New England, Australia

Achieving sustainable production of eggs - Vol 2 017
Animal welfare and sustainability
Edited by: Prof. Julie Roberts, University of New England, Australia

Achieving sustainable cultivation of apples 018
Edited by: Dr Kate Evans, Washington State University, USA

Integrated disease management of wheat and barley 019
Edited by: Prof. Richard Oliver, Curtin University, Australia

Achieving sustainable cultivation of cassava - Vol 1 020
Cultivation techniques
Edited by: Dr Clair Hershey, formerly International Center for Tropical Agriculture (CIAT), Colombia

Achieving sustainable cultivation of cassava - Vol 2 021
Genetics, breeding, pests and diseases
Edited by: Dr Clair Hershey, formerly International Center for Tropical Agriculture (CIAT), Colombia

Achieving sustainable production of sheep 022
Edited by: Prof. Johan Greyling, University of the Free State, South Africa

Achieving sustainable production of pig meat - Vol 1 023
Safety, quality and sustainability
Edited by: Prof. Alan Mathew, Purdue University, USA

Achieving sustainable production of pig meat - Vol 2 024
Animal breeding and nutrition
Edited by: Prof. Julian Wiseman, University of Nottingham, UK

Achieving sustainable production of pig meat - Vol 3 025
Animal health and welfare
Edited by: Prof. Julian Wiseman, University of Nottingham, UK

Achieving sustainable cultivation of potatoes - Vol 1 026
Breeding, nutritional and sensory quality
Edited by: Prof. Gefu Wang-Pruski, Dalhousie University, Canada

Achieving sustainable cultivation of oil palm - Vol 1 027
Introduction, breeding and cultivation techniques
Edited by: Prof. Alain Rival, Center for International Cooperation in Agricultural Research for Development (CIRAD), France

Achieving sustainable cultivation of oil palm - Vol 2 028
Diseases, pests, quality and sustainability
Edited by: Prof. Alain Rival, Center for International Cooperation in Agricultural Research for Development (CIRAD), France

Achieving sustainable cultivation of soybeans - Vol 1 029
Breeding and cultivation techniques
Edited by: Prof. Henry Nguyen, University of Missouri, USA

Achieving sustainable cultivation of soybeans - Vol 2 030
Diseases, pests, food and non-food uses
Edited by: Prof. Henry Nguyen, University of Missouri, USA

Achieving sustainable cultivation of sorghum - Vol 1 031
Genetics, breeding and production techniques
Edited by: Prof. Bill Rooney, Texas A&M University, USA

Achieving sustainable cultivation of sorghum - Vol 2 032
Sorghum utilisation around the world
Edited by: Prof. Bill Rooney, Texas A&M University, USA

Achieving sustainable cultivation of potatoes - Vol 2 033
Production and storage, crop protection and sustainability
Edited by: Dr Stuart Wale, Potato Dynamics Ltd, UK

Achieving sustainable cultivation of mangoes 034
Edited by: Professor Víctor Galán Saúco, Instituto Canario de Investigaciones Agrarias (ICIA), Spain & Dr Ping Lu, Charles Darwin University, Australia

Achieving sustainable cultivation of grain legumes - Vol 1 035
Advances in breeding and cultivation techniques
Edited by: Dr Shoba Sivasankar et al., CGIAR Research Program on Grain Legumes, ICRISAT, India

Achieving sustainable cultivation of grain legumes - Vol 2 036
Improving cultivation of particular grain legumes
Edited by: Dr Shoba Sivasankar et al., CGIAR Research Program on Grain Legumes, ICRISAT, India

Achieving sustainable cultivation of sugarcane - Vol 1 037
Cultivation techniques, quality and sustainability
Edited by: Prof. Philippe Rott, University of Florida, USA

Achieving sustainable cultivation of sugarcane - Vol 2 038
Breeding, pests and diseases
Edited by: Prof. Philippe Rott, University of Florida, USA

Achieving sustainable cultivation of coffee 039
Edited by: Dr Philippe Lashermes, Institut de Recherche pour le Développement (IRD), France

Achieving sustainable cultivation of bananas - Vol 1 040
Cultivation techniques
Edited by: Prof. Gert Kema, Wageningen University, The Netherlands & Prof. André Drenth, University of Queensland, Australia

Global Tea Science 041
Current status and future needs
Edited by: Dr V. S. Sharma, Formerly UPASI Tea Research Institute, India & Dr M. T. Kumudini Gunasekare, Coordinating Secretariat for Science Technology and Innovation (COSTI), Sri Lanka

Integrated weed management 042
Edited by: Emeritus Prof. Rob Zimdahl, Colorado State University, USA

Achieving sustainable cultivation of cocoa - Vol 1 043
Genetics, breeding, cultivation and quality
Edited by: Prof. Pathmanathan Umaharan, Cocoa Research Centre – The University of the West Indies, Trinidad and Tobago

Achieving sustainable cultivation of cocoa - Vol 2 044
Diseases, pests and sustainability
Edited by: Prof. Pathmanathan Umaharan, Cocoa Research Centre – The University of the West Indies, Trinidad and Tobago

Water management for sustainable agriculture 045
Edited by: Prof. Theib Oweis, Formerly ICARDA, Lebanon

Improving organic animal farming 046
Edited by: Dr Mette Vaarst, Aarhus University, Denmark & Dr Stephen Roderick, Duchy College, Cornwall, UK

Improving organic crop cultivation 047
Edited by: Prof. Ulrich Köpke, University of Bonn, Germany

Managing soil health for sustainable agriculture - Vol 1 048
Fundamentals
Edited by: Dr Don Reicosky, USDA-ARS, USA

Managing soil health for sustainable agriculture - Vol 2 049
Monitoring and management
Edited by: Dr Don Reicosky, USDA-ARS, USA

Rice insect pests and their management 050
E. A. Heinrichs, Francis E. Nwilene, Michael J. Stout, Buyung A. R. Hadi & Thais Freitas

Improving grassland and pasture management in temperate agriculture 051
Edited by: Prof. Athole Marshall & Dr Rosemary Collins, University of Aberystwyth, UK

Precision agriculture for sustainability 052
Edited by: Dr John Stafford, Silsoe Solutions, UK

Achieving sustainable cultivation of temperate zone tree fruit and berries – Vol 1 053
Physiology, genetics and cultivation
Edited by: Prof. Gregory Lang, Michigan State University, USA

Achieving sustainable cultivation of temperate zone tree fruit and berries – Vol 2 054
Case studies
Edited by: Prof. Gregory Lang, Michigan State University, USA

Agroforestry for sustainable agriculture 055
Edited by: Prof. María Mosquera-Losada, University of Santiago de Compostela, Spain & Dr Ravi Prabhu, World Agroforestry Centre (ICRAF), Kenya

Achieving sustainable cultivation of tree nuts 056
Edited by: Prof. Ümit Serdar, Ondokuz Mayis University, Turkey & Emeritus Prof. Dennis Fulbright, Michigan State University, USA

Assessing the environmental impact of sustainable agriculture 057
Edited by: Prof. Bo P. Weidema, Aalborg University/2.-0 LCA Consultants, Denmark

Critical issues in plant health: 50 years of research in African agriculture 058
Edited by: Dr. Peter Neuenschwander, IITA & Dr. Manuele Tamò, IITA

Achieving sustainable cultivation of vegetables – Vol 1 059
Physiology, breeding, cultivation and quality
Edited by: Emeritus Prof. George Hochmuth, University of Florida, USA

Achieving sustainable cultivation of vegetables – Vol 2 060
Case studies
Edited by: Emeritus Prof. George Hochmuth, University of Florida, USA

Acknowledgements

We wish to acknowledge the following for their help in reviewing particular chapters:

- Chapter 1: Dr Martin Chilvers, Michigan State University, USA
- Chapter 2: Emeritus Professor Sue Tolin, Virginia Tech, USA; Professor Loren J. Giesler, University of Nebraska-Lincoln, USA; Professor John Hill, Iowa State University, USA
- Chapter 4: Dr Bruce Potter, University of Minnesota, USA
- Chapter 5: Dr Stephen Koenning, North Carolina State University, USA
- Chapter 6: Dr Tom Allen, Mississippi State University, USA
- Chapter 8: Professor Elvira de Mejia, University of Illinois, USA
- Chapter 10: Dr Johann Vollmann, University of Natural Sciences (BOKU), Austria
- Chapter 11: Professor Hans-Henrik Stein, University of Illinois, USA; Dr Jacquie Jacob, University of Kentucky, USA

Introduction

Soybeans are one of the most widely-grown crops in the world. As the world's main source of vegetable protein, they have a wide range of food and non-food uses. Current yields need to increase significantly to meet growing demand but in a way that reduces input use, does not damage the environment and is resilient to climate change. These challenges are addressed in the two volumes of *Achieving sustainable cultivation of soybeans*:

- Volume 1: Breeding and cultivation techniques
- Volume 2: Diseases, pests, food and other uses

Volume 2 reviews advances in understanding and managing the range of diseases and pests that continue to cause significant crop losses. The volume offers detailed coverage of oomycete, fungal, viral and bacterial diseases affecting soybeans. It reviews developments in disease and pest-resistant varieties of soybean as well as integrated pest and weed management. Finally, the volume summarises research on developing the food and non-food uses of soybean, from improving nutritional properties to uses in animal feed and biodiesel.

Part 1 Diseases, pests and weeds

Soybean losses due to diseases of all kinds are estimated to be about 14% of total soybean production. Part 1 discusses fungal, viral and bacterial diseases as well as developments in disease-resistant varieties and integrated pest and weed management. Chapter 1 provides an overview of oomycete and fungal pathogens of soybean. Soybean production faces numerous biotic challenges, leading to significant yield losses each year. The chapter presents cases studies on the oomycete pathogens *Phytophthora sojae* and the *Pythium* species and the true fungi *Cercospora sojina, Phakopsora pachyrhizi, Macrophomina phaseolina* and the *Fusarium* species. Through these case studies, the chapter emphasizes the diversity of the pathogens which affect soybean, including their genetic diversity, different modes of infection and different host ranges, focusing especially on pathogens that affect roots and foliage. The chapter also considers the diverse range of management options, as well as future trends in research which should improve our ability to manage soybean diseases.

Chapter 2 moves from oomcyte and fungal pathogens to consider bacterial and viral diseases affecting soybean production. The chapter reviews bacterial blight, pustule, tan spot and wilt, as well as a few other bacterial diseases affecting soybean. It then discusses in detail some important viral diseases of soybean, namely alfalfa mosaic virus, bean pod mottle virus, soybean mosaic virus, soybean vein necrosis virus, tobacco ringspot virus, and tobacco streak virus. The chapter also considers how the impact of bacterial and viral pathogens can be mitigated through management practices such as crop rotation, utilization of pathogen-free seed, better use of synthetic and non-synthetic pesticides for vector control, and deployment of resistant cultivars. Chapter 3 considers another important group of soybean pests: nematodes. The chapter examines the origins and management of a variety of parasitic nematodes affecting soybean, including soybean cyst nematodes, root-knot nematodes, lesion nematodes, reniform nematodes and lance nematodes. The chapter also reviews nematode resistance in soybean.

Complementing the focus of the first three chapters on specific soybean pests, Chapter 4 concentrates on integrated pest management (IPM) for soybean. The chapter includes a detailed case study on the development of an IPM program for an invasive soybean pest (the case of soybean aphid in the US Midwest). The chapter also examines programmes to manage multiple pests in South America and assesses the barriers to achieving sustainable pest management.

Chapter 5 then reviews recent advances in disease-resistant varieties of soybean. Advances in breeding techniques have facilitated the development of soybean cultivars with broader and more durable resistance, but continued reliance on a small number of major resistance genes remains a concern. Introgression of resistance genes from unadapted germplasm sources with a reduced risk of linkage drag has become more efficient with marker-assisted selection. Advances in DNA sequencing and other technologies have made it possible to identify novel resistance loci and candidate genes. The chapter summarizes what is currently known about resistance to some of the major diseases affecting soybean production, particularly in North America, and genes that condition resistance to those diseases. It explores the possibilities of selection for disease resistance, as well as assessing the contribution of soybeans genomics research to breeding strategies. The chapter also examines the genomics of the pathogens themselves, and focuses particularly on disease resistance from *Glycine soja* and perennial *Glycine* species, as well as discussing transgene-mediated disease resistance.

Complementing Chapter 5, Chapter 6 moves on to discuss advances in pest-resistant varieties of soybean. A number of insect pests have been threatening soybean production in North America, including lepidopteran defoliators, coleopteran defoliators and hemipteran sap-suckers. Among various control methods, host-plant resistance has been of the greatest interest as it is economically and environmentally friendly. The chapter reviews recent advances in research on improving soybean with host-plant resistance in North America, and discuss challenges of developing highly insect-resistant cultivars with competitive yield performance. The chapter discusses attempts to genetically engineer soybean with Bt insecticidal protein genes as well as gene pyramiding in order to gain broad and durable resistances against multiple insects. The chapter also includes a case study on key issues in developing aphid-resistant varieties.

Concluding the first part of the volume, Chapter 7 examines the challenge of integrated weed management in soybean cultivation. Almost all soybean fields have a weed infestation that must be managed for farmers to sustain economic viability. The chapter reviews the practicalities of integrated weed management, which combines plant breeding, cultural, mechanical and chemical practices in suppressing weed populations and minimizing the impacts of weeds on yield. After establishing the importance of a diverse weed management strategy, the chapter examines how farmers set goals for weed management and the importance of focusing on managing the weed seed bank. The chapter then reviews the wide range of management tactics available, and considers interactions between weeds and other pests. Finally, the chapter discusses the factors which limit the adoption of integrated weed management.

Part 2 Food and other uses

The second part of the volume addresses food uses of soybean and other non-food uses. Chapter 8 is devoted to the nutritional, nutraceutical and functional properties of soybeans. Soybeans have desirable nutritional properties, since they contain about twice the protein of other legumes per serving. The chapter examines recent research on the nutritional and nutraceutical efficacy of soybean foods and food or dietary supplement ingredients, including soybean oils, proteins, isoflavones and other compounds for which soybeans are a significant source.

Building on the theme of Chapter 8, Chapter 9 focusses on improving the nutritional value of soybean. Soybean is grown for its high oil and protein concentration in the seed. Numerous studies have been carried out to determine the underlying genetics of soybean seed composition traits and develop molecular markers that can aid in the breeding of soybean cultivars with enhanced nutritional quality. The chapter considers genetics and breeding of soybean to improve the nutritional value of the seed as the end product. The chapter describes the protein content and profile of soybean seeds before considering breeding soybean for a modified fatty acids profile that could extend the shelf life of soybean oil or enhance nutritional and nutraceutical function. Finally, the chapter explores how breeding could improve the content of tocopherols, isoflavones and saponins in soybean seeds, which are known to have important health benefits for humans.

Continuing the focus on the use of soybean as a food source, Chapter 10 considers the problem of allergens in soybean. Soybean protein is one of the most significant sources of food allergens and intolerance, causing both IgE-mediated allergy and food protein induced endocolitis syndrome (FPIES). The chapter first discusses the mechanisms causing allergies and infantile food intolerance. It reviews what we know about adverse food responses, the gut's response to food, the multiple IgE binding proteins in soybean as well as P34, the major soybean allergen. The chapter also discusses how to reduce allergens in soybean plants as well as the particular problem of intolerance in salmonids with implications for aquaculture. The chapter also examines the challenge of mitigating soybean allergens, assessing the potential role of biotechnology and the risk of extrinsically-introduced allergens.

The volume's final chapter, Chapter 11, moves to considering soybean as a source of animal feed, addressing the nutritional considerations for soybean meal use in poultry diets. The poultry feed industry uses about 50% of the soybean meal (SBM) produced in the USA each year. The meal is valuable to the animal feed industry because it is an excellent source of amino acids that would otherwise be lost to the human food chain after the oil has been removed from the bean. The chapter assesses the nutritional content of SBM, as well as its anti-nutritive compounds, and considers the effects of genetic modification on improving SBM as a poultry feed.

Part 1

Diseases, pests and weeds

Chapter 1

Oomycete and fungal pathogens of soybean

Anne E. Dorrance, The Ohio State University, USA

1 Introduction
2 Case studies on diseases caused by Oomycetes
3 Case studies on diseases caused by true fungi: fungal pathogens that affect foliage
4 Case studies on diseases caused by true fungi: fungal pathogens that affect roots
5 Summary
6 Future trends in research
7 Where to look for further information
8 References

1 Introduction

There are numerous biotic challenges to soybean production that contribute to localized and regional yield losses annually. Geographic region, soil type, climate conditions and soybean genetics, alone or in combination, can have a profound influence on which soybean pathogen may have the greatest impact on any given farm in any given year. With both fewer soybean genetic suppliers and seed companies than two decades ago, a cultivar has a greater potential to be grown across a wider range of environmental conditions, and thus has a higher probability of encountering a broader diversity of pathogens.

Added to this are a growing list of soybean pathogens that have been introduced or newly established in geographic regions or are re-emerging pathogens that have recently contributed to substantial yield losses. These include those pathogens that cause disease as a result of changes in climate, which favoured overwintering, for example, *Cercospora sojina* (Cruz and Dorrance, 2009; Zhang and Bradley, 2014), expanded geographic ranges as with *Macrophomina phaseolina* (Lopez Nicora et al., 2016; Sexton et al., 2016), adapted to the deployed genetic resistance as in *Phytophthora sojae* (Dorrance et al., 2016), as well as adapted to fungicides such as *Pythium* spp. and *C. sojina* (Broders et al., 2007a; Standish et al., 2015; Zhang et al., 2012a,b). Additionally, introductions of pathogens into new geographic regions where soybean has no or limited resistance have also occurred such as *Phakopsora pachyrhizi*, which causes soybean rust, in South America, Central America and

http://dx.doi.org/10.19103/AS.2017.0034.15

is now a recurrent problem in the Southern US (Miles et al., 2003; Schneider et al., 2005; Sikora et al., 2014).

The overall soybean yield loss estimates for the United States is collected as part of a long-standing project of the North Central Extension Research Area Committee and Southern Soybean Disease Workers and these are available (1996–2014, 2015) at the following web addresses: http://extension.cropsci.illinois.edu/fieldcrops/diseases/yield_reductions.php and http://cropprotectionnetwork.org/crop-loss-estimates/. This annual survey of soybean pathologists across the United States and Ontario, Canada, now coordinated by Drs. Carl Bradley and Tom Allen, is key to understanding and documenting emerging plant pathogens each year (Allen et al., 2017; Koenning and Wrather, 2010; Wrather and Koenning, 2006; Wrather et al., 2001, 2003). This data was collected from research plots, diagnostic clinic reports and surveys of fields within each state and compiled from reports from producers, crop consultants and extension educators. The losses that were reported during 1996–2014 ranged from 149 to 352 million bushes in 1999 and 2009, respectively (Fig. 1). These losses are subject to the environmental conditions that occurred during each season, in each region, as to which disease may in fact be favoured, with drought or dryer years having the lower reported losses.

There are many types of pathogens that attack soybean with many different approaches to infection, colonization and survival strategies. Necrotrophs are pathogens that obtain nutrition from dead or dying plant materials, for example, *Sclerotinia sclerotiorum* and many of the *Pythium* spp. (Fig. 2a). Biotrophs are plant pathogens that colonize and obtain nutrients from living plant materials, most often through specialized structures such as haustoria that form from *Erysiphe* spp. that cause powdery mildew (Fig. 2b). Downy mildew, soybean rust and powdery mildew are all examples of diseases caused by biotrophs, as these pathogens, *Peronospora manshurica*, *Phakopsora pachyrhizi*, *Erysiphe diffusa* and *Erysiphe glycines*, cannot be cultured and can only live and reproduce on soybean plants. *Phytophthora sojae* is classified as a hemibiotroph, as it colonizes initially in a very similar means with haustoria and excretes proteins into host cells which inhibit plant defences. Following this early infection stage, it switches to a necrotrophic life style.

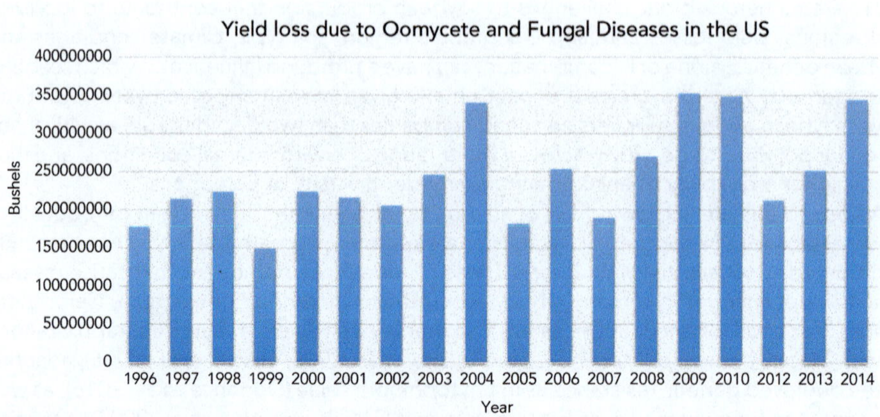

Figure 1 Annual estimated losses from 1996 to 2014 due to Oomycete and fungal pathogens of soybean. Data is a summary of the annual survey of field crop pathologists and can be found at http://extension.cropsci.illinois.edu/fieldcrops/diseases/yield_reductions.php

In addition, some of the diseases that are caused by Oomycete and fungal pathogens have a very limited host range and infect primarily soybean or a few related hosts such as *Peronospora manchurica* and *Phakopsora pachyrhizi*. This is in contrast to pathogens, *Macrophomina phaseolina*, *Pythium ultimum* and *Sclerotinia sclerotiorum*, which all have very large host ranges (>400 host plants). The number of hosts and the type of pathogen (biotroph, hemibiotroph or necrotroph) can greatly influence the type of management approaches that can be implemented.

Water moulds are fungus-like heterotrophs that belong to the superkingdom Chromalveolata (rev. Beakes and Sekimoto, 2009). This group of pathogens has very unique characteristics that separate them from true fungi: the mycelium is coenocytic (lack of cross walls) and composed of cellulose, most of its life is diploid (2n), survival structures are oospores and asexual sporangia produce swimming bi-flagellate zoospores. For soybean, there are necrotrophic Oomycete pathogens which include numerous species of *Pythium*, and the hemibiotroph, *P. sojae* (Kaufman and Gerdeman) and these have all caused significant losses due to stand reductions and root rot. The biotrophic downy mildew, *Peronospora manshurica*, is a widespread foliar disease but losses to this pathogen are limited. Another *Phytophthora* spp., *P. sansomeana*, has been reported in upper Midwest (Hansen et al., 2009) and more recently in China (Tang et al., 2010), and is a pathogen of both corn and soybean (Zelaya-Molina et al., 2010).

True fungi that cause diseases of soybean are very diverse and several are limited to specific geographic regions, primarily due to environmental conditions (Yang and Feng, 2001). For example, aerial blight of soybean caused by *Rhizoctonia solani* AG 1-1A does not occur to date in the upper Midwest, but is common in the Southern US (Yang et al., 1990a,b). *Sclerotinia* stem rot caused by *Sclerotinia sclerotiorum*, another example of a disease that has limited geographic distribution, is favoured by cool temperatures during flowering, and is limited to production regions in the upper Midwest, MN, ND and states near and around the Great Lakes (Grau et al., 2004). Both *Rhizoctonia solani* and *S. sclerotiorum* are necrotophs as are many of the fungi that infect soybean. Most of the fungi that infect soybean belong to phylum Ascomycota, which are in the haploid state for most of their life cycle. They survive or overwinter on host residue as different types of morphological structures, including specialized mycelial masses such as sclerotia, with

Figure 2 Examples of necrotrophs (a) and biotrophs (b) that infect and cause yield losses in soybean. The necrotroph is *Sclerotinia sclerotiorum* with the signs of the pathogen including white fluffy mycelium and black sclerotia. This is compared to a biotroph, *Erysiphe diffusa*, with the signs of the pathogen which are the mycelia covering the top of the leaf.

reproductive structures in the perfect stage (2n) as perithecia and in the imperfect stage (n) as pycnidia or as mycelia embedded in plant tissues. *Phakopsora pachyrhizi*, a biotroph and the only rust fungus that attacks soybean, is a microcyclic rust pathogen that survives through the uredinial stage (n + n) on living or recently detached leaves of host plants (Jurick et al., 2008; Miles et al., 2003).

The diversity of pathogen life styles, life stages, host ranges and genetic composition is indicative of why there is just as much diversity in tactics needed to manage the diseases that they cause on soybean. No one strategy is going to work for all pathogens (Grau et al., 2004; Hartman et al., 2015).

2 Case studies on diseases caused by Oomycetes

2.1 *Phytophthora sojae*

Phytophthora sojae is one of the most studied of all the soybean pathogens with many reviews regarding pathogen diversity, mechanisms of pathogenicity and management (Dorrance and Grünwald, 2009; Dorrance et al., 2009; Grau et al., 2004; Schmitthenner, 1985; Tyler, 2007; Tyler and Gijzen, 2014). This is a soil-borne pathogen, and as with many water moulds infection is highly favoured by high moisture conditions (Fig. 3a). Thus, this is a consistent problem in poorly drained soils throughout the Midwest and Southern regions of the United States, and many regions worldwide including Australia, Brazil, China and Japan to

Figure 3 Examples of Oomycete pathogens of soybean: (a) *Phytophthora* root and stem rot caused by *Phytophthora sojae*, (b) seedling root rot caused by *Pythium* spp., and (c) downy mildew of soybean caused by *Peronospora manshurica*.

name a few (Dorrance and Grünwald, 2009). *P. sojae* has a gene-for-gene interaction with its host and has been successfully managed with the deployment of resistance genes (*Rps*, resistance to *P. sojae*) in commercial cultivars. Twenty-two *Rps* genes have been mapped and more putative sources of additional novel *Rps* genes have been identified (Table 1; Dorrance and Schmitthenner, 2000; Gordon et al., 2009; Matthiesen et al., 2016). However, *P. sojae* populations adapted or already have virulence to many of these *Rps* genes, with the life of a given effective gene ranging from 8 to 20 years (Grau et al., 2004) once it is deployed; thus the constant search for novel *Rps* genes is necessary.

Phytophthora sojae population diversity has primarily focused on pathotype (race), a characteristic which measures the virulence of a population on specific resistance genes (*Rps*) to highlight which resistance genes are effective or ineffective in a region (Dorrance et al., 2016; Grau et al., 2004). The most recent pathotype survey of *P. sojae* isolates recovered during 2011–2014 indicated that most US populations now have virulence to the most commonly deployed *Rps* genes (*Rps1a, Rps1c, Rps1k, Rps3a*), while only a few have high proportion of the population with virulence to *Rps6* and *Rps8* (Dorrance et al., 2016, Table 1). In addition, when the pathotype is compared to older studies, there is a marked increase in complexity in US populations, where the number of ineffective *Rps* genes for each isolate continues to rise (Dorrance et al., 2016; Grau et al., 2004). This is of concern as it is unlikely that these *Rps* genes can be reintroduced into germplasm in the future and the increasing complexity is not hindering pathogenicity within this pathogen population.

Historically, only a few studies explored the population genetic diversity of *P. sojae* in the United States and these both used few isolates (Förster et al., 1994; Meng et al., 1999). A recent genetic analysis of *P. sojae* populations collected in Ohio indicated that there could be greater diversity than previously thought, including evidence for putative heterozygotes indicating that outcrossing maybe occurring at higher frequency than previously thought (Dorrance and Grünwald, 2009). More recently, a broader and deeper sampling of *P. sojae* in four states revealed that there was greater diversity within these populations (Stewart et al., 2016). Interestingly, several distinct populations of *P. sojae* were identified in two neighbouring states, Iowa and South Dakota, as well as among production regions in Ohio (Stewart et al., 2016).

The genome of *P. sojae* was sequenced and initially proposed to have 19 027 predicted gene models and to be 95 Mbp in size (Tyler et al., 2006). Many studies have been completed prior to and following the sequencing of the *P. sojae* genome to identify those genes that are important for *P. sojae* to become established in soybean. The genes identified include elicitin or elicitin like, crinkling and necrosis-inducing protein (CRN), necrosis-inducing peptides (NLP) and RxLR effector families (Anderson et al., 2015; Torto-Alalibo et al., 2007; Tyler and Gijzen, 2014). The RxLR families are very important as all the *Avr* genes cloned to date, those recognized by *Rps* genes in *P. sojae*, are RxLR-type genes (Anderson et al., 2016).

Quantitative disease resistance (QDR) is very important in the *P. sojae*–soybean pathosystem. This has been described as an overall reduction in root rot (smaller lesions, reduced infection frequency and fewer oospores) and with high levels of expression, elimination of the stem rot phase (Mideros et al., 2007). This resistance is effective towards many strains of the pathogen and contributes to yield stability (Dorrance et al., 2003). Two adapted genotypes, Conrad and V71-370, as well as numerous plant introductions have been used to map QDR for *P. sojae* in soybean (Lee et al. 2014; Schneider et al., 2016; Stasko et al., 2016; Tucker et al., 2010). Among these sources of resistance, numerous minor QDR were mapped in Conrad (Burnham et al., 2003;

Table 1 The source, genomic locus and type of soybean resistance genes *(Rps)* that have been fine mapped and cloned to date

Soybean				
Rps	Source	Chrom/ MLG	Cloned	Citation(s)
Rps1a	Blackhawk	3/N		Demirbas et al. 2001 Weng et al., 2001
Rps1b	Sanga L76-2060	3/N	TIR-NBS-LRR	Demirbas et al., 2001 Graham et al., 2002
Rps1c	Arksoy	3/N		Demirbas et al., 2001
Rps1d	PI 103091	3/N		Anderson and Buzzell, 1992
Rps1k	Kingwa	3/N	CC-NBS-LRR	Gao & Bhattacharyya, 2008
Rps2	CNS L76-1988	16/J		Diers et al., 1992 Graham et al., 2002
Rps3a	PI 86972-1 (L83-570)	13/F		Diers et al., 1992
Rps3b	PI 172901	13/F		
Rps3c	PI 340046	13/F		
Rps4	PI 86050	18/G	CC-NBS-LRR	Demirbas et al., 2001 Sandhu et al. 2004
Rps5	PI 91160	18/G		Diers et al., 1992
Rps6	Altona	18/G		Demirbas et al., 2001
Rps7	Harosoy	3/N		Anderson and Buzzell, 1992 Lohnes and Schmitthenner, 1997 Weng et al., 2001
Rps8	PI 399073	13/F		Gordon et al., 2006 Ortega and Dorrance, 2011
RpsYB30	'Youbian 30'	19/L		Zhu et al., 2007
RpsZS18	'Zaoshu 18'	2/D1b		Yao, H, et al. 2010
RpsYD25	'Yudou 25'	3/N		Fan et al., 2009
RpsYu25	'Yudou 25'	3/N		Sun et al., 2011
RpsSN10	'Suinong 10'	13/F		Yu et al., 2010
Rps	'Waseshiroge'	3/N		Sugimoto et al., 2011
Rps9	'Ludou 4'	3/N		Wu et al., 2011
UN2	PI 567139B	16/J		Lin et al., 2013 Li et al., 2016
UN1	PI 567139B	3/N		Lin et al., 2013 Li et al., 2016
Rps10	'Wandou 15'	17/D2	Serine/Threonine protein kinases	Zhang et al. 2013a
RpsYD29	'Yudou 29'	3/N	NBS-LRR	Zhang et al., 2013b

Table 1 *Continued*

		Soybean		
Rps	Source	Chrom/ MLG	Cloned	Citation(s)
RpsJS	Nannong 10-1x	18/G	NBS-LRR	Sun et al. 2014
Rps11	PI 594527	7/M		Ping et al., 2016
Rps12	PI 399036	18/G	NBS-LRR	Sahoo et al. 2017
RpsQ/ Rps9	'Qichadou 1'	3/N	Serine/threonine protein kinase	Li et a., 2017
RpsHN	'Meng8206'	3/N		Niu et al., 2017
RpsWY	'Wayao'	3/N		Cheng, Y et al., 2017

Chrom/MLG: Chromosome and the corresponding molecular linkage group.

Han et al., 2008; Stasko et al., 2016; Wang et al., 2010, 2012) as well as major QDR loci in V71-370 (Tucker et al., 2010), PI427106 and PI427105B (Lee et al., 2014). Saturated soil conditions that occur during flooding events contribute to the development of zoospores. The PI 408105B is a source of resistance for both flooding and QDR for P. sojae. Interestingly, one of the two quantitative trait loci (QTLs) for tolerance to flooding in PI 408105B co-locates to a minor QDR for *P. sojae* (Nguyen et al., 2012).

For regions with high inoculum and favourable conditions for infection, additional disease management strategies have proven useful. Soybeans with QDR are vulnerable to *P. sojae* in the early growth stages (Dorrance et al., 2001). Seed treatments targeting *P. sojae* are used in Ohio and some areas of the Midwest and have helped to protect stand of cultivars even those with high levels of QDR (Dorrance et al., 2003; Dorrance et al., 2009).

2.2 *Pythium* spp.

Numerous species of *Pythium* infect soybean and they predominately affect seed and young seedlings in the early growth stages (Fig. 3b). There is growing evidence of i) the diversity in the species that are present in different fields within and between soybean production regions (Broders et al., 2007, 2009; Coffua et al., 2016; Dorrance et al., 2004; Jiang et al., 2012; Marchand et al., 2014; Murillo-Williams and Pedersen, 2008; Radmer et al., 2017; Rojas et al., 2017a,b; Rizvi and Yang, 1996; Zitnick-Anderson and Nelson, 2015), ii) pathogenicity (Broders et al., 2007; Coffua et al., 2016; Radmer et al., 2017; Rojas et al., 2017a; Zitnick-Anderson and Nelson, 2015) and iii) sensitivity to fungicides (Broders et al., 2007; Dorrance et al., 2004; Matthiesen et al., 2016b; Radmer et al., 2017) that have been recovered from fields with a history of stand establishment issues.

Within Ohio, there are more than 30 different *Pythium* species that cause disease on corn and soybean (Broders et al., 2007, 2009; Dorrance, Balk, Vargas, unpublished data; Eyre, 2016). The region, time and method of sampling can affect which species may be recovered as the environmental factors at the time of infection may have favoured one species of *Pythium* over another. In a large survey in Ohio, during 2007 from baiting soil collected from more than 80 locations, more than 7000 isolates were recovered representing 21 different species including three new species (Broders et al., 2009b, 2012; Ellis et al., 2012). During 2015, more than 15 different species of *Pythium* were recovered from just two fields in Ohio (Eyre, 2016) following very heavy rain events. A greater number of species of *Pythium* (66) were recovered during a broader regional survey of seedling pathogens in 11 Midwestern states, where more than 3000 isolates were recovered over two years (Rojas et al., 2017a,b). In a survey in ND and in the larger survey of the Midwest, greater overall numbers and diversity of species of *Pythium* were recovered during a wet spring compared to a second year where conditions were much drier (Rojas et al., 2017a; Zitnick-Anderson and Nelson, 2015).

Several *Pythium* spp. that infect soybean have been sequenced, including *P. ultimum* var *ultimum*, which was the first to be sequenced (Lévesque et al., 2010). One of the more notable discoveries within the genome of *P. ultimum* var *ultimum* was the absence of RxLR and cutinase gene sequences compared to those of *Phytophthora* species. There were other gene families present in the *Pythium* genome that were associated with pathogenicity including aspartyl protease families, crinklers, glycoside hydrolases, lipases and serine protease inhibitors. These differences between the genes found in the genomes of *Pythium* and *Phytophthora* were thought to relate to the broad host range and necrotrophic abilities of *Pythium* species (Lévesque et al., 2010).

Resistance in soybean to a few *Pythium* spp. has been identified (Balk et al., 2014; Bates et al., 2008; Ellis et al., 2013). A single gene (*Rpa1*) was mapped in the soybean cultivar Archer that conferred resistance to *Pythium aphanidermatum* (Ross et al., 2008). Several QTLs which are associated with resistance to *Pythium irregulare* were mapped in a population of Conrad × Sloan (Stasko et al., 2016) and from PI 424354 in three different populations (Ellis et al., 2013). These QTLs are all minor and contributed 15% or less of the phenotypic variation. Preliminary data also suggest that QTL to one *Pythium* spp. may be different from another (Balk et al., 2014). This is an area where more studies are needed to ensure the incorporation of resistance into high-yielding adapted cultivars.

Differences in geographic locations, environmental conditions and soil factors all contribute to the diversity of *Pythium* spp. that have now been shown to impact soybean (Broders et al., 2009; Rojas et al., 2017b; Zitnick-Anderson et al., 2017). This emphasizes a challenge that for any soybean germplasm or seed treatment to be effective it must have broad activity to a very diverse group of species within the same genus.

3 Case studies on diseases caused by true fungi: fungal pathogens that affect foliage

3.1 *Cercospora sojina*

Cercospora sojina is the causal agent for frogeye leaf spot (Fig. 4a) and has been more commonly associated with yield loss over the past century in the Southern US, and in the past decade the north-central region (Cruz and Dorrance, 2009; Zhang and Bradley, 2014). There are more reports on this pathogen occurring in fields in the north-central region earlier in the season, compared to previous years. One explanation may in part be due to its ability to survive on residue during the winters (Cruz and Dorrance, 2009; Zhang and Bradley, 2014), but also due to the planting of more highly susceptible cultivars as well as changes in the sensitivity to the QoI class of fungicides (Zhang et al., 2012a,b). Overwintering studies were conducted in Ohio during 2007 and conidia were recovered from leaf and stem residue left on the soil surface the following spring (Cruz and Dorrance, 2009). Similar studies were done beginning in the fall of 2008 in Illinois and conidia were recovered up to 24 months when buried at different depths (0, 10 and 20 cm) in the field (Zhang and Bradley, 2014). Fungicides have been used in some regions to manage this foliar pathogen albeit the emergence of resistance to QoI fungicides, first in Tennessee in 2010 (Zhang et al., 2012a,b), then quickly followed in several other states (Zeng et al., 2015) and is now firmly established in southern regions (Standish et al., 2015). The QoI-resistant isolates have a single mutation in cytochrome b gene, codon 143, which is involved in respiration (Zeng et al., 2015) and diagnostic tools are readily available to discern resistance. Both of these results indicate that northern regions should begin to focus on management of this disease, specifically prioritizing deployment of cultivars with resistance to this pathogen.

Cercospora sojina has a gene-for-gene relationship with numerous races identified in several regions (Mian et al., 2008). A resistance gene, *Rcs3*, which is dominant was originally identified in the cultivar 'Davis' and has an immune-type phenotype or interaction with *C. sojina* which is still effective to date (Mian et al., 2009). In addition, twenty new sources of resistance to *C. sojina* that have immune-type response but not the *Rcs3* resistance gene marker haplotype may prove useful in developing new cultivars with high levels of resistance (Mengistu et al., 2011, 2012; Mian et al., 2009).

Population genetic analyses of *C. sojina* isolates collected from Brazil, China and the United States indicate a high level of genetic diversity (Bradley et al., 2012). A second study which examined isolates collected from six fields in Arkansas identified a high probability that mating was occurring in the fields based on a close to 1:1 occurrence of the *MAT1*-1-1 and *MAT1-2* idiomorph mating loci (Kim et al., 2013). This finding could explain in part the high degree of variability that is observed within regions and emphasizes the need to examine QDR loci resistance that slows the rate of development for this polycyclic disease (Bradley et al., 2012).

Figure 4 Examples of fungal pathogens that affect leaves of soybean: (a) frogeye leaf spot caused by *Cercospora sojina* and (b) soybean rust caused by *Phakopsora pachyrhizi*.

3.2 *Phakopsora pachyrhizi*

The movement of this pathogen from its original site in Asia to the Southern hemisphere, Africa, South America and North America has been summarized (Miles et al., 2003; Hartman et al., 2011; Kelly et al., 2015). This rust pathogen contributes annually to losses in parts of South America – 365.5 thousand tons of grain – during 2011/2012 (Godoy et al., 2016). The highest losses recorded for soybean rust in Brazil were 2.67, 2.9 and 3.4 million tons during 2006/2007, 2005/2006 and 2002/2003 production seasons prior to full implementation of management programmes (Godoy et al., 2016). In the United States, losses have been dynamic since it was first identified in 2004 (Kelly et al., 2015), with an estimated yield loss of more than 2.8 million bushels in 2009, 1.0 million in 2013 and less than 1.0 million each year since 2005 (http://extension.cropsci.illinois.edu/fieldcrops/diseases/yield_reductions.php). Interestingly, there were no recorded losses in 2010. As a rust pathogen, it is a biotroph and survives only on living hosts. One potential factor that may contribute to the fact that soybean rust did not become established in northern climates in the United States may be due to freezing temperatures that occur during winters and killing of additional hosts (Sikora et al., 2014). In Brazil, a regulated soybean-free period is thought to have contributed to lower levels of initial inoculum as well as fewer epidemics and yield losses from soybean rust (Godoy et al., 2016).

This rust pathogen exists predominately as urediniospores and no alternate host has been identified. Teliospores are formed at the end of the season but no infection from the basidiospores has been observed. *Phakopsora pachyrhizi* does have a host range of more than 150 species, all through infections from urediniospores. However these hosts are confined to the legume family Fabaceae and subfamily Papilionoideae (Hartman et al., 2011; Kelly et al., 2015). One of these hosts is kudzu (*Pueraria montana*) which was introduced into North America, and some types are very susceptible and provide a suitable overwintering host in the Southern US (Sikora et al., 2014).

Molecular diversity is also high in this rust fungus as measured with AFLPs (Rocha et al., 2015) and SSRs (Anderson et al., 2008; Twizeyimana et al., 2011). Genetic variability in the populations of this microcyclic rust is thought to evolve through somatic hybridizations that occur during asexual reproduction (Vittal et al., 2012). Numerous studies have examined the pathogenic variability, pathotypes, on different sets of soybean differentials (Hartman et al., 2011; Kelly et al., 2015; Paul et al., 2015). In all of the studies both within and across geographic regions, numerous pathotypes (races) were identified. The US isolates had wide differences in aggressiveness as measured by disease severity or density of the uredinia, and these components should be measured to evaluate for resistance (Paul et al., 2015).

As with many biotrophic fungi, there are dominant genes for resistance to *Phakopsora pachyrhizi* and they have been mapped to six loci in the soybean genome with multiple alleles and differentials for this pathogen also available (Paul et al., 2015). There is an immune-type resistance response, but in many of these cases the resistance is characterized by a reddish-brown lesion compared to the susceptible tan lesion. Numerous additional sources of resistance to *Phakopsora pachyrhizi* (*Rpp* genes) have been identified (Miles et al., 2006; Hartman et al., 2011; Paul et al., 2015). However, in many instances, the *Rpp* genes that were deployed were ephemeral, lasting only a few years. Alternatives to immune or single gene-mediated resistance have been suggested, such as partial resistance (Hartman et al., 2011) and potentially non-host or genetically engineered resistance (Godoy et al., 2016; Hartman et al., 2011). There is a need to identify and incorporate more durable resistance to this pathogen as fungicide management continues to become more challenging.

Short-term management solutions have been established with the soybean-free planting periods and improved timing of applications of fungicide (Godoy et al., 2016). However, fungicides are not going to be a long-term answer as the adaptation to fungicides in South America is a troubling development for future management of this disease (Godoy et al., 2016).

4 Case studies on diseases caused by true fungi: fungal pathogens that affect roots

4.1 *Macrophomina phaseolina*

This soil-borne fungal pathogen was previously thought to be a disease of hot, dry climates worldwide, but is emerging as a pathogen in more northern climates (Kauer et al., 2012). *M. phaseolina* has a wide host range, but there are reports of host specialization among isolates (Kaur et al., 2012; Islam et al., 2012; Singh et al., 1990; Su et al., 2001). During the early to mid-2000s there were numerous reports of charcoal rot associated with yield losses on soybean for the first time in northern states in the United States including Iowa (Yang and Navi, 2005), Minnesota (Elaraby, 2003), New York (Cummings and Bergstrom, 2013) and North Dakota (Bradley and del Rio, 2003). In Ohio, *M. phaseolina* was recovered from soils collected from 95% of the soybean and 77% of cornfields sampled during 2013 and 2014 indicating a very wide distribution (Lopez-Nicora et al., 2016). These findings were unexpected, although an earlier survey had identified charcoal rot as a common disease in the southern counties in Ohio (Gearhart et al., 2010). More important than the widespread nature of this pathogen is that isolates of *M. phaseolina* originating from northern regions grew and colonized soybean plants at lower temperatures than isolates from southern regions (Sexton et al., 2016). This indicates that the pathogen population may have adapted to the cooler climates and will impact grain crops in the region.

This fungus is generally thought to be monospecific and haploid in nature (Kaur et al., 2012). The sexual stage has not been identified to date, but this fungus is placed in the ascomycete family Botryosphaeriaceae. There was a high level of genetic diversity among *M. phaseolina* isolates collected from different regions and hosts within the United States (Baird et al., 2010). Due to this high level of genetic diversity, Baird et al. (2010) suggested that the parasexual cycle, and mitotic changes, could be contributing to genetic changes in this fungus, albeit more work should be done. An isolate of *M. phaseolina* collected from jute in Bangladesh has been sequenced (Islam et al., 2012). Approximately 12% of genes within the genome are for plant cell wall–degrading enzymes in addition to those that can inhibit or limit plant defence responses (Islam et al., 2012).

Resistance to *M. phaseolina* has been identified in plant introductions and cultivars, primarily in the maturity group V and later (Mengistu et al., 2007, 2013; Twizeyimana et al., 2012) and in 70 ancestral soybean genotypes (Pawlowski et al., 2014). Sexton et al. (2016) presented a strong case based on pathogenicity assays that evaluations for resistance should utilize geographic-adapted isolates. More surveys from other states and production regions are needed, but the widespread distribution and the adaptation to cooler temperatures indicate that this pathogen will be one to monitor in the future.

4.2 *Fusarium* species

Based on surveys of soybean fields, plants and debris, there is a growing awareness of the wide diversity of *Fusarium* spp., both within and between production regions, which consistently challenge soybeans. Leslie et al. (1990) conducted a snapshot study and collected soybean plants and debris near the end of August from 34 fields in Central and Southern US. In this late season survey, *F. oxysporum*, *F. solani* and *F. equiseti* were most commonly recovered from soybean plants, and these same species along with *F. graminearum* were recovered from soybean debris. More recently, Broders et al. (2007b) primarily recovered *F. graminearum* from symptomatic seedlings, followed by recovery of *F. oxysporum* and five other species from plants collected from problem fields in Ohio for a survey on early seedling death. More importantly, the *F. graminearum* isolates recovered during this study were highly pathogenic on soybean seed compared to the other *Fusarium* species. In a 3-year survey of soybean plants in Iowa, *F. oxysporum*, *F. acuminatum*, *F. graminearum* and *F. solani* were the most frequently recovered species from soybean at the late vegetative and mid-reproductive phases (Díaz et al., 2013b). In a second study, Díaz et al. (2013a) documented the extensive root rot caused by *F. graminearum* in both greenhouse and field microplot studies in Iowa, which clearly showed that this is a very important pathogen of soybean. In addition to *F. graminearum*, they also reported extensive root rot symptoms caused by isolates of *F. virguliforme* (causal agent of sudden death syndrome), *F. proliferatum*, *F. sporotrichioides* and *F. solani*.

There is also diversity among isolates of *F. oxysporum* and *F. commune* in the Midwestern and Southeastern soybean production regions of the United States (Ellis et al., 2014). The *F. oxysporum* species complex from numerous isolates from around the United States were characterized by differences in sequence in the intergenic spacer region of the ribosomal gene, translation elongation factor, the mitochondrial small subunit and the mating-type locus, and were separated into five clades (Ellis et al., 2014). More importantly, the diversity of pathogenicity will have implications in developing diagnostic tests as well as screening germplasm for resistance.

Chitrampalam and Nelson (2016) characterized the *F. solani* species complex associated with isolates that can cause extensive colonization recovered from soybean in North Dakota. These included isolates that were highly pathogenic but also resistant to the most commonly used fungicide, fludioxonil.

Among the *Fusarium* spp. that do infect soybean, *F. virguliforme*, which causes sudden death syndrome, and *F. graminearum* (telemorph: *Gibberella zeae* (Schwein) Petch), which causes seed and root rot (Broders et al., 2007; Xue et al., 2007) as well as pod rot in South America (Pioli et al., 2004), are the most studied. Four different species of *Fusarium* that can cause sudden death syndrome in plants were identified worldwide, but only *F. virguliforme* has been associated with this disease in the United States (Aoki et al., 2003, 2005, 2012; Hartman et al., 2015a; Rupe et al., 1989). Similar to the *Fusarium* spp. listed above, there also exist other types of root-infecting soil-borne fungi. What is unique is the production of one or more toxins that are translocated to the leaves and cause interveinal chlorosis, then necrosis, and early defoliation and plant death in susceptible cultivars. Resistant cultivars have very few symptoms. This host–pathogen system was thoroughly reviewed (Hartman et al., 2015a). Briefly, there is very little variation among isolates of *F. virguliforme* in the United States, except for aggressiveness (Li et al., 2000; Malvick and Bussey, 2008). Numerous QTLs have been mapped and resistant cultivars are actively deployed to manage this disease. However, in a study of four populations where ten QTLs were segregating, five QTLs were identified that may have the greatest effects on the expression of resistance towards *F. virguliforme* (Luckew et al., 2013). In addition, fluopyram, when used as a seed treatment, reduced the foliar symptoms and increased yield in fields in the Midwest region and Ontario, Canada, with a history of sudden death syndrome (Kandel et al., 2016). Based on these results, the combination of a resistant cultivar combined with seed treated with fluopyram is thought to protect yields.

5 Summary

These few case studies emphasize not only the extensive diversity of the pathogens that infect soybean but also modes of infection, host ranges and management options. The population genetics and genomic analysis of these pathogens discussed here and other soybean pathogens as well are in progress. However, many of these studies compared findings with a limited number of strains between regions and not at the depth truly necessary for within-field or regional variation. This is an area that needs focus; larger and deeper sample sizes across and within regions are needed to fully understand the diversity and identity of isolates for screening for resistance and potentially deploying novel resistance approaches which would be effective across the region where a cultivar may be marketed.

In the past two decades, a large number of sources of both R-gene-mediated and quantitative resistance to a few strains of these pathogens were identified. Some have been mapped, and efforts are currently in place to characterize the mechanisms that contribute to the expression of this resistance. With one exception (*Rcs3* for *C. sojina*), R-gene-mediated resistance has been ephemeral, lasting from as few as two to as long as 20 years. Quantitative resistance for necrotrophic, biotrophic and hemibiotrophic pathogens has been identified, but relies on extensive screening in field and greenhouse assays. Fine mapping for QDR loci is far behind the R-gene-mediated traits, but is the backbone of resistance to all pathogens. Better markers, defined number of QDRL to avoid adaptation in the pathogen population as well as a deeper understanding of the role epistasis plays in this resistance are needed.

6 Future trends in research

Novel mapping approaches and screening tools are now available to uncover the mechanisms in crop plants that harness the highest levels of expression of resistance towards a multitude of plant pathogens. This will be especially powerful as soybean cultivars are planted across wider acreages where they could encounter a multitude of biotic and abiotic stressors.

Effectors of plant pathogens, proteins that are secreted during the infection process, have been proposed to be used to screen for resistance in some pathogen–host systems (Anderson et al., 2015; Godoy et al., 2016; Vleeshowers and Oliver, 2014). Identification of key effectors across a large number of isolates can be used not only for resistance but also potentially for diagnostic feature as proposed by Lanubile et al. (2016). Genome-wide association analysis has been used to identify QDRL for several soybean pathogens including *P. sojae* (Schneider et al., 2016). This strategy will identify major QTLs that contribute to large portions of the phenotypic variation. The goal of expedited screening and dense molecular maps increases the likelihood of identifying 'perfect markers', those alleles of genes that are directly tied to the trait, which will be essential to improve the quantitative resistance levels to the multiple pathogens that any given cultivar may encounter.

Based on the number and wide distribution of new and re-emerging pathogens of soybean that have occurred in the past ten years, continual surveys are critical to recognize and adapt cultivars to avoid yield losses. New diagnostic techniques not only to identify the pathogens but also to efficiently check for fungicide sensitivity and pathotype will greatly improve management of soybean diseases.

7 Where to look for further information

There are numerous university, government, soybean commodity and scientific society websites that have a wealth of information related to the cause and management of soybean diseases.

The most recent comprehensive piece is from the American Phytopathological Society with several books that focus on soybean diseases. http://my.apsnet.org/APSStore

- Mueller, D., Wise, K., Sisson, A., Smith, D., Sikora, E., Bradley, C. and Robertson, A (Editors). 2016. *A Farmer's guide to soybean diseases.* APS Press.
- Hartman, G. L., Rupe, J. C., Sikora, E. J., Domier, L. L., Davis, J. A., and Steffey, K. L. (editors) 2015. *Compendium of Soybean Diseases and Pests*, Fifth Edition. APS Press
- Mueller, D., Wise, K., Dufault, N., Bradley, C., and Chilvers, M. 2013. *Fungicides for Field Crops.* APS Press.

The Crop Protection Network (CPN) website (http://cropprotectionnetwork.org/) has numerous fact sheets and crop loss estimates for field crops. These are the results from multi-state Land Grant Universities and International collaborations on studies for management of soybean and corn diseases.

The Plant Management Network (https://www.plantmanagementnetwork.org/) has a series of research articles (*Plant Health Progress*), annual results from seed treatment and fungicide applications towards several diseases (*Plant Disease Management Reports*), and webcasts (*Focus on Soybean, Soybean Fungicide Resistance Hub*).

8 References

Allen, T. W., Bradley, C. A., Sisson, A. J., Byamukama, E., Chilvers, M. I., Coker, C. M., Collins, A. A., Damicone, J. P., Dorrance, A. E., Dufault, N. S., Esker, P. D., Fraske, T. R., Giesler, L. J., Grybauskas, A. P., Hershman, D. E., Hollier, C. A., Isakeit, T., Jardine, D. J., Kelly, H. M., Kemerait, R. C., Kleczewski, N. M., Koenning, S. R., Kurle, J. E., Malvick, D. K., Markell, S. G., Mehl, H. L., Mueller, D. S., Mueller, J. D., Mulrooney, R. P., Nelson, B. D., Newman, M. A., Osborne, L., Overstreet, C., Padgett, G. B., Phipps, P. M., Price, P. P., Sikora, E. J., Smith, D. L., Spurlock, T. N., Tande, C. A., Tenuta, A. U., Wise, K. A. and Wrather, J. A. (2017), 'Soybean yield loss estimates due to diseases in the United States and Ontario, Canada, from 2010 to 2014', *Plant Health Prog.*, 18, 19–27.

Anderson, T. R. and Buzzell, R. I. (1992), 'Inheritance and linkage of the *Rps7* gene for resistance to Phytophthora rot of soybean'. *Plant Dis.*, 76(9), 958–9.

Anderson, R. G., Deb, D., Fedkenheuer, K. and McDowell, J. (2015), 'Recent progress in RXLR effector research', *Mol. Plant-Microbe Interact.*, 28(10), 1063–72.

Anderson, S. J., Stone, C. L., Posada-Buitrago, M. L., Boore, J. L., Neelam, B. A., Stephens, R. M., Luster, D. G., Frederick, R. D. and Pedley, K. F. (2008), 'Development of simple sequence repeat markers for the soybean rust fungus, *Phakopsora pachyrhizi*', *Mol. Ecol. Resour.*, 8(6), 1310–12.

Aoki, T., O'Donnell, K., Homma, Y. and Lattanzi, A. R. (2003), 'Sudden-death syndrome of soybean is caused by two morphologically and phylogenetically distinct species within the *Fusarium solani* species complex- *F. virguliforme* in North America and *F. tucumaniae* in South America', *Mycologia*, 95(4), 660–84.

Aoki, T., O'Donnell, K. and Scandiani, M. M. (2005), 'Sudden death syndrome of soybean in South America is caused by four species of *Fusarium: Fusarium brasiliense* sp. nov., *F. cuneirostrum* sp. nov., *F. tucumaniae*, and *F. virguliforme*', *Mycoscience*, 46(3), 162–83.

Aoki, T., Scandiani, M. M. and O'Donnell, K. (2012), 'Phenotypic, molecular phylogenetic, and pathogenetic characterization *of Fusarium crassistipitatum* sp. nov., a novel soybean sudden death syndrome pathogen from Argentina and Brazil', *Mycoscience*, 53(3), 167–86.

Baird, R. E., Wadl, P. A., Allen, T., McNeill, D., Wang, X., Moulton, J. K., Rinehart, T. A., Abbas, H. K., Shier, T. and Trigiano, R. N. (2010), 'Variability of United States isolates of *Macrophomina phaseolina* based on simple sequence repeats and genus transferability to related genera within Botryosphaeriaceae', *Mycopathologia*, 170(3), 169–80.

Balk, C., Fitzgibbon, T., Huzar Novakowiski, J., Erye, M. and Dorrance, A. E. (2014), 'Assessment of resistance in soybean towards *Pythium ultimum* var. *ultimum* and *Pythium ultimum* var. *sporangiferum*', *Phytopathology*, 104(11), S3.165.

Bates, G. D., Rothrock, C. S. and Rupe J. C. (2008), 'Resistance of soybean cultivar Archer to Pythium damping-off and root rot caused by several *Pythium* spp.', *Plant Dis.*, 92(5), 763–6.

Beakes, G. W. and Sekimoto, S. (2009), 'The evolutionary phylogeny of Oomycetes – insights gained from studies of holocarpic parasites of algae and invertebrates'. In *Oomycete Genetics and Genomics: Diversity, Interactions, and Research Tools* (K. Lamour and S. Kamoun (Eds)). John Wiley & Sons, pp. 1–24.

Bradley, C. A. and del Rio, L. E. (2003), 'First report of charcoal rot on soybean caused by *Macrophomina phaseolina* in North Dakota', *Plant Dis.*, 87(5), 601.

Bradley, C. A., Wood, A., Zhang, G. R., Murray, J. E., Phillips, D. V. and Ming, R. (2012), 'Genetic diversity of *Cercospora sojina* revealed by amplified fragment length polymorphism markers', *Can. J. Plant Pathol.*, 34(3), 410–16.

Broders, K. D., Lipps, P. E., Paul, P. A. and Dorrance, A. E. (2007a), 'Characterization of *Pythium* spp. associated with corn and soybean seed and seedling disease in Ohio', *Plant Dis.*, 91(6), 727–35.

Broders, K. D., Lipps, P. E., Paul, P. A. and Dorrance, A. E. (2007b), 'Evaluation of *Fusarium graminearum* associated with corn and soybean seed and seedling disease in Ohio', *Plant Dis.*, 91(9), 1155–60.

Broders, K. D., Lipps, P. E., Ellis, M. L. and Dorrance, A. E. (2012), 'Erratum: Validation of *Pythium delawarii* as *Pythium delawarense* sp. nov.', *Mycologia*, 104(3), 789.

Broders, K. D., Wallhead, M. W., Austin, G. D., Lipps, P. E., Paul, P. A., Mullen R. W. and Dorrance, A. E. (2009), 'Association of soil chemical and physical properties with *Pythium* species diversity, community composition, and disease incidence', *Phytopathology*, 99(8), 957–67.

Burnham, K. D., Dorrance, A. E., VanToai, T. and St. Martin, S. K. (2003), 'Quantitative trait loci for partial resistance to *Phytophthora sojae* in soybean', *Crop Sci.*, 43(5), 1610–17.

Cheng, Y., Ma, Q., Ren, H., Xia, Q., Song, E., Tan, Z., Li, S., Zhang, G. and Nian, H. (2017), 'Fine mapping of a Phytophthora resistance gene *RpsWY* in soybean (*Glycine max* L.) by high throughput genome-wide sequencing'. *Theor. Appl. Genet.*, 130(5), 1041–51.

Chitrampalam, P. and Nelson, B. (2016), 'Multilocus phylogeny reveals an association of agriculturally important Fusarium solani species complex (FSSC) 11, and clinically important FSSC 5 and FSSC 3 + 4 with soybean roots in the north central United States', *Anonie Van Leeuwenhoek*, 109(2), 335–7.

Coffua, L. S., Veterano, S. T., Clipman, S. J., Mena-Ali, J. I. and Blair, J. E. (2016) 'Characterization of *Pythium* spp. associated with asymptomatic soybean in southeastern Pennsylvania', *Plant Dis.*, 100(9), 1870–9.

Cruz, C. D. and Dorrance, A. E. (2009), 'Characterization and survival of *Cercospora sojina* in Ohio', *Plant Health Prog.*, doi:10.1094/PHP-2009-0512-03-RS.

Cummings, J. A. and Bergstrom, G. C. (2013), 'First report of charcoal rot caused by *Macrophomina phaseolina* in soybean in New York', *Plant Dis.*, 97(7), 1506.

Demirbas, A., Rector, B. G., Lohnes, D. G., Fioritto, R. J., Graef, G. L., Cregan, P. B., Shoemaker, R. C. and Specht, J. E. (2001), 'Simple sequence repeat markers linked to the soybean *Rps* genes for Phytophthora resistance'. *Crop Sci.*, 41(4), 1220–7.

Díaz Arias, M. M., Leandro, L. F. and Munkvold, G. P. (2013a), 'Aggressiveness of Fusarium species and impact of root infection on growth and yield of soybeans', *Phytopathology*, 103(8), 822–32.

Díaz Arias, M. M., Munkvold, G. P., Ellis, M. L. and Leandro, L. F. (2013b), 'Distribution and frequency of *Fusarium* spp. on soybean roots in Iowa', *Plant Dis.*, 97(12), 1557–62.

Diers, B. W., Mansur, L., Imsande, J. and Shoemaker, R. C. (1992), 'Mapping Phytophthora resistance loci in soybean with restriction fragment length polymorphism markers'. *Crop Sci.*, 32(2), 377–83.

Dorrance, A. E., Berry, S. A., Bowen, P. and Lipps, P. E. (2004), 'Characterization of *Pythium* spp. from three Ohio fields for pathogenicity on corn and soybean and metalaxyl sensitivity'. *Plant Health Prog.*, doi:10.1094/PHP-2004-0202-01-RS.

Dorrance, A. E. and Grünwald, N. J. (2009), 'Chapter 10: *Phytophthora sojae*: Diversity among and within populations'. In *Oomycete Genetics and Genomics: Biology Interactions with Plant and Animals and Toolbox* (K. Lamour and S. Kamoun (Eds)). John Wiley & Sons, Inc., pp. 197–212.

Dorrance, A. E., Kurle, J., Robertson, A. E., Bradley, C. A., Giesler, L, Wise, K. and Concibido, V. C. (2016), 'Pathotype diversity of *Phytophthora sojae* in eleven states in the United States', *Plant Dis.*, 100(7), 1429–37.

Dorrance, A. E. and McClure, S. A. (2001), 'Beneficial effects of fungicide seed treatments for soybean cultivars with partial resistance to Phytophthora sojae' *Plant Dis.*, 85(6), 1063–68.

Dorrance, A. E., McClure, S. A. and St.Martin, S. K. (2003), 'Effect of partial resistance on Phytophthora stem rot incidence and yield of soybean in Ohio' *Plant Dis.*, 87(3), 308–12.

Dorrance, A. E. and Schmitthenner, A. F. (2000), 'New sources of resistance to *Phytophthora sojae* in the soybean plant introductions'. *Plant Dis.*, 84(12), 1303–8.

Dorrance, A. E., Robertson, A. E., Cianzo, S., Giesler, L. J., Grau, C. R., Draper, M. A., Tenuta, A. U. and Anderson, T. R. (2009), 'Integrated management strategies for *Phytophthora sojae* combining host resistance and seed treatment', *Plant Dis.*, 93(7), 875–82.

Elaraby, M. E. (2003), 'First report of charcoal rot (*Macrophomina phaseolina*) on soybeans in Minnesota', *Plant Dis.*, 87(2), 202.

Ellis, M. L., Cruz Jimenez, D. R., Leandro, L. F. and Munkvold, G. P. (2014), 'Genotypic and phenotypic characterization of fungi in the *Fusarium oxysporum* complex from soybean roots', *Phytopathology*, 104, 1329–39.

Ellis, M. L., McHale, L. K., Paul, P. A., St. Martin, S. K. and Dorrance, A. E. (2013), 'Soybean germplasm resistant to *Pythium irregulare* and molecular mapping of resistance quantitative trait loci derived from the soybean accession PI 424354', *Crop Sci.*, 53(3), 1008–21.

Ellis, M. L., Wang, H., Paul, P., St. Martin, S. K., McHale, L. and Dorrance, A. E. (2012), 'Identification of soybean genotypes resistant to *Fusarium graminearum* and genetic mapping of resistance quantitative trait loci in the cultivar Conrad', *Crop Sci.*, 52(5), 2224–33.

Eyre, M. (2016), Characterizing the Effect of Increased Phosphorus and Potassium on Seedling Diseases and Sudden Death Syndrome of Soybean in Ohio. Electronic Thesis. The Ohio State University, 171 pp.

Fan, A. Y., Wang, X. M., Fang, X. P., Wu X. F. and Zhu, Z. D. (2009), 'Molecular identification of Phytophthora resistance gene in soybean cultivar Yudou 25'. *Acta Agron. Sin*, 35(10), 1844–50.

Förster, H., Tyler, B. M. and Coffey, M. D. (1994), '*Phytophthora sojae* races have arisen by clonal evolution and by rare outcrosses', *Mol. Plant-Microbe Interact.*, 7(6), 780–91.

Gao, H. and Bhattacharyya, M. K. (2008), 'The soybean-Phytophthora resistance locus *Rps1-k* encompasses coiled coil-nucleotide binding-leucine rich repeat-like genes and repetitive sequences'. *BMC Plant Biol.*, 8, 29.

Gearhart, K., Dugan, D., Grimes, J., Farley, L., Fisher, J., Burskey, C. and Dorrance, A. E. (2010), 'Survey of soybean diseases in the Ohio River Valley Region of Ohio during 2009', *Phytopathology*, 100(6 Supplement), S39.

Godoy, C. V., Seixas, C. D. S, Soares, R. M., Marcelino-Guimarães, F. C., Meyer, M. C. and Costamilan, L. M. (2016), 'Asian soybean rust in Brazil: past, present, and future', *Pesq. Agropec. Bras.*, 51(5): 407–21.

Gordon, S. G., St. Martin, S. K. and Dorrance, A. E. (2006), '*Rps*8 Maps to a resistance gene rich region on soybean molecular linkage group F'. *Crop Sci.*, 46(1), 168–73.

Gordon, S. G., Kowitwanich, K., Pipatpongpinyo, W., St. Martin, S. K. and Dorrance, A. E. (2007), 'Molecular marker analysis of soybean plant introductions with resistance to *Phytophthora sojae*', *Phytopathology*, 97(1), 113–18.

Graham, M. A., Marek, L. F. and Shoemaker, R. C. (2002), 'Organization, expression and evolution of a disease resistance gene cluster in soybean'. *Genetics*, 162(4), 1961–77.

Grau, C. R., Dorrance, A. E., Bond, J. and Russin, J. (2004), 'Chapter 14: Fungal Diseases'. In *Soybeans: Improvement, Production and Uses*, Third Edition (H. R. Boerma and J. E. Specht (co-editors)), Agronomy Monograph No. 16. American Society of Agronomy, Crop Science Society of America, Soil Science Society of America, Madison, WI, pp. 679–763.

Han, Y., Teng, W., Yu, K., Poysa, V., Anderson, T., Qiu, L., Lightfoot, D. A. and Li, W. (2008), 'Mapping QTL tolerance to *Phytophthora sojae* using microsatellite and RAPD/SCAR markers', *Euphytica*, 162(2), 231–9.

Hansen, E. M., Wilcox, W. F., Reeser, P. W. and Sutton, W. (2009), '*Phytophthora rosacearum* and *P. sansomeana*, new species segregates from the *Phytophthora megasperma* 'complex'', *Mycologia*, 101(1), 129–35.

Hartman, G. L., Chang, H. X. and Leandro, L. F. (2015a), 'Research advances and management of soybean sudden death syndrome', *Crop Prot.*, 73, 60–6.

Hartman, G. L., Hill, C. B., Twizeyimana, M., Miles, M. R. and Bandyopadhyay, R. (2011), 'Interaction of soybean and *Phakopsora pachyrhizi*, the cause of soybean rust', *CAB Reviews: Perspectives in Agriculture, Veterinary Science. Nutr. Nat. Resourc.*, 6, 1–13. https://doi.org/10.1079/pavsnnr20116025.

Hartman, G. L., Rupe, J. C., Sikora, E. J., Domier, L. L., Davis, J. A. and Steffey, K. L. (Eds). (2015b), *Compendium of Soybean Diseases and Pests*. APS Press.

Islam, M. S., Haue, M. S., Islam, M. M., Emdad, E. M., Halim, A., Hossen, Q. M. M., Hossain, M. Z., Ahmed, B., Rahim, S., Rahman, M. S., Alam, M. M., Hou, S., Wan, X., Saito, J. A. and Alam, M. (2012), 'Tools to kill: Genome of one of the most destructive plant pathogenic fungi *Macrophomina phaseolina*', *BMC Genomics*, 13, 493.

Jiang, Y. N., Haudenshield, J. S. and Hartman, G. L. (2012), 'Characterization of *Pythium* spp. from soil samples in Illinois', *Can. J. Plant Pathol.*, 34(3), 448–54.

Jurick, W. M., Narváez, D. F., Brennan, M. M., Harmon, C. L., Marois, J. J., Wright, D. L., and Harmon, P. F. (2008), 'Winter survival of the soybean rust pathogen, Phakoposora pachyrhizi, in Florida', *Plant Dis.*, 92(11), 1551–58.

Kandel, Y. R., Wise, K. A., Bradley, C. A., Chilvers, M. I., Tenuta, A. U. and Mueller, D. S. (2016), 'Fungicide and cultivar effects on sudden death syndrome and yield of soybean', *Plant Dis.*, 100(7), 1339–50.

Kaur, S., Dhillon, G. S., Brar, S. K., Vallad, G. E., Chand, R. and Chauhan, V. B. (2012), 'Emerging phytopathogen *Macrophomina phaseolina*: biology, economic importance and current diagnostic trends', *Crit. Rev. Microbiol.*, 38(2), 136–51.

Kelly, H. Y., Dufault, N. S., Walker, D. R., Isard, S. A., Schneider, R. W., Giesler, L. J., Wright, D. L., Marois, J. J. and Hartman, G. L. (2015), 'From select agent to an established pathogen: The response to *Phakopsora pachyrhizi* (Soybean Rust) in North America', *Phytopathology*, 105(7), 905–16.

Kim, H., Newell, A. D., Cota-Sieckmeyer, R. G., Rupe, J. C., Fakhoury, A. M. and Bluhm, B. H. (2013), 'Mating-type distribution and genetic diversity of *Cercospora sojina* populations on soybean from Arkansas: Evidence for potential sexual reproduction', *Phytopathology*, 103(10), 1045–51.

Koenning, S. R. and Wrather, J. A. (2010), 'Suppression of soybean yield potential in the continental United States by plant diseases from 2006 to 2009', *Plant Health Prog.*, doi:10.1094/PHP-2010-122-01-RS.

Lanubile, A., Ellis, M. L., Marocco, A. and Munkvold, G. P. (2016). 'Association of effector Six6 with vascular wilt symptoms caused by *Fusarium oxysporum* on soybean', *Phytopathology*, 106(11), 1404–12.

Lee, S., Mian, R. M. A., Sneller, C. H., Wang, H., Dorrance, A. E. and McHale, L. (2014), 'Joint linkage QTL analysis for partial resistance to *Phytophthora sojae* using six nested inbred populations with heterogenous conditions', *Theor. Appl. Genet.*, 127(2), 429–44.

Leslie, J. F., Pearson, C. A. S., Nelson, P. E. and Toussoun, T. A. (1990), 'Fusarium spp. from corn, sorghum, and soybean in fields in the central and eastern United States', *Phytopathology*, 80(4), 343–50.

Lévesque, C. A., Brouwer, H., Cano, L., Hamilton, J. P., Holt, C., Huitema, E., Raffaele, S., Robideau, G. P., Thines, M., Win, J., Zerillo, M. M., Beakes, G. W., Boore, J. L., Busam, D., Dumas, B., Ferriera, S., Fuerstenberg, S. I., Gachon, C. M. M., Gaulin, E., Govers, F., Grenville-Briggs, L., Horner, N., Hostetler, J., Jiang, R. H. Y., Johnson, J., Krajaejun, T., Lin, H., Meijer, H. J. G., Moore, B., Morris, P., Phuntmart, V., Puiu, D., Shetty, J., Stajich, J. E., Tripathy, S., Wawra, S., van West, P., Whitty, B. R., Coutinho, P. M., Henrissat, B., Martin, F., Thomas, P. D., Tyler, B. M., De Vries, R. P., Kamoun, S., Yandell, M., Tisserat, N. and Buell, C. R. (2010), 'Genome sequence of the necrotrophic plant pathogen *Pythium ultimum* reveals original pathogenicity mechanisms and effector repertoire', *Genome Biol.*, 11, R73.

Li, S., Tam, Y. K. and Hartman, G. L. (2000), 'Molecular differentiation of *Fusarium solani* f. sp *glycines* from other *F. solani* based on mitochondrial small subunit rDNA sequences', *Phytopathology*, 90(5), 491–7.

Li, L., Lin, F., Wang, W., Ping, J., Fitzgerald, J. C., Zhao, M., Li, S., Sun, L., Cai, C. and Ma, J. (2016), Fine mapping and candidate gene analysis of two loci conferring resistance to *Phytophthora sojae* in soybean. *Theor. Appl. Genet.*, 129(12), 2379–86.

Li, Y., Sun, S., Zhong, C., Wang, X., Wu, X. and Zhendong, Z. (2017), Genetic mapping and development of co-segregating markers of *RpsQ*, which provides resistance to *Phytophthora sojae in* soybean. *Theor Appl Genet.*, 130(6), 1223–33.

Lin, F., Zhao, M., Ping, J., Johnson, A., Zhang, B., Abney T. S., Hughes, T. J. and Ma, J. (2013), 'Molecular mapping of two genes conferring resistance to *Phytophthora sojae* in a soybean landrace PI 567139B', *Theor. Appl. Genet.*, 126(8), 2177–85.

Lohnes, D. G. and Schmitthenner, A. F. (1997), 'Position of the Phytophthora resistance gene *Rps7* on the soybean molecular map'. *Crop Sci.*, 37(2), 555–6.

Lopez-Nicora, H. D., Simon, A. C. M., Dossman, B. C., Paul, P. A., Dorrance, A. E., Lindsey, L. E. and Niblack, T. L. (2016), 'Distribution and abundance of *Heterodera glycines* and *Macrophomina phaseolina* in Ohio', *Plant Health Prog.* doi:10.1094/PHP-S-15-0049.

Luckew, A. S., Leandro, L. F., Battacharyya, M. K., Nordman, D. J., Lightfoot, D. A. and Cianzio, S. R. (2013), 'Usefulness of 10 genomic regions in soybean associated with sudden death syndrome resistance', *Theor. Appl. Genet.*, 126(9), 2391–404.

Malvick, D. and Bussey, K. (2008), 'Comparative analysis and characterization of the soybean sudden death syndrome pathogen *Fusarium virguliforme* in the northern United States', *Can. J. Plant Pathol.*, 30(3), 467–76.

Marchand, G., Chen, Y., Berhane, N. A., Wei, L., Lévesque, C. A. and Xue, A. G. (2014), 'Identification of *Pythium* spp. from the rhizosphere of soybeans in Ontario, Canada', *Can. J. Plant Pathol.*, 36(2), 246–51.

Matthiesen, R. L., Abeysekara, N. S., Ruiz-Rojas, J. J., Biyashev, R. M., Saghai Maroof, M. A. and Robertson, A. E. (2016a), 'A method for combining isolates of *Phytophthora sojae* to screen for novel sources of resistance to Phytophthora stem and root rot of soybean', *Plant Dis.*, 100(7), 1424–8.

Matthiesen, R. L., Ahmad, A. A. and Robertson, A. E. (2016b), 'Temperature affects aggressiveness and fungicide sensitivity of four *Pythium* spp. that cause soybean and corn damping-off in Iowa', *Plant Dis.*, 100(3), 583–91.

Meng, X. Q., Shoemaker, R. C. and Yang X. B. (1999), 'Analysis of pathogenicity and genetic variation among *Phytophthora sojae* isolates using RADP', *Mycol. Res.*, 103(2), 173–8.

Mengistu, A., Arelli, P., Bond, J., Nelson, R., Rupe, J., Shannon, G. and Wrather, A. (2013), '*Macrophomina phaseolina* by field screening and laboratory validation', *Plant Health Prog.* doi:10.1094/PHP-2013-0318-01-RS.

Mengistu, A., Bond, J., Mian, R., Nelson, R., Shannon, G. and Wrather, A. (2011), 'Identification of soybean accessions resistant to *Cercospora sojina* by field screening, molecular markers, and phenotyping', *Crop Sci.*, 51(3), 1101–9.

Mengistu, A., Bond, J., Mian, R., Nelson, R., Shannon, G. and Wrather, A. (2012), 'Resistance to frogeye leaf spot in selected soybean accessions in MG I and MG VI', *Plant Health Prog.* doi:10.1094/PHP-2012-0521-02-RS.

Mengistu, A., Ray, J. D., Smith, G. D. and Paris, R. L. (2007), 'Charcoal rot disease assessment of soybean genotypes using colony-forming units index', *Crop Sci.*, 47(6), 2453–61.

Mian, R., Bond, J., Joobeur, T., Mengistu, A., Wiebold, W., Shannon, G. and Wrather, A. (2009), 'Identification of soybean genotypes resistant to *Cercospora sojina* by field screening and molecular markers', *Plant Dis.*, 93(4), 408–11.

Mian, M. A. R., Missaoui, A. M., Walker, D. R., Phillips, D. V. and Boerma, H. R. (2008), 'Frogeye leaf spot of soybean: A review and proposed race designations for isolates of *Cercospora sojina* Hara', *Crop Sci.*, 48(1), 14–24.

Mideros, S., Nita, M. and Dorrance, A. E. (2007), 'Characterization of components of partial resistance, *Rps*2, and root resistance to *Phytophthora sojae* in soybean', *Phytopathology*, 97(5), 655–62.

Miles, M. R., Frederick, R. D. and Hartman, G. L. (2003), 'Soybean Rust: Is the U.S. soybean crop at risk?', *APSnet Features*. Online. doi:10.1094/APSnetFeature-2003-0603.

Miles, M. R., Frederick, R. D. and Hartman, G. L. (2006), 'Evaluation of soybean germplasm for resistance to *Phakopsora pachyrhizi*', *Plant Health Prog.* Online publication.

Murillo-Williams, A. and Pedersen, P. (2008), 'Early incidence of soybean seedling pathogens in Iowa', *Agron. J.*, 100(5), 1481–7.

Nguyen, V. T., Vuong, T. D., VanToai, T., Lee, J. D., Wu, X., Mian, M. A. R., Dorrance, A. E., Shannon, J. G. and Nguyen, H. T. (2012), 'Mapping of quantitative trait loci associated with resistance to *Phytophthora sojae* and flooding tolerance in soybean', *Crop Sci.*, 52(6), 2481–93.

Niu, J., Guo, N., Sun, J., Li, L., Cao, Y., Li, S., Huang, J., Zhao, J., Zhao, T. and Xing, H. (2017), 'Fine mapping of a resistance gene *RpsHN* that controls *Phytophthora sojae* using recombinant inbred lines and secondary populations'. *Front. Plant Sci.*, 8, 538.

Ortega, M. and Dorrance, A. E. (2011), 'Microsporogenesis of *Rps8/rps8* heterozygous soybean lines'. *Euphytica*, 181(1), 77–88.

Paul, C., Frederick, R. D., Hill, C. B., Hartman, G. L. and Walker, D. R. (2015), 'Comparison of pathogenic variation among *Phakopsora pachyrhizi* isolates collected from the United States and international locations, and identification of soybean genotypes resistant to the U.S. isolates', *Plant Dis.*, 99(8), 1059–69.

Pawlowski, M. L., Hill, C. B. and Hartman, G. L. (2014), 'Resistance to charcoal rot identified in ancestral soybean germplasm', *Crop Sci.*, 55(3), 1230–5.

Ping, J., Fitzgerald, J. C., Zhang, C., Lin, F., Bai, Y., Wang, D., Aggarwal, R., Rehman, M., Crasta, O. and Ma, J. (2016), 'Identification and molecular mapping of *Rps11*, a novel gene conferring resistance to *Phytophthora sojae* in soybean', *Theor. Appl Genet.*, 129(2), 445–51.

Pioli, R. N., Mozzoni, L. and Morandi, E. N. (2004), 'First report of pathogenic association between *Fusarium graminearum* and soybean', *Plant Dis.*, 88(2), 220.

Radmer, L., Anderson, G., Malvick, D. M., Rendahl, A., Mallik, A. and Kurle, J. E. (2017) '*Pythium, Phytophthora*, and *Phytopythium* spp. isolated from Minnesota soybean fields, their relative aggressiveness to soybean and corn, and their sensitivity to seed treatment fungicides', *Plant Dis.*, 101(1), 62–72.

Rizvi, S. S. A. and Yang, X. B. (1996), 'Fungi associated with soybean seedling disease in Iowa', *Plant Dis.*, 80(1), 57–60.

Rocha, C. M. L., Vellicce, G. R., García, M. G., Pardo, E. M., Racedo, J., Perera, M. F., de Lucía, A., Gilli, J., Bogado, N., Bonnecarrère, V., German, S., Marcelino, F., Ledesma, F., Reznikov, S., Ploper, L. D., Welin, B. and Castagnaro, A. T. (2015), 'Use of AFLP markers to estimate molecular diversity of *Phakopsora pachyrhizi*', *Electron. J. Biotechnol.*, 18, 6. dx.doi.org/10.1016/j.ejbt.2015.06.007.

Rojas, J. A., Jacobs, J., Napieralski, S., Karaj, B., Bradley, C. A., Chase, T., Esker, P. D., Giesler, L. J., Jardine, D. J., Malvick, D. K., Markell, S. G., Nelson, B. D., Robertson, A. E., Rupe, J. C., Smith, D. L., Sweets, L. E., Tenuta, A. U., Wise, K. A. and Chilvers, M. I. (2017a), 'Oomycete species associated with soybean seedlings in North America – Part I: Identification and pathogenicity characterization', *Phytopathology*, 107(3), 280–92.

Rojas, J. A., Jacobs, J., Napieralski, S., Karaj, B., Bradley, C. A., Chase, T., Esker, P. D., Giesler, L. J., Jardine, D. J., Malvick, D. K., Markell, S. G., Nelson, B. D., Robertson, A. E., Rupe, J. C., Smith, D. L., Sweets, L. E., Tenuta, A. U., Wise, K. A. and Chilvers, M. I. (2017b), 'Oomycete species associated with soybean seedlings in North America – Part II: Diversity and ecology in relation to environmental and edaphic factors', *Phytopathology*, 107(3), 293–304.

Ross, M. L., Rupe, J. C., Chen, P. and Mozzoni, L. A. (2008), 'Inheritance and genetic mapping of resistance to Pythium damping-off caused by *Pythium aphanidermatum* in 'Archer' soybean', *Crop Sci.*, 48(6), 2215–22.

Rupe, J. C. (1989), 'Frequency and pathogenicity of *Fusarium solani* recovered from soybeans with sudden death syndrome', *Plant Dis.*, 73(7), 581–4.

Sahoo, D. K., Abeysekara, N. S., Cianzio, S. R., Robertson, A. E. and Bhattacharyya, M. K. (2017), 'A novel *Phytophthora sojae* resistance *Rps12* gene mapped to a genomic region that contains several *Rps* genes'. *PLoS ONE*, 12(1), e0169950.

Sandhu, D., Gao, H., Cianzo, S. and Bhattacharyya, M. K. (2004), 'Deletion of a disease resistance nucleotide-binding-site leucine-rich-repeat-like sequence is associated with the loss of the Phytophthora resistance gene *Rps4* in soybean'. *Genetics*, 168(4), 2157–67.

Schmitthenner, A. F. (1985), 'Problems and progress in control of Phytophthora root rot of soybean', *Plant Dis.*, 69(4), 362–8.

Schneider, R. W., Hollier, C. A., Whitam, H. K., Palm, M. E., McKemy, J. M., Hernandez, J. R., Levy, L. and DeVries-Paterson, R. (2005), 'First report of soybean rust caused by *Phakopsora pachyrhizi* in the continental United States', *Plant Dis.*, 89(4), 774.

Schneider, R., Rolling, W., Song, Q., Cregan, P., Dorrance, A. and McHale, L. (2016), 'Genome wide association mapping of partial resistance to *Phytophthora sojae* in soybean plant introductions from the Republic of Korea', *BMC Genomics*, 17, 607.

Sexton, Z. F., Hughes, T. J. and Wise, K. A. (2016), 'Analyzing isolate variability of *Macrophomina phaseolina* from a regional perspective', *Crop Prot.*, 81, 9–13.

Sikora, E. J., Allen, T. W., Wise, K. A., Baniecki, J., Bergstrom, G., Bradley, C., Bond, J., Brown-Rytlewski, D., Chilvers, M., Damicone, J., DeWolf, E., Dorrance, A., Dufault, N., Esker, P., Faske, T., Giesler, L., Goldberg, N., Golod, J., Gómez, I. R. G., Grau, C., Grybauskus, A., Franc, G., Hammerschmidt, R., Hartman, G., Henn, A., Hershman, D., Hollier, C., Isakeit, T., Isard, S., Jacobson, B., Jardine, D., Kemerait, B., Koenning, S., Langham, M., Malvick, D., Markell, S., Marois, J. J., Monfort, S., Mueller, D., Mueller, J., Mulrooney, B., Newman, M., Osborne, L., Padgett, G. B., Ruden, B. E., Rupe, J., Schneider, R., Schwartz, H., Shaner, G., Singh, S., Stromberg, E., Sweets, L., Tenuta, A., Vaiciunas, S., Yang, X. B., Young-Kelly, H. and Zidek, J. (2014), 'A coordinated effort to manage soybean rust in North America: A success story in soybean disease monitoring', *Plant Dis.*, 98(8), 864–75.

Singh, S. K., Nene, Y. L. and Reddy, M. V. (1990), 'Influence of cropping systems on *Macrophomina phaseolina* populations in soil', *Plant Dis.*, 74(10), 812–14.

Standish, J. R., Tomoaso-Peterson, M., Allen, T. W., Sabanadzovic, S. and Aboughanem-Sabanadzovic, N. (2015), 'Occurrence of QoI fungicide resistance in *Cercospora sojina* from Mississippi soybean', *Plant Dis.*, 99(10), 1347–52.

Stasko, A. K., Wickramasinghe, D., Nauth, B. J., Acharya, B., Ellis, M. L., Taylor, C. G., McHale, L. and Dorrance, A. E. (2016), 'High density mapping of resistance QTL towards *Phytophthora sojae*,

Pythium irregulare, and *Fusarium graminearum* in the same soybean population', *Crop Sci.*, 56(5), 2476–92.

Stewart, S., Robertson, A. E., Wickramasinghe, D., Draper, M. A., Michel, A. and Dorrance, A. E. (2016), 'Population structure among and within Iowa, Missouri, Ohio, and South Dakota populations of *Phytophthora sojae*', *Plant Dis.*, 100(2), 367–79.

Su, G., Suh, S., Schneider, R. W. and Russin, J. S. (2001), 'Host specialization in the charcoal rot fungus, *Macrophomina phaseolina*', *Phytopathology*, 91(2), 120–6.

Sugimoto, T., Yoshida, S., Kaga. A., Hajika, M., Watanabe, K., Aino, M., Tatsuda, K., Yamamoto, R., Matoh, T., Walker, D. R., Biggs, A. R. and Ishimoto, M. (2011), 'Genetic analysis and identification of DNA markers linked to a novel *Phytophthora sojae* resistance gene in the Japanese soybean cultivar Waseshiroge'. *Euphytica*, 182(1), 133–45.

Sun, S., Wu, X. L., Zhao, J. M., Wang, Y. C., Tang, Q. H., Yu, D. Y., Gai, J. Y. and Xing, H. (2011), 'Characterization and mapping of RpsYu25, a novel resistance gene to *Phytophthora sojae*', *Plant Breed.*, 130(2), 139–43.

Sun, J., Li, L., Zhao, J. Huang, J. Yan, Q., Xing, H. and Guo, N. (2014), 'Genetic analysis and fine mapping of RpsJS, a novel resistance gene to *Phytophthora sojae* in soybean [*Glycine max* (L.) Merr]', *Theor. Appl. Genet.*, 127(4), 913–19.

Tang, Q. H., Gao, F., Li, G. Y., Wang, H., Zheng, X. B. and Wang, Y. C. (2010), 'First report of root rot caused by *Phytophthora sanomeana* on soybean in China', *Plant Dis.*, 94(3), 378.

Torto-Alalibo, T. A., Tripathy, S., Smith, B. M., Arredondo, F. D., Zhou, L., Li, H., Chibucos, M. C., Qutob, D., Gijzen, M., Chunhong, M., Sobral, B. W. S., Waugh, M. E., Mitchell, T. K., Dean, R. A. and Tyler, B. M. (2007), 'Expressed sequence tags from *Phytophthora sojae* reveal genes specific to development and infection', *MPMI*, 20(7), 781–93.

Tucker, D. M., Maroof, M. A. S., Mideros, S., Skoneczka, J. A., Nabati, D. A., Buss, G. R., Hoeschele, I., Tyler, B. M., St. Martin, S. K. and Dorrance, A. E. (2010), 'Mapping quantitative trait loci for partial resistance to *Phytophthora sojae* in a soybean interspecific cross', *Crop Sci.*, 50(2), 628–35.

Twizeyimana, M., Hill, C. B., Pawlowski, M., Paul, C. and Hartman, G. L. (2012), 'A cut-stem inoculation technique to evaluate soybean for resistance to *Macrophomina phaseolina*', *Plant Dis.*, 96(8), 1210–15.

Twizeyimana, M., Ojiambo, P. S., Haudenshield, J. S., Caetano-Anollés, G., Pedley, K. F., Bandyopadhyay, R. and Hartman, G. L. (2011), 'Genetic structure and diversity of *Phakopsora pachyrhizi* isolates from soyabean', *Plant Pathol.*, 60(4), 719–29.

Tyler, B. M. (2007), '*Phytophthora sojae*: root rot pathogen of soybean and model Oomycete', *Mol. Plant Pathol.*, 8(1), 1–8.

Tyler, B. M. and Gijzen, M. (2014), 'The *Phytophthora sojae* genome sequence: Foundation for a revolution'. In *Genomics of Plant-Associated Fungi and Oomycetes: Dicot Pathogens* (R. A. Dean, A. Lichens-Park and C. Kole (Eds)), pp 133–57. DOI:10.1007/978-3-662-44056-8_7.

Tyler, B. M., Tripathy, S., Zhang, X., Dehal, P., Jiang, R. H. Y., Aerts, A., Arredondo, F. D., Baxter, L., Bensasson, D., Beynon, J. L., Chapman, J., Damasceno, C. M. B., Dorrance, A. E., Dou, D., Dickerman, A., Dubchak, I. L., Garbelotto, M., Gijzen, M., Gorgon, S. G., Govers, F., Grunwald, N. J., Huang, W., Ivors, K. L., Jones, R. W., Kamoun, S., Krampis, K., Lamour, K. H., Lee, M., McDonald, W. H., Medina, M., Miejer, H. G. G., Nordber, E. K., Maclean, D. J., Ospina-Giraldo, M. D., Morris, P. F., Phuntumart, V., Putnam, N. H., Rash, S., Rose, J. K. C., Sakihama, Y., Salamov, A. A., Savidor, A., Schuering, C. F., Smith, B. M., Sobral, B. W. S. Terry, A., Torto-Alalibo, T. A., Win, J., Xu, Z., Zhang, H., Grigoriev, I. V., Rokhsar, D. S. and Boore, J. L. (2006), '*Phytophthora* genome sequences uncover evolutionary origins and mechanisms of pathogenesis', *Science* 313, 1261–6.

Urrea, K., Rupe, J. C. and Rothrock, C. S. (2013), 'Effect of fungicide seed treatments, cultivars, and soils on soybean stand establishment', *Plant Dis.*, 97(6), 807–12.

Vittal, R., Yang H. C. and Hartman, G. L. (2012), 'Anastomosis of germ tubes and nuclear migration of nuclei in germ tube networks of the soybean rust pathogen, *Phakopsora pachyrhizi*', *Eur. J. Plant Pathol.*, 132(2), 163–7.

Vleeshouwers, V. G. A. A. and Oliver, R. P. (2014), 'Effectors as tools in disease resistance breeding against biotrophic, hemibiotrophic, and necrotrophic plant pathogens', *Mol. Plant Microbe Interact.*, 27(3), 196–206.

Wang, H., Waller, L., Tripathy, S., St. Martin, S. K., Zhou, L., Krampis, K., Tucker, D. M., Mao, Y., Hoeschele, I., Saghai Maroof, M. A., Tyler, B. M. and Dorrance, A. E. (2010), 'Analysis of genes underlying soybean quantitative trait loci conferring partial resistance to *Phytophthora sojae*', *Plant Genome*, 3(1), 23–40.

Wang, H., Wijeratne, A., Wijeratne, S., Lee, S., Taylor, C., St. Martin, S. K., McHale, L. and Dorrance, A. E. (2012), 'Dissection of two soybean QTL conferring partial resistance to *Phytophthora sojae* through sequence and gene expression analysis', *BMC Genomics*, 13, 428.

Weng, C., Yu, K., Anderson, T. R. and Poysa, V. (2001), 'Mapping genes conferring resistance to Phytophthora root rot of soybean', *Rps1a* and *Rps7*. *J. Hered.*, 92(5), 442–6.

Wrather, J. A. and Koenning, S. R. (2006), 'Estimates of disease effects on soybean yield in the United States 2003 to 2005', *J. Nematol.*, 38(2), 173–80.

Wrather, J. A., Koenning, S. R. and Anderson, T. R. (2003), 'Effect of diseases on soybean yields in the United States and Ontario (1999 to 2002)', *Plant Health Prog.* doi:10.1094/PHP-2003-0325-01-RV.

Wrather, J. A., Stienstra, W. C. and Koenning, S. R. (2001), 'Soybean disease loss estimates for the United States from 1996 to 1998', *Can. J. Plant Pathol.*, 23(2), 122–31.

Wu, X., Zhang, B., Sun, S., Zhao, J., Yang, F., Guo, N., Gai, J. and Xing, H. (2011), 'Identification, genetic analysis and mapping of resistance to *Phytophthora sojae* of Pm28 in soybean', *Agric. Sci. China*, 10(10), 1506–11.

Xue, A. G., Cober, E., Voldeng, H. D., Babcook, C. and Clear, R. M. (2007), 'Evaluation of the pathogenicity of *Fusarium graminearum* and *Fusarium pseudograminearum* on soybean seedlings under controlled conditions', *Can. J. Plant Pathol.*, 29(1), 35–40.

Yang, X. B., Berggren, G. T. and Snow, J. P. (1990a), 'Seedling infection of soybean by isolates of *Rhizoctonia solani* AG-1, causal agent of aerial blight and web blight of soybean', *Plant Dis.*, 74(7), 485–8.

Yang, X. B. and Feng, F. (2001), 'Ranges and diversity of soybean fungal diseases in North America', *Phytopathology*, 91(8), 769–75.

Yang, X. B. and Navi, S. S. (2005), 'First report of charcoal rot epidemics caused by *Macrophomina phaseolina* in soybean in Iowa', *Plant Dis.*, 89(5), 526.

Yang, X. B., Snow, J. P. and Berggren, G. T. (1990b), 'Analysis of epidemics of Rhizoctonia aerial blight of soybean in Louisiana', *Phytopathology*, 80(4), 386–92.

Yao, H., Wang, X., Wu, X., Xiao, Y. and Zhu, Z. (2010), 'Molecular mapping of Phytophthora resistance gene in soybean cultivar Zaoshu18'. *J. Plant Genet. Resour.*, 11(2), 213–17.

Yu, A., Xu, P., Wang, J., Zhang, S., Wu, J., Li, W., Chen, W., Li, N., Fan, S., Wang, X. and Jiang, L. (2010), 'Genetic analysis and SSR mapping of gene resistance to *Phytophthora sojae* race 1 in soybean cv Suinong 10'. *Chinese J. Oil Crop Sci.*, 32 (4), 462–6.

Zelaya-Molina, L. X., Ellis, M. L., Berry, S. A. and Dorrance, A. E. (2010), 'First report of *Phytophthora sansomeana* causing wilting and stunting on corn in Ohio', *Plant Dis.*, 94(1), 125.

Zeng, F., Arnao, E., Zhang, G., Olaya, G., Wullschleger, J., Sierotzki, H., Ming, R., Bluhm, B. H., Bond, J. P., Fakhoury, A. M. and Bradley, C. A. (2015), 'Characterization of quinone outside inhibitor fungicide resistance to *Cercospora sojina* and development of diagnostic tools for its identification', *Plant Dis.*, 99(4), 544–50.

Zhang, G. and Bradley, C. A. (2014), 'Survival of *Cercospora sojina* on soybean leaf debris in Illinois', *Plant Health Prog.*, doi:10.1094/PHP-RS-14-0005.

Zhang, G. R., Newman, M. A. and Bradley, C. A. (2012a), 'First report of the soybean frogeye leaf spot fungus (*Cercospora sojina*) resistant to quinone outside inhibitor fungicides in North America', *Plant Dis.*, 96(5), 767.

Zhang, G., Pedersen, D. K., Phillips, D. V. and Bradley, C. A. (2012b), 'Sensitivity of *Cercospora sojina* isolates to quinone outside inhibitor fungicides', *Crop Prot.*, 40, 63–8.

Zhang, J., Xia, C., Duan, C., Sun, S., Wang, X., Wu, X. and Zhu, Z. (2013a), 'Identification and candidate gene analysis of a novel Phytophthora resistance gene *Rps10* in a Chinese soybean cultivar'. *PLoS ONE*, 8, e69799.

Zhang, J., Xia, C., Wang, X., Duan, C., Sun, S., Wu, X. and Zhu, Z. (2013b), 'Genetic characterization and fine mapping of the novel Phytophthora resistance gene in a Chinese soybean cultivar'. *Theor. Appl. Genet.*, 126(6), 1555–61.

Zhu, Z., Huo, Y. L., Wang, X. M., Huang, J. B. and Wu, X. F. (2007), 'Molecular identification of a novel Phytophthora resistance gene in soybean'. *Acta Agron. Sin*, 33(1), 154–7.

Zitnick-Anderson, K. K. and Nelson, B. D. (2015), 'Identification and pathogenicity of *Pythium* on soybean in North Dakota', *Plant Dis.*, 99(1), 31–8.

Zitnick-Anderson, K. K., Norland, J. E., del Río Mendoza, L. E., Fortuna, A. and Nelson, B. D. (2017), 'Probability models based on soil properties for predicting presence-absence of *Pythium* in soybean roots', *Microb. Ecol.*, doi:10.1007/s00248-017-0958-2.

Chapter 2

Bacterial and viral diseases affecting soybean production

Glen L. Hartman, USDA-ARS and University of Illinois, USA

1 Introduction

2 Bacterial diseases: blight, pustule and tan spot

3 Bacterial diseases: wilts and other bacterial diseases

4 Viral diseases: mosaic and mottle viruses

5 Other viral diseases

6 Summary

7 References

1 Introduction

Worldwide, soybean [*Glycine max* (L.) Merr.] is estimated to be grown on 6% of cultivated land (Hartman et al. 2011). Argentina, Brazil and the United States account for over 80% of the total soybean production (Hartman et al. 2015). In the United States, over 80% of production comes from the North Central Region with the states of Illinois, Iowa and Minnesota producing over 33% of the total production (Hartman et al. 2015).

Microbes such as nitrogen-fixing bacteria or mycorrhizal fungi that infect soybean may be beneficial, but some other microbes cause economic losses in yield (Hartman et al. 2015). Pathogens of soybean include a number species of bacteria, fungi, nematodes, oomycetes, parasitic plants and viruses. Losses due to diseases are estimated to be about 14% of the total soybean production based on estimates from Argentina, Brazil and the United States (Hartman et al. 2015). This chapter focuses mainly on important diseases caused by bacteria and viruses in the United States (Table 1). The general format is to introduce each disease in terms of its general importance, followed by a description of the disease symptoms, the causal agent, disease epidemiology and management. This is followed by a summary section.

http://dx.doi.org/10.19103/AS.2017.0034.16

Table 1 Important bacterial and viral diseases of soybean

Disease	Genus	Species	Resistance[1]
Bacterial blight	*Pseudomonas*	*savastanoi* pv. *glycinea*	Yes
Bacterial pustule	*Xanthomonas*	*axonopodis* pv. *glycines*	Yes
Bacterial tan spot	*Curtobacterium*	*flaccumfaciens* pv. *flaccumfaciens*	No
Bacterial wilt	*Curtobacterium*	*flaccumfaciens* pv. *flaccumfaciens*	No
Bacterial wilt	*Ralstonia*	*solanacearum*	No
Alfalfa mosaic	*Alfamovirus*	*Alfalfa mosaic virus*	Yes
Bean pod mottle	*Comovirus*	*Bean pod mottle virus*	Yes
Soybean mosaic	*Potyvirus*	*Soybean mosaic virus*	Yes
Soybean vein necrosis	*Tospovirus*	*Soybean vein necrosis virus*	No
Tobacco ringspot	*Nepovirus*	*Tobacco ringspot virus*	Yes
Tobacco streak	*Ilarvirus*	*Tobacco streak virus*	Yes

[1] Yes indicates that sources of resistance have been discovered and published. In some cases, the inherence of resistance is known, the genes for resistance are mapped and commercial cultivars are available with this resistance.

2 Bacterial diseases: blight, pustule and tan spot

Bacteria are prokaryotic microbes mostly void of chloroplasts, mitochondria and nuclei (Dworkin et al. 2006). Bacteria are ubiquitous, live in diverse environments and have different types of associations with plants. Epiphytic bacteria occur on the outer surfaces of plants, while endophytes occur inside plant tissues. Most bacteria are saprophytes causing no disease symptoms to plants, while others are symbiotic and help plants grow. However, there is a small group of bacteria that are pathogens, occurring either as plant epiphytes or as endophytes (Janse 2005).

Symptoms of soybean plants infected with bacterial pathogens include chlorosis, necrosis, rots, tumours and wilts. Unlike most pathogenic fungi and nematodes, bacterial pathogens lack active mechanisms for penetrating plant cells. They infect plants through abrasions and injuries caused by arthropod feeding and mechanical damage (rain, wind and cultural practices). In addition, natural plant openings such as flowers, hydathodes, lenticels and stomata allow bacterial cells into the plant (Melotto et al. 2008). Bacterial cells can be disseminated locally or over long distance by aerosols, arthropods, innate mechanical means (clippers, cultivators, etc.) and seeds. Bacterial cells associated with crop debris or seeds may survive for a number of years. Management of bacterial diseases often relies on several different practices that include avoidance, using pathogen-free seeds, resistant cultivars and in some cases antibiotics and antagonists.

2.1 Bacterial blight

Bacterial blight has been identified worldwide, and along with bacterial pustule is one of the most common bacterial diseases of soybeans (Zhao 2015). Bacterial blight symptoms often occur 7–10 days after a precipitation event. The disease is of particular concern in seed generation fields as the bacterium can survive in seed and spread geographically with seed distribution. Yield losses of up to 40% have been reported due to bacterial blight, primarily by reduced green leaf area and premature defoliation (Hwang and Lim 1992; Park and Lim 1986).

Symptoms can be observed on most above-ground plant organs. Infected seeds may have a slight discolouration and when planted, brown lesions on the margins of cotyledons enlarge and turn dark brown, collapsing the infected tissue (Zhao 2015). Young seedlings may be stunted and die if the shoot apex becomes infected. If the seedling infection did not start from infected seed, bacterial cells may enter the plant from neighbouring infected plants or from infested debris. Initial symptoms on young leaves appear as small, angular lesions so that leaves appear water-soaked with yellow halos (Fig. 1). The lesion centres dry out and the colour turns brown to black, and lesions often enlarge and coalesce. The blighted areas tear away, causing the leaves to look tattered. In severe cases, premature defoliation occurs. Lesions also occur on petioles, stems and pods, with pod infection often resulting in seed infection. Stored seeds may appear healthy or shrivelled with a slight discolouration.

The causal bacterium for bacterial blight is *Pseudomonas savastanoi* pv. *glycinea,* a motile, gram-negative rod (1.2–1.5 × 2.3–3.3 μm) with one to several polar flagella (Zhao 2015). Colonies appear white, raised, circular, smooth and glistening on common bacteriological media. Optimal growth occurs at 24–28°C. Strains of the bacterium may be differentiated on soybean genotypes to produce races. The first avirulence gene (*avrA1*) was cloned in 1984 (Staskawicz et al. 1984), and subsequently, several more avirulence genes were identified and characterized (Huynh et al. 1989). Whole-genome sequence analysis of *P. savastanoi* pv. *glycinea* indicated that the pathogen is related to *P. savastanoi* pv. *phaseolicola,* the halo blight pathogen of beans (Qi et al. 2011).

The bacterium survives in infested crop debris and in seeds (Zhao 2015). Disease occurrence is often more common early in the growing season and less common when warmer and drier conditions prevail. The pathogen, disseminated by water droplets during overhead irrigation or rainstorms, survives epiphytically on leaf surfaces and buds and enters the leaves through stomata and wounds. Management techniques include the use of pathogen-free seed, crop rotation with non-hosts to eliminate infested debris, timely overhead irrigation or switching to ground irrigation, the avoidance of mechanical injury when foliage is wet and the use of soybean cultivars with resistance. Four resistance genes (*Rpg1b, Rpg2, Rpg3* and *Rpg4*) may be in use, but their frequency of use in commercial soybean cultivars is not known.

2.2 Bacterial pustule

Bacterial pustule occurs worldwide and often causes yield losses when warm temperatures and frequent precipitation prevail (Hartman 2015b). The disease may cause premature defoliation, which may decrease yield by reducing seed size and number. Yield losses attributed to bacterial pustule can be as high as 20% (Hartman and Wang 1997; Hong et al. 2011; Laviolette et al. 1970).

Symptoms of bacterial pustule begin as minute, pale-green spots with elevated centres on leaf surfaces (Hartman 2015b). As the disease progresses, small, raised, and light in colour pustules form in the centre of lesions, mostly on the lower leaf surfaces (Fig. 1b). Lesions vary in size from small specks to large, irregularly shaped, mottled-brown areas, which develop when smaller lesions coalesce. A severe level of disease often results in some defoliation. Small, reddish-brown, slightly raised spots may develop on pods. Symptoms of bacterial

Figure 1 Important bacterial diseases of soybean: (a) Bacterial blight (caused by *Pseudomonas savastanoi* pv. *glycinea*) leaf symptoms with small angular yellowish lesions; (b) Bacterial pustule (caused by *Xanthomonas axonopodis* pv. *glycines*) leaf symptoms with small reddish-brown lesions; (c) Bacterial tan spot (caused by *Curtobacterium flaccumfaciens* pv. *flaccumfaciens*) leaf symptoms with large necrotic lesions; (d) Bacterial wilt (caused by *Ralstonia solanacearum*) leaf symptoms with interveinal chlorosis and necrosis; (e) Phytoplasma leaf symptoms showing bud proliferation. Courtesy of J. B. Sinclair (1a), R. M. Harveson (1c–d), and M. I. Chilvers (1e). Reproduced, with permission, from Hartman, G. L., et al., eds. 2015. Compendium of Soybean Diseases and Pests. 5th ed. American Phytopathological Society, St. Paul, MN.

pustule differ from those of bacterial blight in that lesions are not water-soaked in the early stages of development and often have distinct pustules in the centre of lesions.

Bacterial pustule is caused by *Xanthomonas axonopodis* pv. *glycines* (Hartman 2015b). The bacterium is a motile, gram-negative rod (0.5–0.9 × 1.4–2.3 μm) with one polar flagellum. Colonies are circular, small and smooth with entire margins that are pale yellow to deep yellow on common bacteriological media. The optimal temperature for growth is 30–33°C. Races of the pathogen exist, and strains from Thailand were grouped into three races based on their interactions with susceptible and resistant soybean cultivars (Anthinuwat at al. 2009). A high-quality genome sequence of a strain from Taiwan was published and a gene cluster that regulates pathogenicity was identified (Weng et al. 2017)

The bacterium survives in seeds, crop residues and the rhizosphere of some plants. Like most other bacteria, this bacterium enters the plant through natural openings and wounds and multiplies intracellularly (Groth and Braun 1989). It spreads via splashing water and windblown rain, and during cultivation when foliage is wet. New infections may occur throughout the growing season when wet or rainy conditions prevail. The disease can be severe in warmer regions and much of the losses are from Asia where races have been reported (Athinuwat at al. 2009). To manage bacterial pustule, maintaining pathogen-free seed fields that produce clean seed is critical. Rotation of soybeans with a non-host crop (e.g. maize) may reduce the survival of the bacterium in infested residue. The first resistant locus, *rxp*, was reported as a single recessive allele from PI548445 (Hartwig and Lehman 1951). Fine mapping narrowed the *rxp* locus to a 33-kb region between two single nucleotide polymorphism (SNP) markers on chromosome (Chr) 17 (Kim et al. 2010). A recent genome-wide association study identified three significant SNPs associated with resistance on Chr 1, 11 and 17 (Chang et al. 2016b). Resistant commercial cultivars include Williams 82 (Bernard and Lindahl 1972) and possibly other commercial soybean cultivars (Goradia et al. 2009). The frequency of using bacterial pustule resistance genes in commercial soybean cultivars is not known.

2.3 Bacterial tan spot

This disease is not as widespread as either bacterial blight or pustule, but may be important in some production regions and in other crops such as dry beans (Harveson 2015). Tan spot was first described in Iowa in 1975 (Dunleavy 1983). This pathogen has been detected on soybean in Brazil, Canada, Russia and the United States. Yield losses of up to 17% have been reported (Harveson 2015).

Tan spot lesions initially are small but may enlarge to cover the entire leaflet. Leaf chlorosis occurs in an oval or elongated pattern along the leaf margins. Leaf chlorosis progresses inward towards the midrib, causing affected tissues to turn brown and dry out (Fig. 1c).

The cause of tan spot is *Curtobacterium flaccumfaciens* pv. *flaccumfaciens*, a flexible coryneform-shaped (0.3–0.5 × 0.6–3.0 μm) aerobic bacterium with one to three lateral or polar flagella (Harveson 2015). Colonies are circular, flat or slightly convex, opaque, smooth, yellow to orange in colour and slightly viscid with entire margins. The pathogen infects many legumes including beans, cowpea, mungbean and peas. Cluster analysis using three molecular typing techniques of 67 isolates of *C. flaccumfaciens* pv. *flaccumfaciens* from multiple crops and sources showed that there were two distinct groups with soybean isolates found in both clusters (Agarkova et al. 2012).

The bacterium survives both internally and externally on seed coats (Harveson 2015). Seeds are the major source of inoculum and for long- and short-distance dispersal of the bacterium. Seed infection rates of up to 30% may occur. Planting infected seeds may cause a systemic infection on the unifoliolate and the first trifoliolate leaves at 25–30°C. Seedling-to-seedling spread is slow, because it is primarily caused by leaf-to-leaf contact during storms. The pathogen is soil-borne, can overwinter on infected plant residues and seeds from previous crops, but its survival is somewhat limited. Management of tan spot is through pathogen-free seed and non-host rotation. Seeds grown in dry climates are usually pathogen-free. Planting early when the soil temperature is cool (below 25°C) may reduce the systemic infection of leaves. There are no characterized genes for resistance to tan spot.

3 Bacterial diseases: wilts and other bacterial diseases

3.1 Bacterial wilts

Two pathogens, *C. flaccumfaciens* pv. *flaccumfaciens* and *Ralstonia solanacearum*, cause bacterial wilt of soybean plants (Harveson 2015). Bacterial wilt is not a common disease of soybean, but the pathogens that incite wilt are reported to affect other crops in most production areas of the world (Harveson 2015; Hayward and Hartman 1994). Since distribution of the disease on soybean is limited, yield loss estimates are not known. However, it is known that single plants infected with wilt can result in plant death.

Symptoms caused by infection of *C. flaccumfaciens* pv. *flaccumfaciens* infected include leaf yellowing, progressive wilting, formation of necrotic leaf lesions with yellow borders, plant stunting and eventual death of the plant (Harveson 2015). Symptoms of *R. solanacearum* infected include foliar chlorosis and dark brown necrotic areas that elongate and form dark borders (Fig. 1d). The necrotic tissues dry and drop from the plant, giving the affected leaves a tattered appearance. In severe cases, plants continue to wilt and die. Infected plants that survive are stunted and relatively non-productive.

Ralstonia solanacearum is a motile, gram-negative rod (0.5 × 1.5 μm) with a polar flagellum (Harveson 2015). Colonies on agar are small, opalescent (becoming brown), irregularly shaped, smooth and mucoid. The bacterium produces different pigments, including pink, red, blue and brown on differential medium. The bacterium affects many crop plants, and has one of the largest host ranges of any bacterial plant pathogen with several races and/or biovars known (Hayward and Hartman 1994).

Wilt caused by *C. flaccumfaciens* pv. *flaccumfaciens* is favoured by temperatures greater than 30°C; optimal pathogen growth in culture occurs at 27–28°C. Plant infection occurs when the pathogen enters the vascular system from the infected seed, the roots of developing seedlings and the wounds on leaves or stems. The *R. solanacearum* bacterium is occasionally seed-borne but survives in soil as a saprophyte. The pathogen becomes more virulent when there is abundant rainfall and added stress on the plant. Management options are very general and include using bacterial-free seed and rotation with non-hosts such as corn and wheat. No commercial cultivars with resistance are known for either of these bacterial wilt pathogens.

3.2 Other bacterial diseases

There are a number of other bacterial diseases that may be commonly found. One of these is wildfire (also known as 'blackfire') caused by *Pseudomonas syringae* pv. *tabaci*, which occurs on cowpea, eggplant, soybean, tobacco and tomato (Hartman 2015c). Symptoms of wildfire include brown necrotic spots on leaves, often surrounded by broad, yellow halos. Restricted spots may occur without sharply delineated chlorotic zones. These spots are dark brown to black, in contrast to the usual light brown, more expansive type of lesion. In damp weather, the lesions enlarge and coalesce, forming large dead areas that become dry and tear away, causing the leaves to have a tattered appearance. In severe cases of wildfire, plants may prematurely defoliate. Pustule lesions caused by *X. axonopodis* pv. *glycines* are natural infection sights for the wildfire bacterium. Not much is known about resistance to the bacterium, but one report showed a differential response to wildfire among some soybean cultivars under delayed planting conditions in Korea (Hong et al. 2012).

Another fairly common bacterial disease is Bacillus seed decay (Hartman 2015a). It causes losses under hot (25–35°C), moist conditions in storage, in the field and in experimental situations. Field losses of nearly 100% were reported in Nigeria when soybean seeds were planted in wet soil at temperatures greater than 30°C. The disease frequently appears on seeds being tested under warm germination conditions (e.g. accelerated ageing tests) as well as under typical germination test conditions. Bacillus seed decay is characterized by a soft decay of seeds with a slimy appearance when seeds are moistened and incubated for a few days.

The Mollicutes, which encompass the Spiroplasmas and Phytoplasmas, cause a number of well-described and not-so-well-described diseases (Mishra 2014). The distinguishing feature is that they lack cell walls. Many of them have very small genomes, and complete genome sequences reveal extensive horizontal gene transfer and genomic plasticity within some species. Their taxonomy is in a state of flux. Mollicutes vary in form, although most have sterols that make the cell membrane somewhat rigid. These pathogens infect plants by arthropod vectors. In some cases, the Mollicutes can be pathogens of arthropods. The Mollicutes in the arthropod vector can either be propagative or non-propagative and move either circulative or non-circulative (Ciancio 2016).

Phytoplasmas commonly occur in the phloem of plants and are transmitted and spread primarily by leafhoppers, plant hoppers and psyllids that feed on phloem (Dermastia et al. 2017). Symptoms caused by these microbes include yellowing, shortening of internodes, phyllody, proliferation of axillary shoots (known as witches' broom), sterility, virescence and reduction in root growth. Phytoplasmas are suspected of causing several soybean diseases, and symptoms include bud proliferation, phyllody and witches' broom. Many of these diseases are seed and leafhopper-transmitted. Phytoplasmas are classified into groups and subgroups based on RFLP analysis of a 1.25-kb PCR-amplified 16S rRNA gene segment using a defined set of 17 restriction enzymes that are consistent with 16S rRNA gene phylogeny (Zhao and Davis 2016). In soybean, the clover proliferation group (16SrVI) and the aster yellows group (16SrI; subgroup B) have been detected in the United States (Chilvers 2015). Symptoms include stunted growth, purple streaks and patches on stems, yellow blotches or marginal necrosis on leaves, stems that remain green even after the leaves and pods have matured, bud proliferation, reduction in the number of pods per node, dull seed coat and relatively dark foliage (Fig. 1e). Specific symptoms have not been linked to a specific phytoplasma. No specific management techniques are known for controlling soybean phytoplasmas, although general practices related to containment and prevention, roguing infected plants and vector control may be useful under certain conditions (Dermastia et al. 2017).

4 Viral diseases: mosaic and mottle viruses

Viruses presumably occur in all living entities including plants. There are more than 1000 viruses described and in soybean, 46 viral diseases have been reported (Tolin 2015b). Viruses contain either RNA or DNA, often surrounded by a protective protein or lipoprotein coat, and live as obligate parasites. Virus macromolecules are isometric (roughly spherical, 20–70 nm in diameter) or elongated rods (rigid or flexuous, up to 2 µm long, 10–20 nm in diameter). Viruses use host machinery to replicate, and are mostly systemic in their hosts. They are transmitted primarily by arthropods that acquire them at feeding and vector them to other hosts. In some cases, viruses are vectored by fungi and nematodes. Viruses are also transmitted by mechanical inoculation, pollination and activities related to cultural practices involved in growing plants. They are often seed-borne and can be seed-transmitted.

Symptoms of viral diseases in plants can range from inconspicuous to severe, including stunting and death. Leaf mottling and mosaics, rigidity, necrosis and misshapen and discoloured seeds are common visual symptoms of virus infection. However, symptoms alone are not enough to complete a diagnosis. To confirm a diagnosis, immunological assays, nucleic acid detection and whole-genome sequencing are used.

The following sections focus on just a few of the viruses reported to infect soybean. Along with this review, other sources of information on additional viruses can be found in an article in the soybean compendium (Hartman et al. 2015) and on soybean virus disease control (Hill and Whitman 2014). The International Committee on Taxonomy of Viruses (ICTV 2014) is the authority for naming viruses.

4.1 *Alfalfa mosaic virus*

The distribution of *Alfalfa mosaic virus* (AMV) is worldwide (Tolin 2015a). The virus has a wide host range infecting plants from 22 dicotyledonous families. AMV infection of soybean was first reported in 1960 and occurs in most soybean-producing states in the United States. In Nebraska, AMV was detected in more than 50% of soybean fields surveyed in 2001 and 2002 (Giesler and Ziems 2006). In Illinois, AMV was detected in 63% of counties intensively sampled over a 3-year period (Hobbs et al. 2010). In Ohio, the virus was not found in 2011 and was only found in 1 out of 22 counties sampled in 2012 (Han et al. 2016). An incidence of 30% or greater of plants showing AMV symptoms was shown to cause yield losses (Mueller and Grau 2007).

Symptoms of AMV include distorted leaves with a bright yellow mosaic pattern. Plants are often stunted, have reduced pod set and produce mottled seeds (Fig. 2).

AMV particles are 18 nm in diameter and 30, 34, 43 and 56 nm in length with a spherical to bacilliform shape (Tolin 2015a). AMV has a tripartite, single-stranded RNA genome along with a fourth subgenomic RNA that encodes the coat protein. They are RNAs encapsidated separately accounting for the four sizes of virus particles.

More than 15 species of aphids transmit the virus, including the soybean aphid (Hill et al. 2001), in a non-persistent manner. AMV is transmitted at a low frequency through soybean seeds. Resistance to AMV was identified and mapped from plant introduction (PI) 153282 (Kopisch-Obuch 2008). Strains of the virus are not known to overcome resistance, although there are no known commercial soybean cultivars available with resistance to AMV.

Figure 2 Important viral diseases of soybean: (a) *Alfalfa mosaic virus*-infected leaf showing yellowing and rugosity. (b) *Bean pod mottle virus*-infected leaf displaying mottling. (c) *Soybean mosaic virus*-infected plant with downward cup-shaped leaves and mosaic. (d) Soybean vein necrosis virus-infected leaf with necrotic lesions associated with major veins. (e) *Tobacco ringspot virus*-infected plants showing bud blight symptoms. (f) *Tobacco streak virus*-infected plant displaying chlorosis and necrosis. Courtesy of C. R. Grau (2a), T. K. Herman (2d), and L. L. Domier (2f). Reproduced, with permission, from Hartman, G. L., et al., eds. 2015. Compendium of Soybean Diseases and Pests. 5th ed. American Phytopathological Society, St. Paul, MN.

4.2 *Bean pod mottle virus*

First noted on soybean in 1958 in the United States, the virus occurs in most production areas in the United States (Domier 2015). Its emergence was considered a threat to soybean production in the United States (Giesler et al. 2002). Surveys indicate the prevalence of *Bean pod mottle virus* (BPMV) in Illinois with 83% and 33% of counties reported as positive for the virus based on two different surveys (Hobbs et al. 2010; Mabry et al. 2003). In Nebraska, 54 and 91% incidences of BPMV were reported in field samples in 2001 and 2002, respectively (Giesler and Ziems 2006). In Ohio, 66% incidence of BPMV was reported in counties sampled in 2011 and 2012 (Han et al. 2016). Yield losses up to 52% were reported when plants were infected early in the season (Hopkins and Mueller 1984), and up to 27% seed losses occurred without considering reductions in seed quality contents and increases in seed mottling (Byamukama et al. 2015).

Symptoms of virus infection are most noticeable on younger leaves as they show green to yellow mottling (Fig. 2b), often become puckered and distorted or develop more advanced chlorotic mottling. Symptoms are less conspicuous during periods of higher temperatures. Symptoms may be enhanced when plants are co-infected with other viruses. Infected plants may show decreased pod formation and reduced seed size, weight and number, and may be predisposed to infection by fungal seed-borne pathogens.

The virus consists of isometric particles approximately 28 nm in diameter and a bipartite, single-stranded RNA genome, and is strongly immunogenic (Domier 2015). BPMV isolates are divided into two subgroups based on nucleotide sequence variation and nucleic acid hybridization assays. Reassortants with the RNAs of the two subgroups typically produce more severe symptoms than one virus from either subgroup (Gu et al. 2002). No strains that overcome resistance have been identified.

The virus is transmitted at low levels through sap, beetles and seeds (Domier 2015). BPMV infects many legume species. Leaf-feeding beetles transmit BPMV. The bean leaf beetle, which efficiently transmits the virus without a latent period for several days, is the primary vector. The source of BPMV includes virus-infected seeds, perennial weeds (panicled tick trefoil serves as a reservoir) and beetles that overwinter in borders of fields and along edges of woods. In a survey of bean leaf beetles in Illinois, beetles tested positive for BPMV in samples from 37 of 41 and 86 of 99 counties assayed in 2000 and 2002, respectively; in addition, western corn rootworm adults tested positive for BPMV in 21 of 21 counties in 2000 and in 20 of 24 counties in 2001 (Mabry et al. 2003). Field-collected western corn rootworm adults transmitted BPMV to potted soybean plants at low rates either directly from BPMV-infected soybean fields or with prior feeding on BPMV-infected plants (Mabry et al. 2003). To manage BPMV, options to control the vectors include delayed planting to reduce the chance of early feeding by overwintering beetles, the application of herbicides to destroy alternative host plants and the use of insecticides either sprayed on foliage or used as a seed treatment to reduce prevalence of beetle vectors (Bradshaw et al. 2008). Commercial soybean cultivars that are BPMV-resistant are not available, but there is tolerance in soybean and resistance from other *Glycine* species (Hill et al. 2007; Zheng et al. 2005; Ziems et al. 2007). Three SNPs on Chr 2, 7 and 9 with a total of four associated LRR-RLK genes may provide insight into location of genes for resistance or tolerance to BPMV (Chang et al. 2016a).

4.3 *Soybean mosaic virus*

First reported in 1921 (Gardner and Kendrick 1921), *Soybean mosaic virus* (SMV) is distributed worldwide (Hill 2015). Early season infection reduces seedling nodulation, pod set, seed size and weight, and increases seed coat mottling and causes a potential increase in *Phomopsis* seed decay (Koning et al. 2001). Soybean surveys for SMV found it infrequently compared to other viruses. In Nebraska, SMV was detected in 0% and 31% of the fields in 2001 and 2002, respectively (Giesler et al. 2006). In Illinois, SMV was detected in 5% of the counties sampled from 2006 to 2008 (Hobbs et al. 2010). In Ohio, no SMV was detected in 24 counties sampled in 2011, and the virus was found in just one county in 2012 (Han et al. 2016). However, when virus infection occurs, yield losses of up to 35% have been documented (Hill et al. 1987; Ross 1977). Yield reductions of plants infected with both SMV and BPMV were found to be greater than reductions caused singly by SMV or BPMV (Calvert and Ghabrial 1983).

Symptoms of SMV infection vary from mild leaf mosaic to severe leaf mosaic, necrosis and abscission (Hill 2015; Fig. 2c). Severely infected plants are stunted and may die early. Symptoms are often specific to the SMV strain and the soybean genotype. Symptoms may be less severe at or over 25°C. Seeds from infected plants often produce seed mottling and hilum bleeding often specific with the interaction of SMV strains and the soybean genotypes (Domier et al. 2011). Additional symptoms or increased severity of symptoms may occur when plants are dually infected with other pathogens.

SMV is a flexuous rod with dimensions of 14 nm diameter × 750 nm length (Hill 2015). The virions are helical and consist of positive-sense, single-stranded RNA. SMV moves bidirectionally in plants and can be detected in all plant tissues including roots and seeds. SMV-infected vascular tissue contains pinwheel-like intracellular inclusion bodies in cross section and bundle-like bodies in longitudinal section. SMV has a wide host range infecting over 42 legume species. Numerous strains of SMV have been identified based on reactions using a set of soybean differentials (Cho and Goodman 1979). There have been a number of additional studies that have reported more strains, some especially unique. One report described a recombinant strain that exhibited different pathogenicity on soybeans and common beans as a result of a recombination event between SMV and *Bean common mosaic virus* (Yang et al. 2014). This strain, SMR-R, occurred in 60% of samples from Chinese soybean fields, and was found to be present in mixed infections with other SMV strains.

The virus is sap and graft transmissible with over 32 aphid species being able to transmit SMV in a non-persistent manner (Hill 2015). In addition to dispersal via aphids, SMV is transmissible through infected seeds (Bowers and Goodman 1979). The level of seed transmission within a given seed lot is often less than 1%, and is dependent on the cultivar and time of infection. In the field, SMV starts from a point source, with secondary spread occurring by aphids. Management of SMV is through several means, including insecticides that control potential aphid vectors, and through the use of clean seed and resistant soybean cultivars. One study indicated that resistant soybean cultivars were better at controlling SMV itself than trying to control vectors transmitting SMV (Pedersen et al. 2007). There are three or more independent loci (*Rsv1, Rsv3* and *Rsv4*) identified and mapped (Hill 2015). Pyramiding sources of resistance are recommended to combat known SMV strains (Saghai et al. 2008). Significant SNPs were found on Chr 13 (*Rsv1*) and Chr 14 (*Rsv3*) with both LRR-RLK and NBS-LRR genes being associated to the SNP markers (Chang et al. 2016b). Almost half (23 of 52 ancestral genotypes tested) of the ancestors of North American lines were resistant to two SMV strains (Wang et al. 2006). However, though available, SMV resistance has not been incorporated into most commercially marketed cultivars available in Illinois (Chawla et al. 2013). In another study, 303 cultivars grown in the mid-south region of the United States were challenged with five SMV strains (Shakiba et al. 2012). The majority did not have any resistance to SMV; most of those that did (74 cultivars) appeared to contain one of the *Rsv1* alleles, with only a few cultivars containing *Rsv3* and *Rsv4* (Shakiba et al. 2012). Thus, although there appears to be some SVM resistance deployed, its use is likely to increase with the use of SNPs that target specific loci such as those used to detect multiple genes in the *Rsv4* locus (Klepadlo et al. 2017).

5 Other viral diseases

5.1 *Soybean vein necrosis virus (SVNV)*

The virus, first reported in Tennessee, USA, in 2008, is now known to be present in all major soybean-producing areas of North America (Zhou et al. 2015), although economic yield losses in commercial soybean fields is not yet known. Based on initial reports, the virus appears to be widespread in the midwestern soybean production area of the United States. A survey from Ohio found SVNV in leaf samples from 5 of 24 counties in 2011 and in 17 of 22 counties in 2012 (Han et al. 2016).

Initial symptoms include thread-shaped clearing along the main veins and some necrotic spots surrounded by chlorosis (Zhou et al. 2015). Severe infection produces purple or dark brown lesions that expand to fill most of the leaf blade (Fig. 2d). Disease symptoms are more evident in the canopy, presumably because the virus vector, the soybean thrips, prefers feeding on newly emerged leaves.

This virus is a tripartite, ambisense RNA genome representing a new lineage in the genus distinct from both the New and the Old World tospoviruses (Zhou et al. 2011). SVNV forms a distinct new clade of tospoviruses. To date, there are no strain variants causing different reactions on soybean genotypes.

SVNV is transmitted by soybean thrips (Zhou and Tzanetakis 2013). Ivyleaf morning glory is a natural, symptomless host that may act as a virus reservoir. SVNV infection may occur at any time during the growing season, but the symptom expression is most visible after flowering. Conditions that favour SVNV are those that promote populations of thrips to survive. There are still some questions about whether SVNV remains localized in leaves or occurs systemically. A recent report indicated that the virus was seed-transmitted (Groves et al. 2016), although another report provided no evidence for seed transmission based on over 2000 seeds assayed (Hajimorad et el. 2015). Management techniques such as controlling thrips, eliminating potential hosts and evaluating soybean genotypes for resistance are currently being validated.

5.2 *Tobacco ringspot virus*

This virus was first described in the United States on soybean in 1946 (Allington 1946) and has since been reported in Australia, Canada, the People's Republic of China and Russia (Hartman and Domier 2015). This virus is less commonly identified than other viruses in surveys. In Illinois, TRSV was found in only one county out of 107 sampled over a period from 2006 to 2008 (Hobbs et al. 2010). In Ohio, the virus was detected in 5 of 24 fields and 6 of 22 fields in 2011 and 2012, respectively (Han et al. 20016). TRSV infection can reduce soybean yields up to 100% as severely infected plants have poor pod set and lack seed (Athow and Laviolette 1961; Crittenden et al. 1966). In general, losses are greatest when young plants are infected or especially when seeds with a high percentage of TRSV are sown.

Symptoms of TRSV include dwarfed rolled leaves with rugose and bronzed blades, ringspots, curving of the terminal bud to form a crook (Fig. 2e) and stunted plants, especially when infected early (Hartman and Domier 2015). Buds may become brown, necrotic and brittle, and are easily broken off of the plant resulting in adventitious leaf and floral buds. Brown streaks are occasionally observed on petioles and large leaf veins. The pith of stems and branches may be brown, first near the nodes and then throughout the stem. Pods generally are underdeveloped or aborted. Those that set before infection often develop dark blotches. Plant maturity may be delayed and plants may remain green and often stunted until harvested or killed by frost.

TRSV consists of a bipartite genome with two single-stranded positive-sense polyadenylated RNA molecules of 7.5 kb (RNA1) and 3.9 kb (RNA2) (Hartman and Domier 2015). These are encapsidated separately in 28-nm icosahedral particles. TRSV is moderately immunogenic, and serological tests are used for diagnosis and to determine virus concentration in infected plants. Strain of the virus exists, although there is no formal race designation (Tu et al. 1986).

The virus causes a systemic infection and moves from infected leaves to stem tips and into roots (Bergeson et al. 1964). Nymphs of *Thrips tabaci* transmit TRSV to soybean at a low level of efficiency (Messieha 1969). The nymphs may retain the virus for at least 14 days after acquisition. The dagger nematode, *Xiphinema americanum,* also transmits TRSV at a low level of efficiency. Seed transmission provides the means for long-range dissemination and carry-over from season to season. Virus-free seed along with the control of thrips, as well as soybean resistance, are all factors in the management of TRSV. The soybean cultivar Young is resistant to TRSV infection (Fasoula et al. 2003) and resistance occurs in soybean (Demski et al. 1999) and *Glycine soja* (Orellana 1981). More recently, a single locus responsible for TRSV sensitivity was reported on Chr 2 with two LLR-RLK genes being located near the locus (Chang et al. 2016a).

5.3 *Tobacco streak virus*

This virus has been found to infect soybean plants in Brazil, India, Iran and the United States (Illinois, Iowa, Kentucky, Ohio Oklahoma and Wisconsin), causing widely varying symptoms (Tolin and Domier 2015). In Wisconsin, the incidence of plants sampled with *Tobacco streak virus* (TSV) was as high as 95% in 2001 (Rabedeaux et al. 2005). In Illinois, TSV was detected in only one county out of 107 counties sampled from 2006 to 2008 (Hobbs et al. 2010). In Ohio, TSV was detected in 3 of 24 and 2 of 22 counties sampled in 2011 and 2012, respectively (Han et al. 2016). In Iowa, TSV was reported in a number of counties in 2013 and was suggested to be an emerging disease in Iowa and other locations (Irizarry et al. 2016). One report from Wisconsin indicated that TSV reduced yields up to 25% and indicated that this was due to plant mortality rather than pod set or seed size reductions (Rabedeaux et al. 2005).

Symptoms of TSV include bud blight (Fig. 2f) and ringspots on leaves (Tolin and Domier 2015). Unlike BPMV and SMV, plants infected with TSV do not result in mottled seeds.

TSV consists of isometric particles, 25–30 nm in diameter, with four segments of positive-sense, single-stranded RNA (Tolin and Domier). TSV has a wide host range with some strain specificity based on hosts (Sharman et al. 2015). Strains that overcome sources of resistance in soybean have not been identified.

Once plants are infected, TSV moves throughout the plant. TSV transmission may be facilitated by multiple thrips species. TSV is also seed-transmitted at a rate as high as 90%. Management measures should emphasize avoiding the introduction of TSV-infected seeds into soybean fields. Resistance to TSV has been reported in at least one soybean cultivar 'Tanner', but it has not been incorporated into adapted cultivars (Wang et al. 2005). Temperature-sensitive resistance in soybean genotypes was also identified (Hobbs et al. 2012).

6 Summary

Soybeans are infected by a number of bacterial and viral pathogens, which can cause some of the most common and important diseases of the crop. Many of these diseases have been researched for decades but others, such as SVNV, are considered as emerging diseases. For emerging diseases, more research is often needed to understand disease impact on soybean yield, associated vector relationships, sources of host resistance and suitable management schemes. The importance of bacterial and viral pathogens of soybean often change because of discoveries of new pathogenic strains, new pathogens or new vectors. This results in new challenges for disease management.

The interaction of soybean and its pathogens and their vectors is dynamic, and factors such as climate and human activity often lead to geographical differences in terms of the importance of any one bacterial and viral disease of soybean. In addition, the expansion of the soybean crop itself to new areas of production in Brazil, Canada and some countries in Africa may lead to an increase in the occurrence of bacterial and viral diseases as new strains of the pathogens or vectors move from other hosts to soybean.

In addition to the changing geographic range of soybean production, about one-third of commercial soybean cultivars on the market each year in the United States are newly released cultivars. Many of these have had a short, often geographically focused, testing cycle and may not contain the necessary disease resistance genes or selection for tolerance. This could lead to increased susceptibility to bacterial and viral pathogens. In addition to the rapid overturn of soybean cultivars, there are seasonal changes within and among years that affect bacterial and viral pathogens and their vectors resulting in a dynamic system of change. Thus, disease management practices need to be dynamic as well. Specific disease management techniques vary some by disease and region; however, good disease management practices start with healthy seeds free of bacterial and viral pathogens. Other management techniques include planting soybeans in rotation with crops that are not hosts of soybean bacterial and viral pathogens, controlling vectors that disperse soybean bacterial and viral pathogens and using disease resistance. The latter is dependent on many factors including the ability to find useable sources of disease resistance that either will be or has been incorporated into commercial cultivars. The incorporated resistance needs to be effective against current strains of bacterial and viral pathogens and also durable through time to produce sustainable results.

Future advances in agricultural technology will include increased or enhanced pathogen resistance in soybean by using small RNAs to regulate gene expression. This has been shown to provide immunity against viruses and was used to control AMV, BPMV and SMV to simultaneous infection of the three viruses (Zhang et al. 2011). In addition to genetically modifying soybean, development of novel biological products, including microorganisms and their metabolites that will control pathogens and their vectors, and enhance plant growth or stimulate innate resistance, will further increase disease management strategies. A combined effort of disease management techniques, through both traditional and biotechnological approaches, will provide the tools needed to reduce the importance of diseases of soybean caused by bacterial and viral pathogens.

7 References

Agarkova, I. V., Lambrecht, P. A., Vidaver, A. K. and Harveson, R. M. 2012. Genetic diversity among *Curtobacterium flaccumfaciens* pv. *flaccumfaciens* populations in the American high plains. *Can. J. Microbiol.* 58:788–801.

Allington, W. B. 1946. Bud blight of soybean caused by the tobacco ring-spot virus. *Phytopathology* 36:319–22.

Athinuwat, D., Prathuangwong, S., Cursino, L. and Burr, T. 2009. *Xanthomonas axonopodis* pv. *glycines* soybean cultivar virulence specificity is determined by *avrBs3* homolog *avrXg1*. *Phytopathology* 99:996–1004.

Athow, K. L. and Laviolette, F. A. 1961. The relation of seed-transmitted tobacco ringspot virus to soybean yield. *Phytopathology* 51:341–2.

Bergeson, G. B., Athow, K. L., Laviolette, F. A. and Thomasine, S. M. 1964. Transmission, movement, and vector relationships of tobacco ringspot virus in soybean. *Phytopathology* 54:723–8.
Bernard, R. L. and Lindahl, D. A. 1972. Registration of Williams soybean. *Crop Sci.* 12:716.
Bowers Jr., G. R. and Goodman, R. M. 1979. Soybean mosaic virus: Infection of soybean seed parts and seed transmission. *Phytopathology* 69:569–72.
Bradshaw, J. D., Rice, M. E. and Hill, J. H. 2008. Evaluation of management strategies for bean leaf beetles (Coleoptera: Chrysomelidae) and *Bean pod mottle virus* (Comoviridae) in soybean. *J. Econ. Entomol.* 101:1211–27.
Byamukama, E., Robertson, A. E. and Nutter, Jr., F. W. 2015. Bean pod mottle virus time of infection influences soybean yield, yield components, and quality. *Plant Dis.* 99:1026–32.
Calvert, L. A. and Ghabrial, S. A. 1983. Enhancement by soybean mosaic virus of bean pod mottle titer in doubly infected soybean. *Phytopathology* 73:992–7.
Chang, H.-X., Brown, P., Lipka, A., Domier, L. L. and Hartman, G. L. 2016a. Genome-wide association and genomic prediction identifies associated loci and predicts the sensitivity of *Tobacco ringspot virus* in soybean plant introductions. *BMC Genomics* 17:153. DOI:10.1186/s12864-12016-12487-12867.
Chang, H.-X., Lipka, A., Domier, L. L. and Hartman, G. L. 2016b. Characterization of disease resistance loci in the USDA soybean germplasm collection using genome-wide associations. *Phytopathology* 106. dx.doi.org/10.1094/PHYTO-01-16-0042-FI.
Chawla, S., Bowen, C. R., Slaminko, T. L., Hobbs, H. A. and Hartman, G. L. 2013. A public program to evaluate commerical soybean cultivars for pathogen and pest resistance. *Plant Dis.* 97:568–97.
Chilvers, M. I. 2015. Clover proliferation and aster yellows. In: G. L. Hartman, J. C. Rupe, E. F. Sikora, L. L. Domier, J. A. Davis and K. L. Steffey (Eds), *Compendium of Soybean Diseases and Pests*. American Phytopathological Society, St. Paul, pp. 27–9.
Cho, E.-K. and Goodman, R. M. 1979. Strains of soybean mosaic virus: Classification based on virulence in resistant soybean cultivars. *Phytopathology* 69:467–70.
Ciancio, A. 2016. Traveling bacteria: Vectors. In: A. Ciancio (Ed.), *Invertebrate Bacteriology: Function, Evolution and Biological Ties*. Springer Science+Business Media B.V., Dordrecht, pp. 145–83.
Crittenden, H. W., Hastings, K. M. and Moore, G. M. 1966. Soybean losses caused by tobacco ringspot virus. *Plant Dis. Rep.* 50:910–13.
Dermastia, M., Beraccini, A., Constable, F. and Mehle, N. 2017. Phytoplasmas – dangerous and intriguing bacteria. In: M. Dermastia et al. (Eds), *Grapevine Yellows Diseases and Their Phytoplasma Agents*. SpringerBriefs in Agriculture. Springer Nature, Cham, Switzerland. DOI:10.1007/978-3-319-50648-7_1.
Domier, L. L. 2015. Bean pod mottle virus. In: G. L. Hartman, J. C. Rupe, E. F. Sikora, L. L. Domier, J. A. Davis and K. L. Steffey (Eds), *Compendium of Soybean Diseases and Pests*. American Phytopathological Society, St. Paul, pp. 117–19.
Domier, L. L., Hobbs, H. A., McCoppin, N. K., Bowen, C. R., Steinlage, T. A., Chang, S., Wang, Y. and Hartman, G. L. 2011. Multiple loci condition seed transmission of *Soybean mosaic virus* (SMV) and SMV-induced seed coat mottling in soybean. *Phytopathology* 101:750–6.
Dunleavy, J. M. 1983. Bacterial tan spot, a new foliar disease of soybeans. *Crop Sci.* 23:473–6.
Dworkin, M., Falkow, S., Rosenberg, E., Schleifer, K. H. and Stackebrandt, E. (Eds). 2006. *The Prokaryotes: A Handbook on the Biology of Bacteria*. 3rd ed. Springer Press, New York.
Fasoula, V. A., Harris, D. K., Bailey, M. A., Phillips, D. V. and Boerma, H. R. 2003. Identification, mapping, and confirmation of a soybean gene for bud blight resistance. *Crop Sci.* 43:1754–9.
Gardner, M. and Kendrick, J. B. 1921. Soybean mosaic. *J. Agric. Res.* 22:111–14.
Giesler, L. J., Ghabrial, S. A., Hunt, T. E. and Hill, J. H. 2002. *Bean pod mottle virus:* A threat to U.S. soybean production. *Plant Dis.* 86:1280–9.
Giesler, L. J. and Ziems, A. D. 2006. Incidence of *Alfalfa mosaic virus, Bean pod mottle virus,* and *Soybean mosaic virus* in Nebraska soybean fields. *Plant Health Progress* doi:10.1094/PHP-2006-0424-01-HM.

Goradia, L., Hartman, G. L. and Daniel, S. L. 2009. Evaluation of glyphosate-resistant soybean cultivars for resistance to bacterial pustule. *Eur. J. Plant Path.* 124:331–5.

Groth, D. E. and Braun, E. J. 1989. Survival, seed transmission, and epiphytic development of *Xanthomonas campestris* pv. *glycines* in the north-central United States. *Plant Dis.* 73:326–30.

Groves, C., German, T., Dasgupta, R., Mueller, D. and Smith, D. L. 2016. Seed tranmission of soybean vein necrosis virus: The first tospovirus implicated in seed transmission. *PLoS ONE* 11(1): e0147342.

Gu, H.-C., Clark, A. J., de Sá, P. B., Pfeiffer, T. W., Tolin, S. and Ghabrial, S. A. 2002. Diversity among isolates of *Bean pod mottle virus*. *Phytopathology* 92:446–52.

Hajimorad M. R., Halter, M. C., Wang, Y., Staton, M. E. and Hershman, D. E. 2015. Evaluation of seed transmissibility of Soybean vein necrosis-associated virus in two soybean cultivars grown under field conditions. *J. Plant Pathol. Microb.* 6:278. DOI:10.4172/2157-7471.1000278.

Han, J., Domier, L. L., Cassone, B. J., Dorrance, A. and Feng, Q. 2016. Assessment of common soybean-infecting virusesin Ohio, USA, through multi-site sampling and high-throuput sequencing. *Plant Health Progress* 17: 133–40.

Hartman, G. L. 2015a. Bacillus seed decay. In: G. L. Hartman, J. C. Rupe, E. F. Sikora, L. L. Domier, J. A. Davis and K. L. Steffey (Eds), *Compendium of Soybean Diseases and Pests*. American Phytopathological Society, St. Paul, pp. 24–5.

Hartman, G. L. 2015b. Bacterial pustule. In: G. L. Hartman, J. C. Rupe, E. F. Sikora, L. L. Domier, J. A. Davis and K. L. Steffey (Eds), *Compendium of Soybean Diseases and Pests*. American Phytopathological Society, St. Paul, pp. 19–20.

Hartman, G. L. 2015c. Wildfire. In: G. L. Hartman, J. C. Rupe, E. F. Sikora, L. L. Domier, J. A. Davis and K. L. Steffey (Eds), *Compendium of Soybean Diseases and Pests*. American Phytopathological Society, St. Paul, pp. 20–1.

Hartman, G. L. and Domier, L. L. 2015. Tobacco ringspot virus. In G. L. Hartman, J. C. Rupe, E. F. Sikora, L. L. Domier, J. A. Davis and K. L. Steffey (Eds), *Compendium of Soybean Diseases and Pests*. American Phytopathological Society, St. Paul, pp. 121–2.

Hartman, G. L., Rupe, J. C., Sikora, E. F., Domier, L. L., Davis, J. A. and Steffey, K. L. (Eds). 2015. *Compendium of Soybean Diseases and Pests*. American Phytopathological Society, St. Paul.

Hartman, G. L. and Wang, T. C. 1997. Bacterial pustule development and associated yield loss on nine soybean lines, pp. 194–6. In: V, B. Napompeth (Ed.) *Proceeding World Soybean Research Conference*. Kasetsart University Press, Bangkok, Thailand.

Hartman, G. L., West, E. and Herman, T. 2011. Crops that feed the world 2. Soybean-worldwide production, use, and constraints caused by pathogens and pests. *Food Security* 3:5–17.

Hartwig, E. E. and Lehman, S. G. 1951. Inheritance of resistance to the bacterial pustule disease in soybeans. *Agron. J.* 43:226–9.

Harveson, R. M. 2015. Bacterial wilt and bacterial tan spot. In: G. L. Hartman, J. C. Rupe, E. F. Sikora, L. L. Domier, J. A. Davis and K. L. Steffey (Eds), *Compendium of Soybean Diseases and Pests*. American Phytopathological Society, St. Paul, pp. 21–4.

Hayward, A. C. and Hartman, G. L. (Eds). 1994. *Bacterial Wilt: The Disease and Its Causative Agent Pseudomonas Solanacearum*. CABI, Wallingford, Oxfordshire, UK.

Hill, J. H. 2015. Soybean mosaic virus. In: G. L. Hartman, J. C. Rupe, E. F. Sikora, L. L. Domier, J. A. Davis and K. L. Steffey (Eds), *Compendium of Soybean Diseases and Pests*. American Phytopathological Society, St. Paul, pp. 114–17.

Hill, J. H., Alleman, R., Hogg, D. B. and Grau, C. R. 2001. First report of transmission of *Soybean mosaic virus* and *Alfalfa mosaic virus* by *Aphis glycines* in the New World. *Plant Dis.* 85:561.

Hill, J. H., Bailey, T. B., Benner, H. I., Tachibana, H. and Durand, D. P. 1987. Soybean mosaic virus: Effects of primary disease incidence on yield and seed quality. *Plant Dis.* 71:237–9.

Hill, J. H., Koval, N. C., Gaska, J. M. and Grau, C. R. 2007. Identification of field tolerance to *Bean pod mottle* and *Soybean mosaic viruses* in soybean. *Crop Sci.* 47:212–18.

Hill, J. H. and Whitham, S. A. 2014. Control of virus diseases in soybeans. *Adv. Virus Res.* 90:355–90.

Hobbs, H. A., Jossey, S., Wang, Y., Hartman, G. L. and Domier, L. L. 2012. Diverse soybean accessions identified with temperature-sensitive resistance to Tobacco Streak Virus. *Crop Sci.* 52: 738–44.

Hobbs, H. A., Herman, T. K., Slaminko, Wang, Y., Nguyen, B. T., McCoppin, N. K., Domier, L. L. and Hartman, G. L. 2010. Occurrences of soybean viruses, fungal diseases, and pests in Ilinois soybean rust sentinel plots. *Plant Health Progress* Online: DOI:10.1094/PHP-2010-0827-1001-BR.

Hong, S., Kim, Y., Jee, H., Shim, C., Kim, M. Park, J., Han, E. and Kee, B. 2011. Influence of disease severity of bacterial pustule caused by *Xanthomonas axonopodis* pv. *glycines* on soybean yield. *Res. Plant Dis.* 17:317–25.

Hong, J. K., Sung, C. H., Kim, D. K., Yun, H.-T., Jung, W. and Kim, K. D. 2012. Differential effect of delayed planting on soybean cultivars varying in susceptibility to bacterial pustule and wildfire in Korea. *Crop Prot.* 42:244–9.

Hopkins, J. D. and Mueller, A. J. 1984. Effect of bean pod mottle virus on soybean yield. *J. Econ. Entomol.* 77:943–7.

Huynh, T. V., Dahlbeck, D. and Staskawicz, B. J. 1989. Bacterial blight of soybean: Regulation of a pathogen gene determining host cultivar specificity. *Science* 245:1374–80.

Hwang, I. and Lim, S. M. 1992. Effects of individual and multiple infections with three bacterial pathogens on disease severity and yield of soybeans. *Plant Dis.* 76:195–8.

International Committee on Taxonomy of Viruses (ICTV). 2014. *Virus Taxonomy.* ICTV, Virology Division–IUMS. Available online at www.ictvonline.org.

Irizarry, M. D., Groves, C. L., Elmore, M. G., Bradley, C. A., Dasgupta, R., German, T. L. and et al. 2016. Re-emergence of Tobacco streak virus infecting soybean in the United States and Canada. *Plant Health Progress* 17:92–4.

Janse, J. D. 2005. *Phytobacteriology, Principles and Practice.* CABI Publishing. Oxford, UK.

Kim, D. H., Kim, K. H., Van, K., Kim, M. Y. and Lee, S. H. 2010. Fine mapping of a resistance gene to bacterial leaf pustule in soybean. *Theor. Appl. Genet.* 120:1443–50.

Klepadlo, M., Chen, P., Shi, A., Mason, R. E., Korth, K. L. and Srivastava, V. 2017. Single nucleotide polymorphism markers for rapid detection of the *Rsv4* locus for soybean mosaic virus resistance in diverse germplasm. *Mol. Breed.* DOI:10.1007/s22032-016-0595-3.

Koning, G., TeKrony, D. M., Pfeiffer, T. W. and Ghabrial, S. A. 2001. Infection of soybean with *Soybean mosaic virus* increases susceptibility to *Phomopsis* spp. seed infection. *Crop Sci.* 41:1850–6.

Kopisch-Obuch, F. J., Koval, N. C., Mueller, E. M., Paine, C., Grau, C. R. and Diers, B. W. 2008. Inheritance of resistance to *Alfalfa mosaic virus* in soybean PI 153282. *Crop Sci.* 48:933–40.

Laviolette, F. A., Athow, K. L., Probst, A. H. and Wilcox, J. R. 1970. Effect of bacterial pustule on yield of soybeans. *Crop Sci.* 10:150–1.

Mabry, T. R., Hobbs, H. A., Steinlage, T. A., Johnson, B. B., Pedersen, W. L., Spencer, J. L. and et al. 2003. Distribution of leaf-feeding beetles and *Bean pod mottle virus* (BPMV) in Illinois and transmission of BPMV in soybean. *Plant Dis.* 87:1221–5.

Melotto, M., Underwood, W. and He, S. Y. 2008. Role of stomata in plant innate immunity and foliar bacterial diseases. *Annu. Rev. Phytopathol.* 46:101–22.

Messieha, M. 1969. Transmission of tobacco ringspot virus by thrips. *Phytopathology* 59:943–5.

Mishra, S. 2014. Mollicutes and Plant Diseases. Discovery Publishing, New Delhi, India.

Mueller, E. E. and Grau, C. R. 2007. Seasonal progression, symptom development, and yield effects of *Alfalfa mosaic virus* epidemics on soybean in Wisconsin. *Plant Dis.* 91:266–72.

Orellana, R. G. 1981. Resistance to bud blight in introductions from the germ plasm of wild soybean. *Plant Dis.* 65:594–5.

Park, E. W. and Lim, S. M. 1986. Effects of bacterial blight on soybean yield. *Plant Dis.* 70:214–17.

Pedersen, P., Grau, C., Cullen, E., Koval, N. and Hill, J. H. 2007. Potential for integrated management of soybean virus disease. *Plant Dis.* 91:1255–9.

Qi, M., Wang, D., Bradley, C. and Zhao, Y. F. 2011. Genome sequence analyses of *Pseudomonas savastanoi* pv. *glycinea* and *in silico* subtractive hybridization-based comparative genomics with nine phytopathogenic pseudomonads. *PLoS ONE* 6:e16451.

Rabedeaux, P. F., J. M. Gaska, N. C. Kurtzweil and C. R. Grau. 2005. Seasonal progression and agronomic impact of *Tobacco streak virus* on soybean in Wisconsin. *Plant Dis.* 89:391–6.

Ross, J. P. 1977. Effect of aphid-transmitted soybean mosaic virus on yields of closely related resistant and susceptible soybean lines. *Crop Sci.* 17:869–72.

Saghai Maroof, M. A., Jeong, S. C., Gunduz, I., Tucker, D. M., Buss, G. R. and Tolin, S. A. 2008. Pyramiding of *Soybean mosaic virus* resistance genes by marker-assisted selection. *Crop Sci.* 48:517–26.

Shakiba, E., Chen, P., Gergerich, R., Li, S., Dombek, D. and Shi, A. 2012. Reactions of commercial soybean cultivars from the mid South to *Soybean mosaic virus*. *Crop Sci.* 52:1990–7.

Sharman, M., Thomas, J. E. and Persley, D. M. 2015. Natural host range, thrips and seed transmission of distinct *Tobacco streak virus* strains in Queensland, Australia. *Ann. Appl. Biol.* 167:197–207.

Staskawicz, B. J., Dahlbeck, D. and Keen, N. T. 1984. Cloned avirulence gene of *Pseudomonas syringae* pv. *glycinea* determines race-specific incompatibility on *Glycine max*. Merr. *Proc. Natl. Acad. Sci. U.S.A.* 81:6024–8.

Tolin, S. A. 2015a. Alfalfa mosaic virus. In: G. L. Hartman, J. C. Rupe, E. F. Sikora, L. L. Domier, J. A. Davis and K. L. Steffey (Eds), *Compendium of Soybean Diseases and Pests*. American Phytopathological Society, St. Paul, pp. 106–7.

Tolin, S. A. 2015b. Diseases caused by viruses. In: G. L. Hartman, J. C. Rupe, E. F. Sikora, L. L. Domier, J. A. Davis and K. L. Steffey (Eds), *Compendium of Soybean Diseases and Pests*. American Phytopathological Society, St. Paul, pp. 106–7.

Tolin, S. A. and Domier, L. L. 2015. Tobacco streak virus. In: G. L. Hartman, J. C. Rupe, E. F. Sikora, L. L. Domier, J. A. Davis and K. L. Steffey (Eds), *Compendium of Soybean Diseases and Pests*. American Phytopathological Society, St. Paul, pp. 109–10.

Tu, J. C. 1986. Strains of tobacco ringspot virus isolated from soybean (*Glycine max*) in southwestern Ontario (Canada). *Can. J. Plant Sci.* 66:491–8.

Wang, Y., Hobbs, H. A., Bowen, C. R., Bernard, R. L., Hill, C. B., Haudenshield, J. S., Domier, L. L. and Hartman, G. L. 2006. Evaluation of soybean cultivars, 'Williams' isogenic lines, and other selected soybean lines for resistance to two *Soybean mosaic virus* strains. *Crop Sci.* 46: 2649–53.

Wang, Y., Hobbs, H. A., Hill, C. B., Domier, L. L., Hartman, G. L. and Nelson, R. L. 2005. Evaluation of ancestral lines of U.S. soybean cultivars for resistance to four soybean viruses. *Crop Sci.* 45:639–44.

Weng S.-F., Luo, A.-C., Lin, C.-J. and Tseng, T.-T. 2017. High-quality genome sequence of *Xanthomonas axonopodis* pv. *glycines* strain 12609 isolated in Taiwan. *Genome Announc.* 5:e01695-16. https://doi.org/10.1128/genomeA.01695-16.

Yang, Y., Lin, J., Zheng, G., Zhang, M. and Zhi, H. (2014). Recombinant soybean mosaic virus is prevalent in Chinese soybean fields. *Arch. Virol.* 159:1793–6.

Zhang, X., Sato, S., Ye, X., Dorrance, A. E., Morris, T. J., Clemente, T. E. and Qu, F. 2011. Robust RNAi-based resistance to mixed infection of three viruses in soybean plants expressing separate short hairpins from a single transgene. *Phytopathology* 101:1264–9.

Zhao, Y. F. 2015. Bacterial blight. In: G. L. Hartman, J. C. Rupe, E. F. Sikora, L. L. Domier, J. A. Davis and K. L. Steffey (Eds), *Compendium of Soybean Diseases and Pests*. American Phytopathological Society, St. Paul, pp. 18–19.

Zhao, Y. and Davis, R. E. 2016. Criteria for phytoplasma 16Sr group/subgroup delineation and the need of a platform for proper registration of new groups and subgroups. *Int. J. Sys. Evol. Microbiol.* 66:2121–3.

Zheng, C. M., Chen, P. Y., Hymowitz, T. and Gergerich, R. 2005. Evaluation of *Glycine* species for resistance to *Bean pod mottle virus*. *Crop Prot.* 24:49–56.

Zhou, J., Domier, L. L., Sabanadzovic, S. and Tzanetakis, I. E. 2015. Soybean vein necrosis virus. In: G. L. Hartman, J. C. Rupe, E. F. Sikora, L. L. Domier, J. A. Davis and K. L. Steffey (Eds), *Compendium of Soybean Diseases and Pests*. American Phytopathological Society, St. Paul, pp. 124–5.

Zhou, J., Kantartzi, S. K., Wen, R.-H., Newman, M., Hajimorad, M. R., Rupe, J. and Tzanetakis, I. 2011. Molecular characterization of a new tospovirus infecting soybean. *Virus Genes* 43:289–95.

Zhou, J. and Tzanetakis, I. E. 2013. Epidemiology of soybean vein necrosis-associated virus. *Phytopathology* 103:966–71.

Ziems, A. D., Giesler, L. J., Graef, G. L., Redinbaugh, M. G., Vacha, J. L., Berry, S., Madden, L. V. and Dorrance, A. E. 2007. Response of soybean cultivars to *Bean pod mottle virus* infection. *Plant Dis.* 91:719–26.

Chapter 3

Nematode pathogens of soybean

T. L. Niblack and H. D. Lopez-Nicora, The Ohio State University, USA

1 Introduction
2 General nematode biology
3 Soybean cyst nematodes
4 Root-knot nematodes, *Meloidogyne* spp.
5 Lesion nematodes, *Pratylenchus* spp.
6 Reniform nematodes
7 Lance nematodes, *Hoplolaimus* spp.
8 Other nematodes
9 Conclusion
10 Where to look for further information
11 References

1 Introduction

In general, plant-parasitic nematodes that may detrimentally affect soybean production are microscopic roundworms that feed on or are in plant roots. They are obligate parasites, meaning that they cannot use any other food source and require a living, reasonably healthy (not necessarily economically productive) host in order to survive and reproduce. As such, they rarely kill their hosts. They exist in virtually all soils that support plant growth and comprise an important component of the soil biota and its ecosystem services.

The presence of plant-parasitic nematodes in fields planted with soybean (*Glycine max*) is to be expected. In the literature, there are reports of more than 100 species of plant-parasitic nematodes associated with soybean (Schmitt and Noel, 1984) with more than 50 species documented in the host database, 'Nemabase' (Ferris et al., 2017). Of these, only a few are widely considered to be economically important soybean pathogens. Worldwide, in the countries that produced 97% of the global soybean supply in 2006, only two nematode categories were listed among the disease-causing agents responsible for significant soybean yield losses (Wrather et al., 2010). They estimated that nematodes caused yield loss of 9.9 million metric tonnes, second only to that caused by soybean rust (13.2 million metric tonnes). The soybean cyst nematode (*Heterodera glycines*) alone was considered responsible for most of the loss due to nematodes (7.2 million metric tonnes).

http://dx.doi.org/10.19103/AS.2017.0034.18

The nematode categories, 'root-knot' and 'other nematodes', include nematodes such as lance, lesion, reniform, sting and stubby-root nematodes.

Whether a particular nematode is considered to have global significance is, of course, unimportant to an individual farmer facing a nematode problem. The purpose of this chapter is to not only to provide basic information on the known soybean pathogens, but also suggest some other considerations that may be pertinent. For example, many plant-parasitic nematodes can cause direct yield loss if subject to soybean monoculture combined with susceptible rotation crops over time. The indicators of such a situation are unusually high population densities of the nematode, compared with those more normally observed, combined with unsatisfactory or declining yields. Soybeans have only been produced in some regions for a relatively short period of time, and the potential for nematode adaptation to soybean continues as production expands or intensifies. In addition, soybean yield loss may be a result of the additive effects of several nematode species. The presence of more than one pathogenic species in a field is not uncommon (Eisenback, 1993; Herman et al., 1988; Niblack et al., 1986) and has rarely been well investigated. Finally, as soil-borne organisms are in a close association with roots, there is a high likelihood that disease interactions with other soil-borne pathogens are occurring and being overlooked (although some are well known).

2 General nematode biology

The specifics of nematode development should be understood because they may affect the choices of protocols used for extraction and diagnosis (Hallman and Viaene, 2013); however, there are general commonalities. For example, all plant-parasitic nematodes are vermiform (worm-shaped) at some point in their life cycles, all have a form of protrusible stylet (Fig. 1), which allows them to penetrate and feed on plant cells, and all reproduce by means of eggs (Fig. 2). The eggs undergo all stages of embryonic development within a pliable, translucent egg shell (Fig. 2). The final stage of nematode development within the egg is the juvenile stage. The juveniles may emerge from the eggs as first-stage (J1) or second-stage (J2), depending on the nematode, but in either event, the development from J1 to J2 is punctuated by a moult of the cuticle. Each additional stage of development

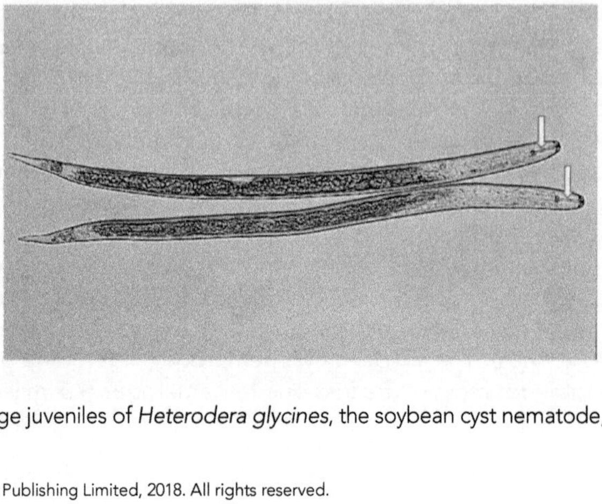

Figure 1 Second-stage juveniles of *Heterodera glycines*, the soybean cyst nematode, with stylets (arrows).

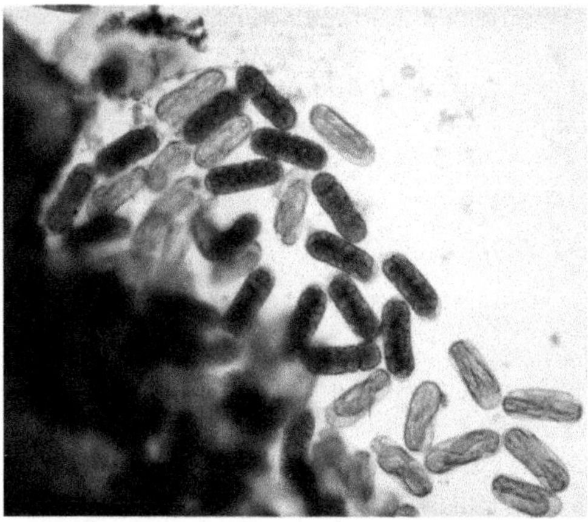

Figure 2 Eggs of *Heterodera glycines*, the soybean cyst nematode, at various stages of embryonic development.

is likewise punctuated by a moult: J2 nematodes develop through two additional stages, J3 and J4, before becoming adults. The mode of nematode reproduction may be sexual or asexual, but this is usually unimportant for the prediction of disease potential. In some species, males may not feed although they appear morphologically to be capable of feeding, and in other cases, males may be incapable of feeding due to lack or reduction of a stylet or other structures needed for feeding.

The infective stages are those responsible for the primary infection and may be limited to one life stage or may characterize all life stages, depending on the nematode species. For species that produce easily extractable eggs, such as the cyst nematodes, the number of eggs per unit volume of soil represents the best measure of disease potential, combined with host susceptibility and other factors. For most species, the number of juveniles or adults is the only direct measure of nematode population density and disease potential.

Nematodes described as 'migratory' retain their vermiform shape throughout the life cycle. For those described as 'sedentary', the females lose their somatic musculature and develop into sac-like shapes that are incapable of relocation, whereas the males (if present) are vermiform and motile.

Nematode feeding may be described as endo- or ectoparasitic, or some combination thereof. In general, the endoparasites feed from within plant tissue compared with ectoparasites, which feed from the outside. Endoparasites as a group are more damaging to plants.

The total length of the life cycle from egg to egg for the most economically important soybean-pathogenic nematodes is similar, around 4 to 6 weeks per generation in the field. Host suitability for a given nematode is often gauged through its rate of reproduction during a season or experiment, expressed as a reproduction factor (RF). The RF is calculated as the final nematode population density (Pf) present at the end of a field season or experiment, divided by the initial population density (Pi). An RF >1 indicates that nematode reproduction took place. RF values, along with other measures, are useful to compare host suitabilities of soybean cultivars, alternative hosts or treatment effects (Barker et al., 1986; Mueller and Noling, 1996).

The plant-parasitic nematodes known to cause disease and yield loss in soybean have been the subjects of several reviews (Hartman et al., 2015; Niblack et al., 2004; Riggs and Niblack, 1993; Schmitt and Noel, 1984). Only the nematodes considered to be of widespread economic significance will be covered in this chapter.

3 Soybean cyst nematodes

H. glycines, the soybean cyst nematode, is the most economically important pathogen of soybean in the United States and Canada (Allen et al., 2017; Koenning and Wrather, 2010; Tylka and Marett, 2014; Wang et al., 2003). Worldwide, *H. glycines* is responsible for significant losses incurred by soybean growers every year (Niblack et al., 2006; Wrather et al., 2001). *H. glycines* has a broad host range (Poromarto and Nelson, 2010; Poromarto et al., 2015); however, soybean is the major agricultural host (Venkatesh et al., 2000). Although there are various potential management options, research has demonstrated the limitations of each.

3.1 Origin and distribution

The origin of *H. glycines* remains uncertain. Evidence indicates that the pathogen had its origin in Asia, where soybean was grown as a native crop (Hymowitz, 1970, 2004; Riggs, 2004).

Farmers in China gave the name of 'fire-burned seedling' to a disease in soybean, but it was not until 1899 that a report from northeastern China indicated that this disease was caused by *H. glycines* (Liu et al., 1997). If soybean has its origins in China, where farmers long ago identified the 'fire-burned seedling' disease, and the disease was later proven to be caused by *H. glycines*, it seems likely that China is the origin of this pathogen and that both pathogen and crop evolved together (Riggs, 2004). There was controversy over the ancestral origins of *H. glycines* populations in the United States (Noel, 1992). Overlapping morphological characteristics with other *Heterodera* spp. generated confusion in early classification of *H. glycines* (Ichinohe, 1952). For example, in 1951, *H. glycines* was first identified as *Heterodera goettingiana* (Goffart, 1951), and was not described as *H. glycines* until 1952 (Ichinohe, 1952).

Soon after the first report of *H. glycines* in North Carolina (Winstead et al., 1955), the pathogen was detected in several other states (Noel, 1992). The introduction, through different entry points, of infested soil from China that was used as a source of *Bradyrhizobium* inoculant (Hymowitz, 1990) was a probable source of *H. glycines* infestation and rapid dispersion throughout soybean-growing areas in the United States (Riggs, 2004). Local movement of infested soil aided by wind, flood and field run-off in combination with machinery traffic contributed to dissemination of the nematode. In addition, the ability of *H. glycines'* eggs to survive in cysts through the digestive tracts of migratory birds allowed for further spread of the nematode (Epps, 1971; Smith et al., 1992).

Currently, *H. glycines* is found in major soybean production regions in North America (Tylka and Marett, 2014) with a steady increase in reports of new fields infested with this nematode. Moreover, *H. glycines* is present in most countries where soybean is produced (Riggs, 2004). As soybean-producing areas expand, so does *H. glycines*, placing soybean production in infested fields at risk.

3.2 *Heterodera glycines* biology

H. glycines is an obligate endoparasite. The cyst for which the nematode is named is the environmentally resistant remains of the female nematode, which contains viable eggs. The infective unit of *H. glycines* is second-stage juvenile (J2) (Niblack et al., 1986; Wrather and Anand, 1988). The protective cyst and the ability of the eggs to become dormant enable the nematode to survive environmental extremes and be passively dispersed to diverse environments (Koenning, 2004).

Similar to most plant-parasitic nematodes, *H. glycines* undergoes four juvenile stages through moulting (or ecdysis). Each moult results in a subsequent juvenile stage until the fourth moult results in the adult stage. The life cycle of *H. glycines* has been described in detail elsewhere (Koenning, 2004; Lauritis et al., 1983; Niblack, 2005; Niblack et al., 2006). Lauritis et al. (1983) characterized the life cycle of *H. glycines* in monoxenic culture on soybean roots at 25 °C and is a good reference for the *H. glycines* life cycle and duration of each life stage under controlled environmental conditions. Similar life cycle characterization during soybean-growing seasons and under different field conditions would greatly complement our knowledge of this nematode's life cycle.

The life cycle of *H. glycines* (Fig. 3) begins with the egg, in which embryonic development results in a first-stage juvenile (J1) inside the egg. The first moult to the second-stage juvenile (J2) takes place within the eggshell. At this point, through enzymatic and/or physical processes, the nematode hatches. The hatching process is influenced by several factors such as soil temperature, presence of a host (and root exudates) and time (Yen et al., 1995). Very little, however, is known and understood of the hatching mechanisms, and research is still needed on the topic.

The newly hatched J2 is exposed to the soil environment, and successful completion of its life cycle depends on its ability to localize, penetrate and initiate a feeding site (syncytium) in a compatible (susceptible) host root. The infective juvenile locates roots,

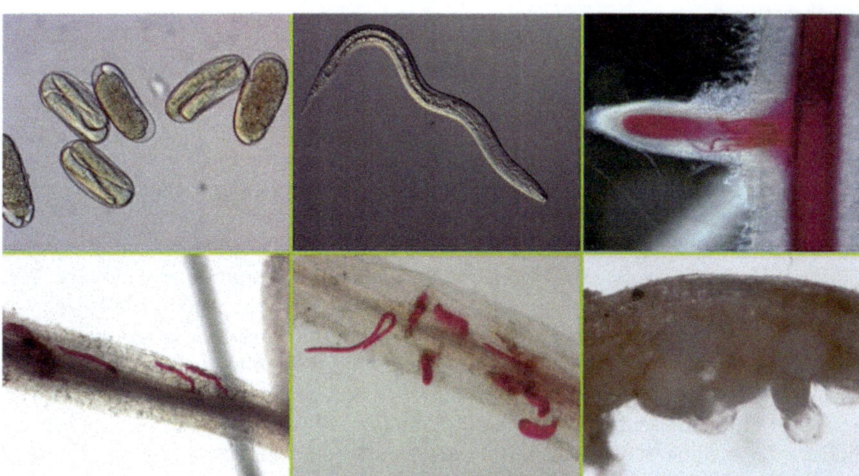

Figure 3 Life cycle collage of *Heterodera glycines*, the soybean cyst nematode, first row from upper left: eggs, second-stage juvenile (J2), infected root; second row from lower left: J2 initiating feeding sites, multiple juvenile stages and male in infected root, females and egg masses on root surface (photos courtesy of Ursula Reuter-Carlson).

presumably through chemolocation (Dusenbery, 1987; Perry, 1996; von Mende et al., 1998), and penetrates the root through the epidermis. With the aid of the stylet and the production of cellulases and other enzymes (Davis et al., 2004; Niblack, 2005; Niblack et al., 2006), the J2 penetrates and migrates intercellularly towards the vascular cylinder.

The major damage at this stage is a physical damage to the root, though more evidence indicates that the nematode secretes effector proteins produced in the oesophageal glands that suppress the plant's defence response during initial infection (Goverse and Smant, 2014; Haegeman et al., 2012; Mantelin et al., 2015). *H. glycines* will equally infect non-hosts and susceptible or resistant hosts. The ability to initiate a feeding site, however, will determine the compatibility between the nematode and the host.

Upon its arrival to the vascular system, the J2 will probe cells (cortical parenchyma, endodermal or pericycle cells) in the periphery of the vascular cylinder, using its protractible stylet, without damaging the plasma membrane of these cells. The precise location of the syncytium can be influenced by the available water content in the soil and plant (Johnson et al., 1993). The nematode will secrete effector proteins that will reprogramme these cells (Gardner et al., 2015). Currently, characterizing the nematode effector proteins and understanding their functions in the host–nematode interaction is one of the most exciting research areas with its potential for new management approaches.

Compatible cells, referred to as initial syncytial cells, undergo tremendous physiological changes and recruit adjacent cells into a multinucleate feeding site (syncytium), a metabolic sink that functions as a transfer cell to redirect plant photosynthetic products to the nematodes. The initiation and maintenance of the syncytium are the main difference between a compatible and incompatible host reaction to *H. glycines* parasitism. For detailed and fascinating descriptions of the physiological and molecular changes that take place in the initiation, formation and development of syncytia induced by *H. glycines*, the reader should consult Noel (2004) and Sobczak and Golinowski (2009, 2011). A J2 that successfully initiates a syncytium loses motility, begins feeding and continues growing.

The now sedentary J2 undergoes subsequent moults to J3, J4 and adulthood. Sexual differentiation becomes apparent at J3 (Raski, 1950) and at J4, males and females are evidently dimorphic. Under controlled environmental conditions, males regain their vermiform shape and exit the root at approximately 8 to 9 days after root infection (Lauritis et al., 1983). Sedentary females continue to grow and the posterior portion of their bodies eventually protrudes from the root with the anterior part remaining embedded. Spheroid, white females are visible 9 to 10 days after penetration (Lauritis et al., 1983).

The ratio of male to female is approximately 1:1, but this can be skewed by several conditions such as host resistance or environmental factors (Colgrove and Niblack, 2005). Sexual reproduction (amphimixis) is obligate in *H. glycines* (Triantaphyllou and Hirschmann, 1962, 1964). Males are attracted to females, likely due to the release of sex pheromones (Jaffe et al., 1989). After the female is inseminated, she begins to lay fertilized eggs in an external gelatinous matrix, and later to retain the eggs inside the body (Ichinohe, 1952). The range of egg production per female ranges from 40 to more than 600 (Sipes et al., 1992); however, a general average is 200 eggs per female (Niblack, 2005).

There is room for more research in the final portion of the *H. glycines* life cycle. Areas for further research include: a better understanding of sex differentiation and the mating process; male to female ratios under diverse environmental conditions; the composition and function of the gelatinous matrix (Lopez-Nicora and Niblack, 2012) and the ability to hybridize with other *Heterodera* spp. (Miller, 1988, 1989; Potter and Fox, 1965).

3.3 Population dynamics and their impact on soybean

Several researchers have demonstrated that the initial population density (Pi) of soil-borne pathogen is the main factor that relates to yield reduction and it is, therefore, used in modelling and estimating yield losses (Barker and Olthof, 1976; Barker et al., 2004; Ferris, 1987; Francl and Dropkin, 1986; Gilligan, 1983; Niblack et al., 2006; Schmitt and Ferris, 1998). *H. glycines* Pi is commonly quantified as number of eggs per unit of soil at planting. It is crucial though to remember that both the host (i.e., soybean yield) and the pathogen (i.e., *H. glycines* Pi) are subject to environmental conditions. Therefore, a *H. glycines* damage threshold will vary from one field (environment) to another. Once the nematode is detected in a field, population levels and virulence profile are important items to inform management decisions.

The spatial distribution of *H. glycines* is aggregated in the soil (Francl, 1986; Gavassoni et al., 2001, 2007). Consequently, the effects of this pathogen on soybean yield are not spatially uniform in infested fields. This spatial autocorrelation violates the independence assumption of several statistical models (Kutner et al., 2004). Spatial statistical analysis, therefore, could be used to account for the aggregated effect of *H. glycines* on soybean yield in naturally infested fields (Hughes, 1988; Liu et al., 2014, 2015).

Classic symptoms such as stunted plants with chlorotic (yellowing) foliage may be observed only when *H. glycines* population densities are high. These symptoms occur most often in fields with nutrient deficiencies, in interactions with other pathogens (mainly with fungal pathogens) and/or under stressful environmental conditions (such as high temperature and drought). Often, however, *H. glycines* infection causes no visible symptoms (Niblack, 2005; Wang et al., 2003). The fact that significant yield reduction (15 to 30%) may take place with absolutely no above-ground symptoms (Wang et al., 2003) is one of the main reasons fields must be sampled for the presence and abundance of *H. glycines*. The quality of the diagnosis is dependent on the method and timing of sampling (Niblack and Tylka, 2008).

Nematode infection will cause both physical (root invasion) and physiological (formation of syncytia acting as metabolic sinks) damages to soybean plants, stressing them and ultimately reducing their yield. If stressful environmental conditions are present, *H. glycines* infection may enhance the symptoms of these stressful factors. For example, nutrient deficiency symptoms increase with elevated *H. glycines* population density, which makes the symptoms more noticeable (Melakeberhan, 2004; Smith et al., 2001). Amending the nutrient deficiency will reduce visible symptoms, but the elevated levels of *H. glycines* will continue to impact soybean performance with no visible symptoms.

Another environmental factor that can affect *H. glycines* population densities and consequently soybean production is soil pH (Pedersen et al., 2010; Rogovska et al., 2009; Tylka et al., 1998). A positive correlation between *H. glycines* population densities and soil pH revealed that in infested fields, more alkaline soils supported higher *H. glycines* levels and, therefore, lower soybean yield. The mechanism involved in this relationship is not known.

3.4 Biotic disease interactions

Disease interactions between *H. glycines* and soil-borne (mainly fungal and fungal-like) pathogens include *Calonectria crotalariae* (= *ilicicola*) (Overstreet and McGawley, 1988; Overstreet et al., 1990), *Diaporthe phaseolorum* (Russin et al., 1989), *Fusarium* spp.

(Brzostowski et al., 2014; Gao et al., 2006; Hartman et al., 1995; Marburger et al., 2014; McLean and Lawrence, 1993a, 1993b; Melgar et al., 1994; Roy et al., 1989; Rupe et al., 1997, 1999; Westphal et al., 2014; Xing and Westphal, 2006), *Macrophomina phaseolina* (Francl et al., 1988; Todd, 1993; Todd et al., 1987; Winkler et al., 1994), *Phialophora gregata* (Niblack et al., 1992; Tabor et al., 2003, 2006; Tubajika et al., 1994), *Phytophthora sojae* (Adeniji et al., 1975; Kaitany et al., 2000) and *Rhizoctonia solani* (Dave, 1975; Frohning, 2013). The interaction studies between *H. glycines* and these pathogens described the effect on not only the host but also the reproduction of either pathogen involved in the disease complex.

There are additional studies that evaluated the effect of insects (soybean aphids) on the above-ground parts of plants in the presence of cyst nematodes infecting root systems (Heeren et al., 2012; McCarville et al., 2012, 2014). Understanding the intricate network of interaction that takes place in the field will help improve crop pest management; however, interaction studies can yield results that are difficult to interpret. The reader should refer to Barker and McGawley (1998), Bond and Wrather (2004) and Khan (1993) for additional details on nematode interactions with other pathogens.

3.5 Genetic diversity

Breeding for resistance to *H. glycines* will be a perpetual process. The Red Queen hypothesis is well illustrated in the co-evolution of soybean and *H. glycines*. (Alice was told by the Red Queen in *Alice in Wonderland*: 'Now, here, you see, it takes all the running you can do, to keep in the same place. If you want to get somewhere else, you must run at least twice as fast as that!' (Carroll, 1872).) Although resistance to pathogens exists, efforts to develop resistant plants must continue as the pathogens adapt (Clay and Kover, 1996). *H. glycines* populations exhibit diversity for virulence (Niblack et al., 2002) both among (Niblack et al., 1993; Rao-Arelli et al., 1992; Sikora and Noel, 1991) and within populations (Colgrove et al., 2002; Zhang et al., 1998).

Soon after *H. glycines* was first reported in the United States in 1954 (Winsted et al. 1955), adaptation to resistant cultivars was observed (Ross, 1962) and the 'arms race' to develop new soybean cultivars resistant to the more virulent *H. glycines* populations started (Brim and Ross, 1966). The widespread use of soybean cultivars with two main sources of resistance to *H. glycines* (i.e., 'Peking', plant introduction [PI] 548402, and PI 88788), however, exerted selection pressure on the nematode, which resulted in a shift of pathogen virulence and adaptation to these cultivars in major soybean-producing regions of the United States (Acharya et al., 2016; Faghihi et al., 2010; Hershman et al., 2008; Mitchum et al., 2007; Niblack et al., 2008; Zheng and Chen, 2006). To describe the observed variability in *H. glycines* virulence, the race scheme (Golden et al., 1970) and later the *H. glycines* (HG) Type (Niblack et al., 2002) were developed. An adequate understanding of the purpose, information produced and limitations of either a race or an HG-Type test is important for integrated pest management and a good explanation can be found in Tylka (2016).

Attempts to measure genetic variation in *H. glycines* populations started in the 1960s (Golden et al., 1970) with the race test, which included four soybean differentials, a susceptible soybean check ('Lee') and the description of four *H. glycines* races. In 1988 (Riggs and Schmitt, 1988), the 16 'theoretical' *H. glycines* races were described based on the four original soybean differentials (Golden et al., 1970). The race of a *H. glycines* population was determined through a bioassay. Soybean differentials would be infested with a specific *H. glycines* population and, after the completion of the nematode life cycle, *H. glycines* females were collected from the soybean differentials and the susceptibility checked. Female (or virulence) Index (FI) values were obtained for each indicator (differential) line by dividing the average number of *H. glycines* females counted on the tested line by the average number of females counted on the susceptible standard and multiplying this value by 100. If the specific *H. glycines* population produced an FI greater than or equal to 10, then that population was considered able to reproduce on that soybean differential and a 'yes' or '+' was assigned to that differential. If the FI was below 10, then the tested line received a 'no' or a '–' because the nematode population could not reproduce over 10% compared with the susceptible check and would not, in theory, be detrimental in the field (Niblack et al., 2006).

This initial attempt to describe the genetic variation in *H. glycines* populations soon saw its limitations (Niblack and Riggs, 2004; Niblack et al., 2002, 2006). In 2002, a revised classification scheme for genetically diverse *H. glycines* populations was proposed: the HG Type (Niblack et al., 2002). The HG Type, which describes the virulence profile of a *H. glycines* population, is determined through a bioassay conducted under standardized conditions in a greenhouse (Niblack et al., 2002, 2009). Seven soybean lines, used in breeding programmes as sources of resistance, were included as differentials in the HG-Type test. These lines are labelled 1 through 7 according to their chronological public release (Table 1). The HG-Type test allows the addition of new sources of resistance as these are identified without the virulence profile of the nematode increasing exponentially as it would with the race test. The FI value is obtained in a manner identical to that described for the race test in the previous paragraph.

The virulence profile of a *H. glycines* population is defined based on the ability to produce (or not) an FI above or equal to 10 on any (or none) of the seven differentials. For example, if the FI for a *H. glycines* population did not surpass 10 on any of the differentials, then that population is considered an HG Type 0 (formerly race 3). If the nematode can produce an FI above 10 on PI 88788 (differential #2), for example, this population would be assigned an HG Type 2 (formerly race 1). Notice that an HG Type 2.5.7 (currently the most prevalent *H. glycines* virulence profile in the United States) would still be considered race 1 in the previous classification scheme. The HG Type provides additional information about the virulence profile of a nematode population in terms of aggressiveness, since the FI produced on each differential must be reported (Niblack et al., 2002).

The HG Type is a useful resource to make soybean cultivar recommendations to growers; however, it does not explain all the variations observed between and within *H. glycines* populations. Based on the FI, resistance to *H. glycines* ranges from highly resistant to susceptible though a spectrum of moderate and low resistance (Niblack, 2005; Schmitt and Shannon, 1992). Variability in the levels of resistance from soybean lines with the same source of resistance and variability in the aggressiveness of different *H. glycines* populations with the same assigned HG Type are commonly observed in greenhouse bioassays and, for this reason, results should be accompanied by yield data in field trials (Niblack, 2005; Niblack et al., 2006). Current research on resistance to *H. glycines*, however, has contributed to better understanding of the unexplained variation observed in greenhouse bioassays (Brucker et al., 2005; Yu et al., 2016).

Analogous to the FI observed in greenhouse assays, *H. glycines* RF, a ratio of the nematode final (Pf) to initial (Pi) population, is obtained from field trials. The RF is used

Table 1 Soybean indicator lines used for the determination of *Heterodera glycines* (HG) Type (Niblack et al., 2002)

Indicator number	Line[a]	Comment
1	PI 548402	Along with several other PIs, also known as 'Peking'; included in original race test (Golden et al., 1970)
2	PI 88788	Source of resistance in most *H. glycines*-resistant cultivars; included in original race test (Golden et al., 1970)
3	PI 90763	Included in original race test (Golden et al., 1970)
4	PI 437654	Source of resistance in 'Hartwig' and other highly resistant cultivars
5	PI 209332	Resistance is similar to that in PI 88788 and PI 548316 (Colgrove and Niblack, 2008)
6	PI 89772	Resistance is similar to that in PI 548402 (Colgrove and Niblack, 2008)
7	PI 548316	Also known as 'Cloud'; resistance is similar to that in PI 88788 and PI 209332 (Colgrove and Niblack, 2008)

[a] PI = plant introduction, accession numbers from U.S. National Plant Germplasm System.

to assess the *H. glycines* population density at the end of the season in relation to the presence of a specific soybean cultivar. For example, an RF below 1 indicates a reduction in *H. glycines* population levels under a specific soybean cultivar. An RF above 1 will indicate that the soybean cultivar favoured the reproduction of the nematode and consequently increased its population density. Complementing FI information generated in greenhouse bioassay with RF from field studies is important because environmental factors, which play a crucial role in the disease triangle (Agrios, 2005), are at work in the field and not under the controlled conditions of the greenhouse. Additionally, information on soybean performance (i.e., yield) related to *H. glycines* resistance levels can most accurately be obtained in field trials.

3.6 Resistance to *Heterodera glycines*

The use of *H. glycines*-resistant soybean cultivars remains the best-recommended management method to reduce the effect of the nematode on soybean production where rotation to a non-host is not economically feasible (Niblack, 2005). Resistance to *H. glycines* in soybean is quantitative and was described as a complex and multigenic process where major and minor genes were involved in conferring resistance to *H. glycines* (Shannon et al., 2004). The primary *H. glycines* resistance genes were reported from PI 548402 (i.e., 'Peking') and were named *rhg1* (resistance to H. g*lycines*), *rhg2*, *rhg3* (Caldwell et al., 1960) and *Rhg4* (Matson and Williams, 1965); additionally, *Rhg5* was reported from PI 88788 (Rao-Arelli et al., 1994). Genes *Rhg4* and *Rhg5* were originally described as dominant, while the other three fit a recessive gene inheritance model.

Concibido et al. (2004) reviewed the immense effort that took place over a decade to map quantitative trait loci (QTLs) and genes responsible to confer resistance to *H. glycines*. Two major QTLs from different sources conferring the greatest resistance to *H. glycines* are mapped to regions containing *rhg1* and *Rhg4*. Not surprisingly, these are the two major genes found in most soybean cultivars with resistance to *H. glycines* in North America

(Concibido et al., 2004). For example, the *rhg1* locus was mapped to a major QTL in several sources of resistance (Kim et al., 2010), such as PI 437654 (Webb et al., 1995), PI 209332 (Concibido et al., 1996), Peking, PI 90763, PI 88788 (Concibido et al., 1997; Glover et al., 2004), PI 89772 (Yue et al., 2001) and PI 404198A (Guo et al., 2006).

In 2010, the genome of soybean 'Williams 82' was sequenced, annotated and released (Schmutz et al., 2010). Fine mapping of *rhg1-b* allele (name given by the Soybean Genetics Committee to the *rhg1* allele derived from PI 88788) placed it in an interval of 67 kb in the Williams 82 genome sequence (Kim et al., 2010). Rather than having a recessive gene action, the *rhg1-b* derived from PI 88788 has an additive or incomplete dominance gene action (Kim et al., 2010). Previous research had reportedly identified and cloned the *rhg1* and *Rhg4* loci derived from Peking, both with leucine-rich repeat (LRR) and receptor-like kinase characteristics (Hauge et al., 2001; Lightfoot and Meksem, 2002; Ruben et al., 2006). The fine mapping of *rhg1-b* did not include the reportedly cloned *rhg1* (Kim et al., 2010). Further experiments demonstrated that the LRR-kinase genes at the *rhg1* and *Rhg4* are not involved in *H. glycines* resistance (Liu et al., 2012; Melito et al., 2010).

A major breakthrough in *H. glycines* resistance research took place in 2012 (Cook et al., 2012; Liu et al., 2012). Resistance to *H. glycines* derived from PI 88788 was not due to the presence of *rhg1-b,* but to the copy number in tandem of a 31-kb segment which included three different genes, all of which were involved in resistance (Cook et al., 2012). Additionally, the gene at the *Rhg4* locus was map based, cloned from soybean lines with resistance to *H. glycines* derived from Peking (Liu et al., 2012). Recent research showed that in addition to *Rhg1* copy number, the type (specific sequence) and interaction with the *Rhg4* gene define the degree of *H. glycines* resistance in soybean (Yu et al., 2016). The 'non-classical' genes involved in resistance to *H. glycines* and the fact that *rhg1-b* requires several copy numbers of a specific segment are helping researchers revisit previous results and understand many of the inconsistencies and variability observed in the *H. glycines*–soybean interaction (Brucker et al., 2005). The reader should consult Mitchum (2016) and Kim et al. (2016) for more detailed updates on the status of soybean resistance to *H. glycines.*

3.7 Management

Once *H. glycines* becomes established in a field, eradication is very unlikely. A universal damage threshold for *H. glycines* is difficult (if not impossible) to obtain because the biology of the nematode and its effect on soybean performance are influenced by environmental factors. An action threshold, therefore, is preferred (Niblack et al., 2006). Consequently, the first step in managing *H. glycines* is periodic soil sampling (Niblack and Chen, 2004; Niblack and Tylka, 2008; Young, 1998). Detection of this nematode in a field will then determine what management strategies need to be implemented.

Quantification of *H. glycines* population density from infested fields is the next step. The main objective in quantifying *H. glycines* population density (e.g., eggs per 100 cm^3 soil) is to estimate the damage potential of the nematode. For this purpose, it is recommended to obtain the nematode initial population density (Pi) at planting. Pi of soil-borne pathogen is the main factor that relates to yield reduction and it is, therefore, used in modelling and estimating yield losses (Barker and Olthof, 1976; Barker et al., 2004; Ferris, 1987; Gilligan, 1983; Niblack et al., 2006; Schmitt and Ferris, 1998).

Additionally, periodic evaluation of *H. glycines* Pi from a field will indicate if the applied management strategies have been successful in reducing the nematode population density.

Finally, identifying and understanding (Tylka, 2016) the HG Type of the *H. glycines* population from a particular field will aid in selecting soybean cultivars resistant to that specific *H. glycines* population. As previously mentioned, shifts in virulence and adaptation to resistant cultivars have been reported (Acharya et al., 2016; Faghihi et al., 2010; Hershman et al., 2008; Mitchum et al., 2007; Niblack et al., 2008; Zheng et al., 2006); therefore, using cultivars with different sources of resistance may help reduce the nematode population densities while avoiding shifts in virulence (Niblack, 2005). Ultimately, once *H. glycines* have been detected in a field, keeping the nematode levels down is an easier task than trying to bring an elevated population density to non-damaging levels.

Management of soil-borne pathogens responsible for monocyclic diseases generally involves reduction of initial inoculum (Agrios, 2005). The use of soybean cultivars resistant to *H. glycines* and non-host crop rotation remain the best methods to reduce the impact of this pathogen on soybean production (Niblack and Chen, 2004; Niblack, 2005). Adequate use of resistant cultivars and crop rotation in an integrated management plan resulted in the 'multiple rotation' scheme (Niblack, 2005; Niblack et al., 2006; Niblack and Tylka, 2008): rotate with an *H. glycines* non-host crop, rotate with *H. glycines*-resistant soybean cultivars and rotate soybean cultivars with different sources of resistance. In addition, other agronomic practices (i.e., planting date, tillage, good soil fertility, adequate weed management, etc.) integrated with the 'multiple rotation' scheme ultimately enhance the efficacy of *H. glycines* management.

Each component of the 'multiple rotation' scheme has a specific role in the integrated *H. glycines* management. Rotation with non-host crop is an effective way to reduce *H. glycines* population density (Niblack, 2005; Niblack et al., 2006; Niblack and Chen, 2004). In regions where the soil does not freeze during winter, the use of a non-host crop can reduce *H. glycines* levels to below damage threshold in one (Koenning et al., 1993) or two years (Francl and Dropkin, 1986). In regions where *H. glycines* overwinters in frozen soil, *H. glycines* population densities obtained at any point during the overwintering period (September to April) do not differ (Chen et al., 2001; Jackson et al., 2005) and additional years of non-host cropping sequence are required to lower *H. glycines* population densities (Chen et al., 2001; Miller et al., 2006). Furthermore, Jackson et al. (2005) demonstrated that the viability and infectivity of the nematode were not affected when soybean was planted after the non-host crops commonly used in crop rotation (e.g., oat, canola, sesame, corn, sorghum and red clover).

Corn, which is commonly used in crop rotation to manage *H. glycines*, was classified among the least effective non-host crops in reducing *H. glycines* population density (Miller et al., 2006). Usually, however, planting more effective non-host crops is not an option where farmers are unwilling to rotate out of corn for economic reasons. Other studies, for example, demonstrated that winter wheat in double crop sequence both reduced *H. glycines* population densities and affected its viability (Baird and Bernard, 1984; Hershman and Bachi, 1995; Long and Todd, 2001); however, the effect of wheat on *H. glycines* viability was not consistent (Koenning and Anand, 1991; Miller et al., 2006). The mechanisms by which crop rotation affects *H. glycines* population densities are not well understood, and it is an area for more research. There is a re-emerging interest in evaluating cover crops as non-host crops and biological control agents which, through root exudates, can produce toxic compounds affecting the viability and infective ability of *H. glycines* (Jackson et al., 2005; Miller et al., 2006; Niblack and Chen, 2004; Warnke et al., 2008).

Two important components of the 'multiple rotation' management are to rotate *H. glycines*-resistant soybean cultivars and to rotate sources of resistance. We have previously (see previous sections) discussed the importance of using resistant cultivars and rotating sources of resistance in an integrated *H. glycines* management system. Soybean cultivars resistant to *H. glycines* reduce population density and yield more than susceptible cultivars; however, yield may still be negatively related to Pi in *H. glycines*-resistant soybean cultivars. Development and deployment of resistant cultivars must incorporate the most recent knowledge of resistance mechanisms to *H. glycines* (Cook et al., 2012; Kim et al., 2010; Lee et al., 2015; Yu et al., 2016). Copy number and type of *H. glycines* resistance genes must be taken into consideration when new breeding lines are developed, and this information should be provided to growers for better decision making.

Finally, there are several agronomic practices that can enhance *H. glycines* management and they do so by alleviating plant stress. Examples of these practices can be found in Niblack and Chen (2004) and involve management of fall weeds (Venkatesh et al., 2000; Poromarto et al., 2010, 2015), adequate soil fertility (Melakeberhan, 2004; Pedersen et al., 2010; Rogovska et al., 2009; Smith et al., 2001) and tillage (Perez-Hernandez and Giesler, 2014; Westphal et al., 2009), among others. The use of chemical treatment has not shown efficacy in reducing *H. glycines* population density in the past (Smith et al., 1991). Biocontrol is also an option which remains largely unexplored. Recently, however, new products are available in the market and are being tested in different environments to evaluate their efficacy and consistency (Beeman and Tylka, 2017; Broderick, 2016; Musil, 2016; Zaworski, 2014).

Integration of different management techniques such as the ones described earlier would lead to sustainable management of soybean with regard to *H. glycines* (and surely other biotic and abiotic soybean problems). The main objective of integrated pest management is to alter and/or disrupt, through different agricultural practices, the disease (biology) cycle of the pathogen and consequently reduce its impact on the crop. A thorough understanding of the biology of the pathogen is, therefore, necessary. The previous statement forces us to face the gaps in our knowledge of the biology of *H. glycines*.

We do not fully understand the hatching process of *H. glycines* eggs and processes involved in overwintering and viability. We do not know the composition of the gelatinous matrix where females lay their eggs, which serves to protect the offspring. Recent knowledge gained on the mechanisms of resistance to *H. glycines* must now be applied in breeding programmes and that knowledge extended to the growers for use in management decision-making. We know very little about the genetic structure of *H. glycines*. Identification and characterization of several reported effector proteins involved in parasitism are exciting research areas, and more are there to be uncovered soon.

4 Root-knot nematodes, *Meloidogyne* spp.

Meloidogyne is the most damaging genus of plant-parasitic nematodes worldwide, causing yield losses and crop failures on a wide array of cultivated plants. Root-knot nematodes are so-named because their feeding induces galling that makes the roots appear knotted (Fig. 4). Feeding by root-knot nematodes not only diverts resources from the shoot to support nematode growth and development, but also interferes with nutrient and water transport by the roots. Root galling increases with increases in soil population

Figure 4 Galls on soybean roots, the symptom of root-knot (*Meloidogyne* sp.) nematode infection (photo courtesy of Jason Bond).

density of the nematodes. Above-ground symptoms appear as infection levels increase and are visible as unthrifty growth, wilting and chlorosis (Fig. 5). An excellent resource for practitioners desiring more information about root-knot nematodes and the disease they cause can be found in Perry et al. (2009).

Root-knot nematodes have been recognized as soybean parasites since the late 1800s (Franck, 1882), but were not recognized as widespread until the 1940s (Atkinson, 1944). Several species of *Meloidogyne* are known to cause root-knot disease of soybean (Ferris et al., 2017; Niblack et al. 2004). Of these, only *M. incognita*, *M. arenaria* and *M. javanica* are widespread in economic importance. This is not to say that other *Meloidogyne* species are not emerging (Bernard and Jennings, 1997; Elling, 2013; Koenning, 2015) or important in more temperate regions, such as *M. hapla* in North America.

Figure 5 Above ground symptoms of root-knot (*Meloidogyne* sp.) nematode infection (photo courtesy of Jason Bond).

4.1 Life cycle and disease cycle

The infective stage of this nematode is the second-stage juvenile (J2). J2s are difficult for untrained observers to identify in soil extracts at dissecting microscope magnifications because of their small size (250 to <500 μm body length) and very fine stylets (Fig. 6). J2s invade roots near the tips and migrate to the vascular tissue, where they induce permanent feeding sites known as giant cells. As the nematodes develop, they lose somatic musculature and become sedentary. The giant cells become enlarged and multinucleate, and each giant cell system becomes a metabolic sink that supports nematode development and reproduction (Gommers and Dropkin, 1977). The giant cell system may also disrupt the vascular tissue, impairing its function and resulting in reduced water and nutrient transport. Cortical cells near the giant cells become enlarged by means of hyperplasia and hypertrophy (increases in cell number and size, respectively). The root-knots or galls that are the result of this induced growth may coalesce when infection levels are high, and each gall may contain numerous females.

J2s infect roots regardless of the resistance status of the host plant. Resistance in soybean often results in the death of the J2 during the feeding site establishment phase. In susceptible plants, J2s progress through the J3 and J4 life stages, and usually to adulthood as females, completely embedded within gall tissue. Reproduction is usually apomictic, and males are few in number, depending on the species, environmental conditions and population density. The life cycle may take 4 to 6 weeks, and each female can produce 300 or more eggs (Koenning, 2015). The potential for population density increase is, therefore, very high during a single growing season.

Meloidogyne species survive periods when hosts are not present either as J2 in the soil or as eggs within a gelatinous matrix produced by the female at the root surface. J2s may emerge from the eggs at any time the soil temperature is above freezing, depending on the species. The population density in a field declines during the overwinter period and during rotations with non-hosts.

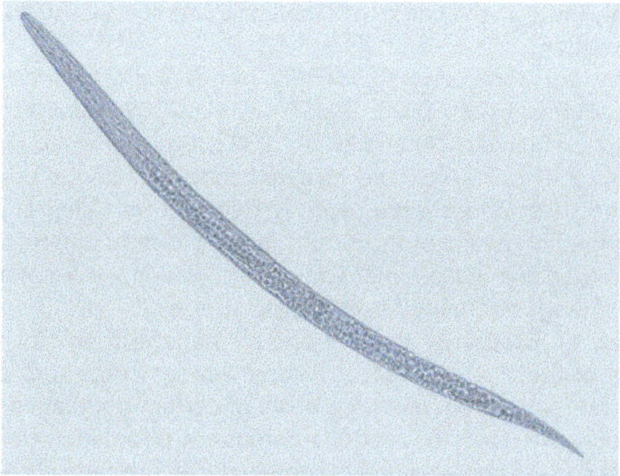

Figure 6 Root-knot (*Meloidogyne* sp.) nematode J2. Note the delicate stylet, which makes the nematode difficult to distinguish under low magnification (photo courtesy of Zak Ralston).

4.2 Diagnosis

The above-ground symptoms of root-knot disease are non-specific and range from subtle to severe chlorosis and stunting. Confirming the presence of *Meloidogyne* in a soybean field is not difficult; the root galls are distinctive and unlikely to be mistaken for other problems. Observation of root galling in association with above-ground symptoms is usually sufficient to diagnose the cause of the disease. The J2 can readily be extracted from soil by means of standard techniques with care to choose the proper sieve aperture size (van Bezooijen, 2006).

Identification of the species present in a given field is essential for management, not only because genetic resistance in soybean is somewhat species-specific, but also because more than one species may be found in a field and each species has a very wide host range that would affect choices of rotation crops for management. Unfortunately, identification of *Meloidogyne* species is not a simple task. The low-tech option for species identification is a differential host test (Hunt and Handoo, 2009), which can take 60 days or more. Species can be identified by trained personnel by means of preparation and examination of several female perineal patterns (Hunt and Handoo, 2009), an exacting and somewhat subjective process. Polymerase chain reaction analysis and other molecular methods may also be used (Adam et al., 2007; Blok and Powers, 2009) and are required for the confirmation of species identified by other methods.

Races of various *Meloidogyne* species have been described (Koenning, 2015), but the differential host tests for race identification do not include soybean and may have no significance for soybean production in itself. Race designations for local populations may be important for cropping system decisions (Fourie et al., 2006).

4.3 Management

Root-knot nematode infestations are a challenge for management in soybean. There are no economical or environmentally-friendly nematicides available, although nematicide research and development continues (Yu and Potter, 2008). Sustainable management can be achieved primarily through the use of resistant or tolerant soybean cultivars, along with non- or poor-host rotations.

Soybean breeders are continuing to identify genes for resistance to *Meloidogyne* species (Carpentieri-Pipolo et al., 2005; Shearin et al., 2009; Yates et al., 2010) and to develop useful resistant cultivars (Anand et al., 2002; Boerma et al., 2011; Chen et al., 2004; Shannon et al., 2007a, 2007b), and research continues on the uses of resistance in areas where root-knot nematodes have recently been found (Allen et al., 2005; Kruger et al., 2008). Sanitation, by means of thorough cleaning of equipment before it is moved from one field to another, and vigilant management of weeds are recommended practices due to the wide host ranges of many *Meloidogyne* species.

Success in the use of rotation or cover crops to manage root-knot disease depends on prior knowledge of the species present. Graminaceous crops such as grain sorghum (*Sorghum bicolor*) and maize (*Zea mays*) are hosts of certain populations of *M. incognita*, but in short-term rotations, each may reduce nematode population densities effectively. Rotation crops must fit into the economic framework of the farm as well; for example, many non-host crops may be highly effective, but if the crops are not marketable, they represent losses to the farmers as well as to local food security. For example, velvet bean

(*Mucuna pruriens*) and bahiagrass (*Paspalum notatum*) were both highly effective at reducing unidentified *Meloidogyne* species and the soybean cyst nematode, *H. glycines*, and at increasing subsequent soybean yields in field trials (Weaver et al., 1998), but the limited economic benefit from these crops may reduce their consideration for rotations in some areas.

Many different approaches to management of root-knot nematodes have been explored, and a common theme is that these approaches may be partially effective and should be incorporated in integrated management schemes. The results seem to be somewhat site-specific, and most would require some research to determine their applicability in a given situation. Some examples include the use of organic amendments (Everts et al., 2006; Zasada et al., 2008); cover crops, green manure crops and other plant additives (Adekunle, 2011; Zasada et al., 2003); biofumigation (Masler et al., 2010; Zasada et al., 2009); biological agents or products (Adekunle and Akinsanmi, 2005; Meyer et al., 2004; Nitao et al., 2001; Nitao et al., 2002; Siddiqui et al., 2005; Xia et al., 2011) and combinations of approaches (El-Sherif and Ismail, 2009; Everts et al., 2006),

5 Lesion nematodes, *Pratylenchus* spp.

Pratylenchus species are commonly known as lesion or root-lesion nematodes. They are usually classified as migratory endoparasites, although they may also feed ectoparasitically on soybean (Rebois and Huettel, 1986). Lesion nematodes directly damage soybean roots by feeding within the root cortex, causing necrotic lesions that may coalesce as nematode population densities increase within the root, resulting in extensive necrosis (Acosta and Malek, 1981) that may easily be mistaken for (or occur simultaneously with) damage caused by root-rotting microorganisms.

Soybean is a host for a number of *Pratylenchus* species (Ferris et al., 2017), including but not limited to *P. agilis, P. alleni, P. brachyurus, P. coffeae, P. crenatus, P. hexincisus, P. neglectus, P. penetrans, P. scribneri, P. thornei, P. vulnus* and *P. zeae* (Acosta and Malek, 1979, 1981; Acosta et al., 1979; Briar et al., 2007; Dickerson, 1979; Owen et al., 2014; Rebois and Golden, 1985; Schmitt and Barker, 1981; Wartman and Bernard, 1985; Zirakparvar, 1980, 1982). Pathogenicity to soybean varies by geographic region, perhaps due to differences in temperature optima, and even by isolate (Machado et al., 2007; Silva and Inomoto, 2002).

Lesion nematodes reproduce either parthenogenically or sexually, depending on the species. During a growing season, the generation time may be from four to eight weeks (Loof, 1991). Lesion nematodes are vermiform throughout their life cycle after hatching, and their development follows the typical pattern of hatching at the J2 stage, undergoing three additional juvenile stages before the adult stage with accompanying moults between stages. All stages are infective and can be extracted from soil and roots.

5.1 Diagnosis

Lesion nematodes may reduce yields without causing notable symptoms above ground. As with most root-parasitic nematodes, symptoms of severe infestations include chlorosis and stunting.

Diagnosis of lesion nematodes as the cause of soybean disease is not straightforward. The optimum time for sampling is generally later in the season (MacGuidwin and Bender, 2012). Sampling requires not only taking samples from greater depths than are normally taken in soybean (i.e., 15 to 20 cm) (Owen et al., 2014; Pudasaini et al., 2006), but also extracting worms from both soil and roots (MacGuidwin and Bender, 2012). In soils with certain characteristics, such as hard pans or high clay content, deep sampling may not be possible, but these characteristics also limit root penetration. Depth of sampling should be determined on site.

Standard techniques are sufficient to yield reasonable estimates of lesion nematode population densities (Hallman and Viaene, 2013; MacGuidwin and Bender, 2012; van Bezooijen, 2006). Soil and root samples from soybean are likely to contain multiple species of plant-parasitic nematodes, and highly likely to contain *Pratylenchus* as this is one of the more common genera. Consultation with a diagnostician trained in nematode identification is recommended especially if *Pratylenchus* is the predominant genus in a sample, as the species differ in pathogenicity and host range (Rebois and Golden, 1985; Zirakparvar, 1980).

Pratylenchus spp. are distinguished from other common soybean-parasitic nematodes by their short but robust stylet, strong cephalic sclerotization and a conoid to bluntly rounded tail (Fig. 7). The females have a single ovary outstretched anteriorly from the vulva, which may be found located in 70% or more of the total body length measured from the head (Castillo and Vovlas, 2007). Identification of species is based on a number of morphological characteristics, including lip and tail morphology, stylet length and cuticular features, among others (Castillo and Vovlas, 2007). Molecular confirmation of species can usually be accomplished with standard protocols.

Because *Pratylenchus* is a common genus in agricultural soils, its presence alone is not evidence of potential economic injury to soybean. Damage thresholds have been established for some species, but because the host–parasite relationship depends on factors that vary by location, threshold numbers are not generalizable. A given *Pratylenchus* species may be of concern when the species is adapted to soybean and the population density exceeds those of other plant-parasitic nematodes in a sample.

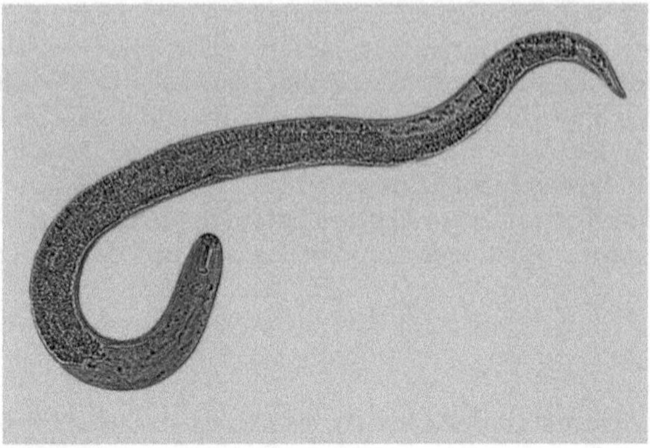

Figure 7 Lesion (*Pratylenchus* sp.) nematode female (photo courtesy of Abasola Simon).

5.2 Management

Lesion nematodes are quite common in most row crop production systems. As noted earlier, there are differences within and between species for adaptation to soybean and potential for damage; identification of the species present in a particular field is critical because *Pratylenchus* spp. as a group have broad host ranges. Cropping system and weed management decisions must include consideration of the host range.

Numerous studies have addressed lesion nematode management tactics as alternatives to nematicides treatment. Examples include the use of cover crops (Kimpinski et al., 2000; Pedersen and Rodriguez-Kabana, 1991) and alternative crops (Johnson et al., 1998; Koenning et al., 1985; Kratochvil et al., 2004; Owen et al., 2014), manipulation of planting date (Koenning et al., 1985) and/or additions of amendments (Cohen et al., 2005; Kratochvil et al., 2004). All these studies have in common that their conclusions are mostly not generalizable, i.e., they are site or region specific for biologic or economic reasons. Research on and practice of management of *Pratylenchus* species will need to take these issues into account.

6 Reniform nematodes

Rotylenchulus reniformis, the reniform nematode, is one of only a few species described in the genus and is the most economically important among them, primarily due to its pathogenicity on cotton (*Gossypium hirsutum*). *R. reniformis* has a worldwide distribution in tropical, subtropical and adjacent temperate regions (McGawley and Overstreet, 2015; Nakasono, 2004; Niblack, *personal observation*) and can parasitize more than 300 plant species, including soybean and other crop species (Davis and Webster, 2005; Robinson et al., 1997).

The symptoms of *R. reniformis* infection on soybean are similar to those caused by other root-pathogenic nematodes and include stunting, chlorosis, unthrifty growth and the production of empty seed pods (McGawley and Overstreet, 2015). Yield losses of 30% to 60% are possible in individual fields (McGawley and Overstreet, 2015).

The presence of many small, soil-covered, sticky masses on the roots of symptomatic plants, along with necrosis of the root system, is an indication of *R. reniformis* infection, but there is no visible galling. Diagnosis of this nematode, of course, requires observations of females in soil extracts.

6.1 Life cycle and disease cycle

Reniform nematode population densities can become very high when susceptible crops are grown. Robinson (2007) reported that 60 to 490 pre-parasitic (immature) females per cm^3 soil have been observed. Immature *R. reniformis* females are small hoplolaimoid nematodes with body lengths of 340 to 420 μm and a stylet with posteriorly sloping knobs (Siddiqi 1972) (Fig. 8). Males, which may or may not be required for reproduction (Nakasono, 2004), have reduced stylets and oesophageal structures (Fig. 8) and remain vermiform and outside the root. During egg production, the female produces a gelatinous matrix that may completely cover her posterior region, into which she deposits 59 to 72 eggs (Rebois, 1973) although much higher egg production has been reported (Robinson, 2007).

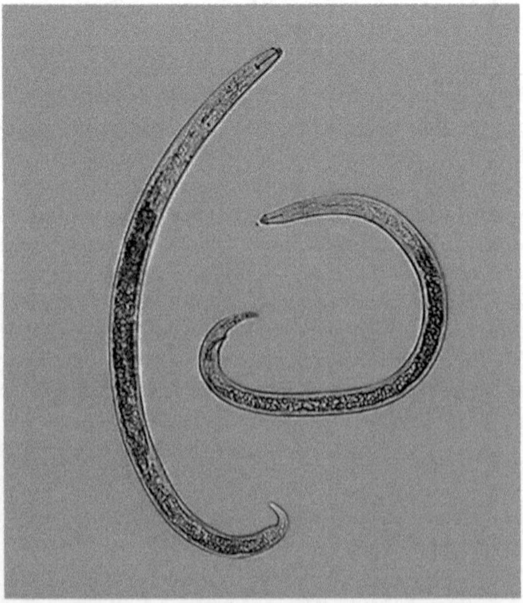

Figure 8 Pre-parasitic female (left) and male *Rotylenchulus reniformis* (photo courtesy of Jonathon D. Eisenback).

R. reniformis is a sedentary semi-endoparasitic pathogen that is unique in several ways compared with the other major sedentary parasites of soybean, root-knot and cyst nematodes. As with the others, the J2 emerges from the egg when conditions are conducive, but rather than infecting a root, the J2 undergoes the juvenile moults rapidly without shedding the cuticle and without feeding. Retention of the cuticle during development may be related to this nematode's remarkable ability to survive in dry soil (Robinson, 2007). Whereas root-knot and cyst nematodes infect at the J2 stage and become enlarged after initiating feeding sites and as they develop embedded within host tissue, a pre-parasitic vermiform female reniform nematode is the infective stage (Fig. 8). The female penetrates the epidermis and disrupts cortical root tissue in a path more perpendicular than parallel to the axis of the vascular cylinder as do root-knot and cyst J2. The reniform female induces formation of a feeding site beginning with a single endodermal cell, called a nurse cell. As the feeding site develops, the nurse cell is joined by numerous contiguous pericycle cells into a syncytial system with the characteristics of high metabolic activity (Rebois et al., 1975) seen in root-knot-nematode-induced giant cells and cyst-nematode-induced syncytia. The presence of many reniform-nematode-induced syncytial systems in a root can seriously impair root function, contributing to disease development. As the syncytium develops, the posterior portion of the female becomes enlarged and takes on the kidney shape for which the nematode is named (Fig. 9). She is a 'semi-endoparasite' because her head remains embedded in root tissue. The life cycle of *R. reniformis* can be completed on soybean in less than three weeks (Rebois, 1973), which contributes to potentially large population increases in a single growing season.

Figure 9 Female *Rotylenchulus reniformis* protruding from an infected root (photo courtesy of Jonathon D. Eisenback).

Population densities can be high at much greater depths than those at which soil samples are typically taken (Westphal et al., 2004). The vermiform stages are easily extracted from soil with standard techniques (Hallman and Viaene, 2013; van Bezooijen, 2006). This nematode's wide host range, short life cycle, ability to survive desiccation, adaptation to soils with high silt or clay content and deep soil distribution contribute to its successful spread and threat to susceptible crops.

6.2 Management

Nematicides are either unavailable or cost-prohibitive for use on soybean. The available alternatives for management are resistance or tolerance, soil amendments, cover crops and crop rotations (Gaur and Perry, 1991; Stetina et al., 2014). The use of resistant soybean cultivars in cropping systems is advisable and relatively easy to implement, but injudicious use of resistance will likely result in adaptation of *R. reniformis* to resistance (Leach et al., 2012) similar to that which occurs with other nematode-pathogen-resistant host systems.

The host range of *R. reniformis* makes the use of cover crops and crop rotations problematic, and recommendations for the use of these approaches will be region specific (Guertal et al., 1997). Likewise, the host range of this nematode is such that weed management is essential in infested fields (Davis and Webster, 2005; Robinson et al., 1997) even during non-host growing seasons. Crops that are poor or non-hosts are corn, grain sorghum, peanut and rice (Niblack et al., 2004). McGawley and Overstreet (1995) reported that wheat and oat may be useful in rotations for reniform nematode management as well, but only if the fields were left fallow during the remaining growing season after grain harvest.

A wide variety of seed treatments involving nematicidal or nemastatic compounds or organisms, as well as a newly invigorated interest in biological control, are evident in a number of regional and commodity reports, along with research reports in publications such as *Plant Disease Management Reports* (see link in references). An isolate of the parasitic bacterium *Pasteuria* may have utility for management of *R. reniformis* (Schmidt et al., 2010) in addition to its applications for the management of other plant-pathogenic nematodes.

7 Lance nematodes, *Hoplolaimus* spp.

Hoplolaimus species are known commonly as lance nematodes. These large (1 to 2 mm), robust, vermiform nematodes have a strong cephalic structure with very distinct stylets (40 to 50 μm; Fig. 10) and rounded tails (Fig. 11). Females have two ovaries extending in opposite directions from a vulva that is located mid-body, 50 to 60% of the body length from the anterior end. Among the species that have been reported as parasites of soybean, *Hoplolaimus columbus* is clearly the most pathogenic and may cause up to 48% yield suppression in susceptible, intolerant soybean cultivars (Noe, 1993). *Hoplolaimus galeatus* has been widely reported in association with soybean in the United States, but there is much less evidence of its pathogenicity. In addition, recent phylogenetic analysis of *Hoplolaimus* species showed that *H. galeatus* has been over-reported and that populations identified morphometrically as *H. galeatus* are likely to be *H. stephanus* (Holguin et al., 2015). *H. magnistylus*, another species very similar morphologically to *H. galeatus*, was recently reported on soybean (Donald et al., 2013), but its pathogenicity is not established. Few other reports of lance nematodes pathogenic to soybean exist.

Hoplolaimus species are generally classified as migratory endoparasites, but may also feed ectoparasitically. They are vermiform in all life stages and infective from the J2 to adult stage. *H. columbus* can complete its life cycle in about 7 weeks (Schmitt and Noel, 1984), slightly longer than the generation times of other endoparasitic nematodes. This nematode's endoparasitic feeding is on cortical, endodermal and phloem cells (Fortuner, 1991), which results in collapse of the cells and formation of necrotic lesions. Secondary root growth may be reduced. Above-ground symptoms are those typical of nematode injury: stunting and chlorosis, with a reduction in the number of pods produced.

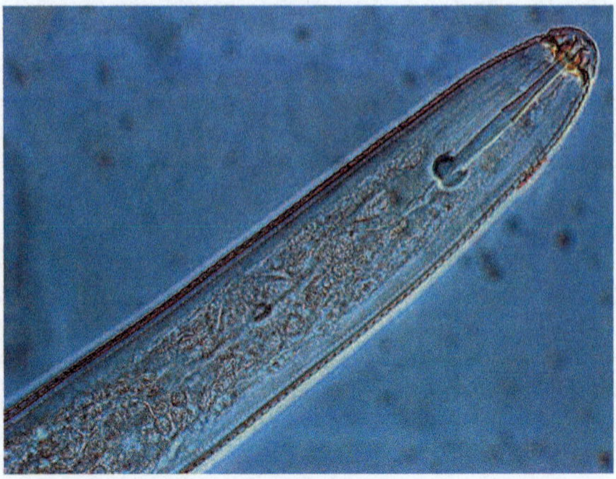

Figure 10 Oesophageal region of a lance (*Hoplolaimus* sp.) nematode (photo courtesy of Tesfamariam Mekete).

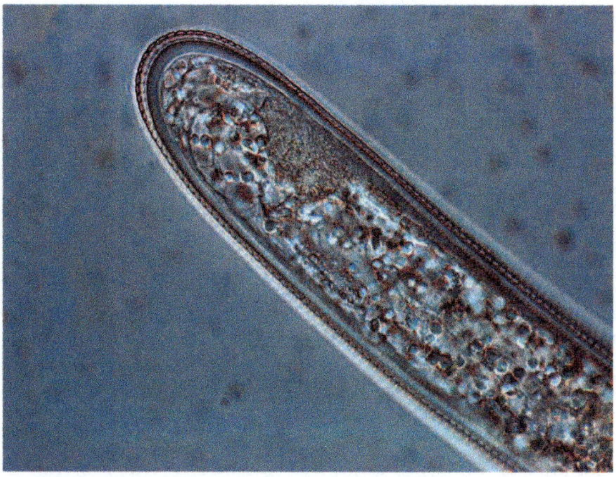

Figure 11 Tail of a lance (*Hoplolaimus* sp.) nematode (photo courtesy of Tesfamariam Mekete).

7.1 Management

Hoplolaimus columbus can reduce soybean yield 10% at a Pi as low as 4 worms/100 cm^3 soil (Noe, 1993) up to 100 to 200 worms/100 cm^3 soil (Schmitt and Bailey, 1990), but no determinations of threshold densities have been made for other *Hoplolaimus* species on soybean. Threshold densities are affected by so many factors that they are not generalizable. *Hoplolaimus* individuals appear commonly in soil samples collected from agronomic crops in the United States; thus, identification of species may be important when population densities are equivalent to or higher than the values reported for *H. columbus*. The dual assays recommended for extraction of *Pratylenchus* (MacGuidwin and Bender, 2012) are also useful for *Hoplolaimus*.

Hoplolaimus species have wide host ranges, which limits the utility of crop rotation for sustainable management. Cropping systems and their impact on nematode populations and yield are affected by many factors, including soil types, fertilizer amendments, organic amendments and tillage (Noe et al., 1991) and may, therefore, be considered site-specific strategies. Although resistance to *Hoplolaimus* is not available, several soybean cultivars have shown high levels of tolerance in the field (Koenning, 2002; Mueller et al., 1988; Nyczepir and Lewis, 1979; Schmitt and Imbriani, 1987). Soybean yields have also been increased by subsoiling in soils with hard pans that limit root penetration (Koenning, 2002). Weed management must be considered in fields infested with high levels of *Hoplolaimus* spp. due to their host ranges.

8 Other nematodes

In some areas, nematodes other than the major economically important pathogens may be of concern. The nematode host database (Ferris et al., 2017) lists about 60 species associated with soybean production. In the list *Diseases of Soybeans* compiled by Lim and Chamberlain (2001), there are 12 common-name groups of nematode genera and species included as soybean disease-causing agents. In addition to the ones already covered in this chapter are pin (*Paratylenchus* spp.), ring (*Criconemella ornata*), sheath (*Hemicycliophora* spp.), spiral (*Helicotylenchus* spp.), sting (*Belonolaimus gracilis* and *Belonolaimus longicaudatus*), stubby-root (*Paratrichodorus minor*) and stunt (*Quinisulcius acutus* and *Tylenchorhynchus* spp.) nematodes. Limited information is available on the specific soybean–nematode relationships for these species, but some may be found in reviews (e.g., Hartman et al., 2015; Niblack et al., 2004) and on websites. Nematode species with wide host ranges may have the opportunity to adapt to soybean as production expands into new areas (Fourie et al., 2001).

Readers may find of interest some recent reports of nematode parasites of soybean, such as that of a stunt nematode, *Merlinius brevidens* (Owen et al., 2014), a pin nematode, *Paratylenchus neoamblycephalus* (Ankrom et al., 2017) and a root-knot nematode, *Meloidogyne graminicola* (Long et al., 2017). These are fairly common species for which clear evidence of pathogenicity to soybean is still lacking. The sting nematode, *B. longicaudatus*, a well-established soybean pathogen, was recently reported from a new location (Handoo et al., 2010), and a new species of needle nematode, *Longidorus glycines*, was described (Ye and Robbins, 2004). It is also interesting that *Helicotylenchus* spp. were included in the Lim and Chamberlain (2001) list, because although spiral nematodes are common in soybean fields and may cause small lesions on roots (Orbin, 1973), these migratory semi-endoparasites are not thought to cause serious diseases in soybean (Johnson et al, 1998; Pedersen and Rodriguez-Kabana, 1991).

Although nematodes occur in polyspecific assemblages, studies have not been common on nematode–nematode interactions in the field or in controlled environments. Studying disease interactions among obligate parasites is fraught with technical difficulties, not excluding the long life cycles (compared with bacterial or fungal pathogens). Studies including more than one nematode species, therefore, tend to focus on population dynamics, with some exceptions (Johnson et al., 1998; Guy and Lewis, 1987; Lawn and Noel, 1986, 1990; Lim and Castillo, 1978; McGawley and Chapman, 1983; McSorley and Gallaher 1993, 1994; McSorley and Dickson, 1989, 1993, 1994; Melakeberhan (1998); Melakeberhan and Dey (2003); Pedersen and Rodriguez-Kabana, 1991; Rodriguez-Kabana and Thurlow, 1980; Todd, 1991).

9 Conclusion

Soybean supports nematode assemblages that are highly likely to contain one or more potentially pathogenic nematode species, which may or may not cause sufficient disease to result in measurable yield loss (depending on cultivar susceptibility, cropping system soil and other environmental factors) or warrant control measures. The simple presence of a pathogenic nematode does not by itself signal a nematode problem. Exceptions include the soybean cyst nematode and those which are damaging at low population densities, such as sting and needle nematodes. As we have noted several times, for most nematodes, population densities significantly higher than those of other species present signals a problem. Once that observation is made, species identification is essential as the key to development of sustainable cropping systems and integrated pest management strategies. As such, involvement of personnel with specialized training in nematode species identification is crucial.

Particular attention should be paid to sampling for nematodes, as the commonly recommended 15 to 20 cm depth of sampling may be insufficient for accurate assessments

(Pudasaini et al., 2006; McSorley and Dickson, 1990). Extraction of nematodes from soil and roots for direct observations can be accomplished with relatively low-tech equipment (Hallman and Viaene, 2013; van Bezooijen, 2006).

Assessment of sedentary endoparasites may require special procedures (Perry et al., 2009; Niblack et al., 2009) and root population densities of migratory endoparasites may be important depending on time of sampling (MacGuidwin and Bender, 2012).

The use of host resistance is the most useful and flexible tactic for nematode management.

Good resistance is available for the soybean cyst nematode, some root-knot nematode species and the reniform nematode, but resistant soybean cultivars adapted for any particular region may be unavailable. Resistance is not available for important migratory endoparasites, including lance and lesion nematodes (although tolerance may be an option), or for migratory ectoparasites.

The available options beyond host resistance vary by nematode and geographical region. The general categories of options are also affected by economic considerations. For example, if the use of a rotation crop is functionally equivalent to a bare fallow treatment (i.e., non-income-producing) or is impossible to implement due to lack of proper equipment, then it has little chance of being adopted.

For additional thoughts on the options for nematode management, as well as needs for research in this area, the reader is referred to several excellent reviews on the subject, including Barker and Koenning (1998), Bridge (1996), Duncan and Noling (1998) and Stirling (2008). Some readers may be interested to find more information and references regarding the important roles of nematodes in crop production in Briar et al. (2007), Neher and Olson (1999) and Wang and McSorley (2005).

10 Where to look for further information

Food and Agriculture Organization of the United Nations (2017), Commonwealth Institute of Helminthology, *C. I. H. Descriptions of Plant Parasitic Nematodes.* CABI. http://www.fao.org/faostat/en/#home

Hallman, J. and Viaene, N. (2013) 'PM 7/119 (1) Nematode extraction', *OEPP/EPPO Bulletin* 43(3), 471–95, DOI: 10.1111/epp.12077.

Hartman, G., Rupe, J., Sikora, E., Domier, L, Davis, J. and Steffey, K., eds., (2015), *Compendium of Soybean Diseases and Pests 5th Edition*, APS Press.

Mai, W. F., McMullen, P. G., Lyon, H. H. and Loeffler, K. (1996), *Plant-parasitic Nematodes, A Pictorial Key to Genera*, Cornell University Press.

Perry, R. N., Moens, M. and Starr, J. L., eds. (2009), *Root-knot Nematodes*, CABI.

Plant Disease Management Reports, American Phytopathological Society, http://www.plantmanagementnetwork.org/pub/trial/search.aspx

Schmitt, D. P., Wrather, J. A. and Riggs, R. D. (eds.) (2004), *Biology and Management of the Soybean Cyst Nematode*, 2nd ed. Schmitt & Associates of Marceline, Marceline, MO.

Sharma, S. B. (ed.) (1998), *The Cyst Nematodes*. Kluwer Academic Publishers.

van Bezooijen, J. (2006), Methods And Techniques For Nematology, https://www.wur.nl/upload_mm/4/e/3/f9618ac5-ac20-41e6-9cf1- c556b15b9fa7_MethodsandTechniquesforNematology.pdf

11 References

Acharya, K., Tande, C., and Byamukama, E. (2016), 'Determination of *Heterodera glycines* virulence phenotypes occurring in South Dakota', *Plant Disease,* 100(11), 2281–6.

Acosta, N., Edwards, D.I., and Malek, R.B. (1979), 'Susceptibility of soybean cultivars to Pratylenchus scribneri', *Journal of Agriculture of the University of Puerto Rico* 63, 103–10.

Acosta, N., and Malek, R. B. (1979), 'Influence of temperature on population development of eight species of *Pratylenchus* on soybean', *Journal of Nematology* 11, 22932.

Acosta, N., and Malek, R.B. (1981), 'Symptomatology and histopathology of soybean roots infected by *Pratylenchus scribneri* and *Pratylenchus alleni*', *Journal of Nematology* 13, 612.

Adam, M. A. M., Phillips, M. S. and Blok, V. C. (2007), 'Molecular diagnostic key for identification of single juveniles of seven common and economically important species of root-knot nematode (*Meloidogyne* spp.)', *Plant Pathology*, 56, 190–7.

Adekunle, O. (2011), 'Amendment of soil with African marigold and sunn hemp for management of *Meloidogyne incognita* in selected legumes', *Crop Protection* 30, 1392–5.

Adekunle, O., and Akinsanmi, O. (2005), 'Bioactivity of *Fusarium oxysporum* f. sp. *glycines* and *Sclerotium rolfsii* filtrates on egg hatching, survival and infectivity of juveniles of *Meloidogyne incognita* race 2', *Australian Journal of Experimental Agriculture* 45, 99–102.

Adeniji, M. O., Edwards, D. I., Sinclair, J. B., and Malek, R. B. (1975), 'Interrelationship of *Heterodera glycines* and *Phytophthora megasperma* var. *sojae* in soybeans', *Phytopathology*, 65, 722–5.

Agrios, G. N. (2005), *Plant Pathology, 5th ed.*, Elsevier Academic Press.

Allen, T. W., Bradley, C. A., Sisson, A. J., Byamukama, E., Chilvers, M. I., Coker, C. M., Collins, A. A., Damicone, J. P., Dorrance, A. E., Dufault, N. S., Esker, P. D., Faske, T. R., Giesler, L. J., Grybauskas, A. P., Hershman, D. E., Hollier, C. A., Isakeit, T., Jardine, D. J., Kelly, H. M., Kemerait, R. C., Kleczewski, N. M., Koenning, S. R., Kurle, J. E., Malvick, D. K., Markell, S. G., Mehl, H. L., Mueller, D. S., Mueller, J. D., Mulrooney, R. P., Nelson, B. D., Newman, M. A., Osborne, L., Overstreet, C., Padgett, G. B., Phipps, P. M., Price, P. P., Sikora, E. J., Smith, D. L., Spurlock, T. N., Tande, C. A., Tenuta, A. U., Wise, K. A., and Wrather, J. A. (2017), 'Soybean yield loss estimates due to diseases in the United States and Ontario, Canada, from 2010 to 2014', *Plant Health Progress* 18, 19–27. http://dx.doi.org/10.1094/PHP-RS-16-0066.

Allen, J., Schmidt, M., and Bond, J. (2005), 'Incidence of *Meloidogyne incognita* and development of resistant soybean germplasm in Illinois', *Plant Health Progress.* doi:10.1094/PHP-2005-0606-01-RS.

Anand, S., Sleper, D., Young, L., Arelli, P., Shannon, J., and Wrather, J. (2002), 'Registration of S96-2692 soybean germplasm line resistant to three soybean nematodes', *Crop Science* 42, 1753–4.

Ankrom, K. E., Lopez-Nicora, H. D., Niblack, T. L., and Lindsey, L. E. (2017), 'First report of a pin nematode (*Paratylenchus neoamblycephalus*) from soybean in Ohio', *Plant Disease* 101(7), 1330. (doi:10.1094/PDIS-12-16-1728-PDN).

Atkinson, R. E. (1944), 'Diseases of soybeans and peanuts in the Carolinas in 1943', *Plant Disease Reporter Supplement* 148, 254–9.

Baird, S. M., and Bernard, E. C. (1984), 'Nematode population and community dynamics in soybean wheat cropping and tillage regimes', *Journal of Nematology* 16, 379–86.

Barker, K. R., and Koenning, S. R. (1998) 'Developing sustainable systems for nematode management', *Annual Review of Phytopathology* 36, 165–205.

Barker, K. R., Koenning, S. R., and Schmitt, D. P. (2004), 'Population density based management', in *Biology and Management of Soybean Cyst Nematode*, 2nd *ed.*, D. P. Schmitt, J. A. Wrather, and R. D. Riggs (eds.), pp. 89–110. Schmitt & Associates of Marceline, Marceline, MO.

Barker, K. R., and McGawley, E. C. (1998), 'Interrelations with other microorganisms and pests', in *The Cyst Nematodes*, S. B. Sharma (ed.), pp. 266–92. Chapman & Hall.

Barker, K. R., and Olthof, T. H. A. (1976), 'Relationships between nematode population densities and crop responses', *Annual Review of Phytopathology* 14, 327–53.

Barker, K. R., Townshend, J. K., Bird, G. W., Thomason, I. J., and Dickson, D. W. (1986), 'Determining nematode population responses to control agents' in Hickey, K. D. (ed.), *Methods for evaluating pesticides for control of plant pathogens*, pp. 283–96. American Phytopathological Society Press.
Beeman, A., and Tylka, G. L. (2017), 'Assessing the effects of Ilevo and Votivo seed treatments on reproduction, hatching, motility, and root penetration of the soybean cyst nematode, *Heterodera glycines*', *Plant Disease* (In press/First look-posted 9/7/17). doi:10.1094/PDIS-04-17-0585-RE
Bernard, E., and Jennings, P. (1997), 'Host range and distribution of the clover root-knot nematode, *Meloidogyne trifoliophila*', *Journal of Nematology* 29(4S), 662–72.
Blok, V. C., and Powers, T. O. (2009), 'Biochemical and molecular identification', in Perry, R. N., Moens, M., and Starr, J. L. (eds.), *Root-knot Nematodes*, pp. 98–118. CABI Publishing.
Boerma, H., Phillips, D., Walker, D., Missaoui, A., Monteros, M., Ha, B., and Wood, E. (2011), 'Registration of Asian Soybean Rust–resistant soybean germplasm G01-PR16', *Journal of Plant Registrations* 5, 118–22.
Bond, J., and Wrather, J. A. (2004), 'Interactions with other plant pathogens and pests', in *Biology and Management of Soybean Cyst Nematode, 2nd ed.*, D. P. Schmitt, J. A. Wrather, and R. D. Riggs (eds.), pp. 111–29. Schmitt & Associates of Marceline, Marceline, MO.
Briar, S. S., Stinner, D., Miller, S. A., Grewal, P. S., and Somasekhar, N. (2007), 'Soil nematode community, organic matter, microbial biomass and nitrogen dynamics in field plots transitioning from conventional to organic management', *Applied Soil Ecology* 37, 256–66.
Bridge, J. (1996), 'Nematode management in sustainable and subsistence agriculture', *Annual Review of Phytopathology* 34, 201–25.
Brim, C. A., and Ross, J. P. (1966), 'Registration of Pickett Soybeans (Reg. No. 52)', *Crop Science* 6(3), 305.
Broderick, K. C. (2016), 'Diversity and virulence of soybean cyst nematode (*Heterodera glycines* Ichinohe) in Nebraska', Theses, Dissertations, and Student Research in Agronomy and Horticulture. Paper 109. University of Nebraska-Lincoln http://digitalcommons.unl.edu/agronhortdiss/109 (accessed 18 April 2017).
Brucker, E., Carlson, S., Wright, E., Niblack, T., and Diers, B. (2005), 'Rhg1 alleles from soybean PI 437654 and PI 88788 respond differentially to isolates of *Heterodera glycines* in the greenhouse', *Theoretical and Applied Genetics* 111(1), 44–9.
Brzostowski, L. F., Schapaugh, W. T., Rzodkiewicz, P. A., Todd, T. C., and Little, C. R. (2014), 'Effect of host resistance to *Fusarium virguiliforme* and *Heterodera glycines* on sudden death syndrome disease severity and soybean yield', *Plant Health Progress*. doi:10.1094/PHP-RS-13-0100.
Caldwell, B. E., Brim, C. A., and Ross, J. P. (1960), 'Inheritance of resistance to soybean cyst nematode, *Heterodera glycines*', *Agronomy Journal* 52(11), 635–636.
Carpentieri-Pipolo, V., Souza, A., Kikuchi, A., Mandarino, J., and Carrao-Panizzi, M. (2005), 'Association of isoflavonoids with the incompatible response of soybean roots to *Meloidogyne incognita* race 3', *Nematropica* 35, 103–10.
Carroll, L. (1872), *Through the Looking-Glass and What Alice Found There*. Macmillan.
Castillo, P., and Vovlas, N. (2007), *Pratylenchus (Nematoda:Pratylenchidae) Diagnosis, Biology, Pathogenicity and Management*, Koninklijke Brill NV, Leiden, the Netherlands.
Chen, P., Riggs, R., Rupe, J., and Sneller, C. (2004), 'Registration of 'Ozark' soybean', *Crop Science* 44, 1872–3.
Chen, S., Porter, P. M., Reese, C. D., and Stienstra, W. C. (2001), 'Crop sequence effects on soybean cyst nematode and soybean and corn yields', *Crop Science* 41(6), 1843–9.
Clay, K., and Kover, P. X. (1996), 'The Red Queen hypothesis and plant/pathogen interactions', *Annual Review of Phytopathology* 34(1), 29–50.
Cohen, M. F., Mazzola, M., and Yamasaki, H. (2005), '*Brassica napus* seed meal soil amendment modifies microbial community structure, nitric oxide production and incidence of *Rhizoctonia* root rot', *Soil Biology and Biochemistry* 37, 1215–27.
Colgrove, A. L., and Niblack, T. L. (2005), 'The differential effect of resistant soybean on adult sex ratios of *Heterodera glycines*', *Journal of Nematology* 37(2), 161–7.

Colgrove, A. L., and Niblack, T. L. (2008), 'Correlation of Female Indices from virulence assays on inbred lines and field populations of *Heterodera glycines*', *Journal of Nematology* 40(1):39–45.

Colgrove, A. L., Smith, G. S., Wrather, J. A., Heinz, R. D., and Niblack, T. L. (2002), 'Lack of predictable race shift in *Heterodera glycines*-infested field plots', *Plant Disease* 86(10), 1101–8.

Concibido, V. C., Denny, R. L., Lange, D. A., Orf, J. H., and Young, N. D. (1996), 'RFLP mapping and marker-assisted selection of soybean cyst nematode resistance in PI 209332', *Crop Science* 36(6), 1643– 1650.

Concibido, V. C., Lange, D. A., Denny, R. L., Orf, J. H., and Young, N. D. (1997), 'Genome mapping of soybean cyst nematode resistance genes in 'Peking', PI 90763, and PI 88788 using DNA markers', *Crop Science* 37(1), 258–264.

Concibido, V. C., Diers, B. W., and Arelli, P. R. (2004), 'A decade of QTL mapping for cyst nematode resistance in soybean', *Crop Science* 44(4), 1121–31.

Cook, D. E., Lee, T. G., Guo, X., Melito, S., Wang, K., Bayless, A., Wang, J., Hughes, T. J., Willis, D. K., Clemente, T. E., Diers, B.W., Jiang, J., Hudson, M. E. , and Bent, A. F. (2012), 'Copy number variation of multiple genes at *Rhg1* mediates nematode resistance in soybean', *Science* 338(6111), 1206–9.

Dave, G. S. (1975), 'Interrelationships of *Rhizoctonia solani* with *Heterodera glycines*, *Pratylenchus scribneri* and *Tylenchorhynchus martini* on 'Clark 63' soybeans', PhD thesis, University of Illinois, Urbana, Illinois.

Davis, E. L., R. S. Hussey, and Baum, T. J. (2004), 'Getting to the roots of parasitism by nematodes', *Trends in Parasitology* 20(3), 134–41.

Davis, R. F., and Webster, T. M. (2005), 'Relative host status of selected weeds and crops for *Meloidogyne incognita* and *Rotylenchulus reniformis*', *Journal of Cotton Science* 9, 41–46.

Dickerson, O. J. (1979), 'The effects of temperature on *Pratylenchus scribneri* and *Pratylenchus alleni* populations on soybeans and tomatoes', *Journal of Nematology* 11, 23–26.

Donald, P., Holguin, C. M., and Agudelo, P. (2013), 'First report of lance nematode (*Hoplolaimus magnistylus*) on corn, soybean, and cotton in Tennessee', *Plant Disease* 97, 1389, http://dx.doi.org/10.1094/PDIS-09-12-0881-PDN.

Duncan, L. W., and Noling, J. W. (1998), 'Agricultural sustainability and nematode integrated pest management', in Plant and Nematode Interactions, *Agronomy Monograph* No. 36, pp. 251–87. American Society of Agronomy, Crop Science Society of America, Soil Science Society of America.

Dusenbery, D. B. (1987), 'Prospects for exploiting sensory stimuli in nematode control', in *Vistas on Nematology*, J. A. Veech and D. W. Dickson (eds.), pp. 131–5. Society of Nematologists, Inc.

Eisenback, J. D. (1993), 'Interactions between nematodes in cohabitance', in *Nematode Interactions*, Khan, M. W. (ed.), pp. 134–4. Chapman & Hall. DOI:10.1007/978-94-011-1488-2_7.

Elling, A. (2013), 'Major emerging problems with minor *Meloidogyne* species', *Phytopathology* 103, 1092–102.

El-Sherif, A., and Ismail, A. (2009), 'Integrated management of *Meloidogyne incognita* infecting soybean by certain organic amendments, *Bacillus thuringiensis*, *Trichoderma harzianum* and oxamyl with reference to NPK and total chlorophyll status', *Plant Pathology Journal* 8, 159–64.

Epps, J. M. (1971), 'Recovery of soybean cyst nematodes (*Heterodera glycines*) from the digestive tract of blackbirds', *Journal of Nematology* 3(4), 417–19.

Everts, K., Armentrout, D., Gallagher, L., Sardanelli, S., and Kratochvil, R. (2006), 'Root-knot and root-lesion nematode suppression by cover crops, poultry litter, and poultry litter compost', *Plant Disease* 90, 487–92.

Faghihi, J., Donald, P. A., Noel, G., Welacky, T. W., and Ferris, V. R. (2010), 'Soybean resistance to field populations of *Heterodera glycines* in selected geographic areas', *Plant Health Progress* doi:10.1094/PHP-2010-0426-01-RS.

Ferris, H. (1987), 'Components and techniques of integrated pest management threshold determinations for soilborne pathogens', *Plant Disease* 71, 452–55.

Ferris, H., Caswell-Chen, E. P., and Westerdahl, B. B. 'NEMABASE A Database of the Host Status of Plants to Nematodes', accessed January, 2017, http://plpnemweb.ucdavis.edu/nemaplex/Nemabase2010/Nemabase%20Search%20Menu.htm

Fortuner, R. (1991), 'The Hoplolaiminae', in Nickle, W. R. (ed.), *Manual of Agricultural Nematology*, pp. 669–719. Marcel Dekker.

Fourie, H., Loots, G.C., and McDonald, A.H. (2001), 'Plant-parasitic nematodes in field crops in South Africa. 6. Soybean', *Nematology* 3, 447–54.

Fourie, H., McDonald, A. H., and De Waele, D. (2006), 'Host suitability of South African and foreign soybean cultivars to *Meloidogyne incognita* race 2', *South African Journal of Plant and Soil* 23(2), 132–7.

Franck, A. B. (1882), 'Gallender *Anguillula raciciola* Greff on *Soja hispida, Medicago sativa, Lactuca sativa,* und *Pirus communis*', *Verhandlungen des Botanischen Vereins der Provinz Brandenburg* 23:45–55.

Francl, L. J. (1986), 'Spatial analysis of *Heterodera glycines* populations in field plots,', *Journal of Nematology* 18(2), 183–9.

Francl, L. J., and Dropkin, V. H. (1986), '*Heterodera glycines* population-dynamics and relation of initial population to soybean yield', *Plant Disease* 70(8), 791–5.

Francl, L. J., Wyllie, T. D., and Rosenbrock, S. M. (1988), 'Influence of crop rotation on population density of *Macrophomina phaseolina* in soil infested with *Heterodora glycines*', *Plant Disease* 72(9), 760–4.

Frohning, J. R. (2013), 'Evaluation of *Rhizoctonia solani - Heterodera glycines* interactions on soybean', MSc thesis, University of Illinois, Urbana, Illinois.

Gao, X., Jackson, T. A, Hartman, G. L., and Niblack, T. L. (2006), 'Interactions between the soybean cyst nematode and *Fusarium solani* f. sp. *glycines* based on greenhouse factorial experiments', *Phytopathology* 96(12), 1409–15.

Gardner, M., Verma, A., and Mitchum, M. G. (2015), 'Emerging roles of cyst nematode effectors in exploiting plant cellular processes', in C. Escobar and C. Fenoll (eds.), *Advances in Botanical Research Vol. 73, Plant Nematode Interactions: A View on Compatible Interrelationships. Chapter 11*, Elsevier. pp. 259–91.

Gaur, H. S., and Perry, R. N. (1991). 'The biology and control of the plant parasitic nematode *Rotylenchulus reniformis*', *Agricultural Zoology Reviews* 4, 177–212.

Gavassoni, W. L., Tylka, G. L., and Munkvold, G. P. (2001), 'Relationships between tillage and spatial patterns of *Heterodera glycines*' *Phytopathology* 91(6), 534–45.

Gavassoni, W. L., Tylka, G. L., and Munkvold, G. P. (2007), 'Effects of tillage practices on dissemination and spatial patterns of *Heterodera glycines* and soybean yield', *Plant Disease* 91(8), 973–8.

Gilligan, C. A. (1983), 'Modeling of soilborne pathogens', *Annual Review of Phytopathology* 21(1), 45–64.

Glover, K. D., Wang, D., Arelli, P. R., Carlson, S. R., Cianzio, S. R., and Diers, B. W. (2004), 'Near isogenic lines confirm a soybean cyst nematode resistance gene from PI 88788 on linkage group J', *Crop Science* 44(3), 936–941.

Goffart, H. (1951) in: *Nematoden der Kulturpflanzen Europas*, p. 144. Parey-Verlag, Berlin.

Golden, A. M., Epps, J. M., Riggs, R. D., Duclos, L. A., Fox, J. A., and Bernard, R. L. (1970), 'Terminology and identity of infraspecific forms of the soybean cyst nematode (*Heterodera glycines*)', *Plant Disease Reporter* 54(7), 544–546.

Gommers, F. J., and Dropkin, V. H. (1977), 'Quantitative histochemistry of nematode-induced transfer cells', *Phytopathology* 67, 869–73.

Goverse, A., and Smant, G. (2014), 'The activation and suppression of plant innate immunity by parasitic nematodes', *Annual Review of Phytopathology* 52, 243–65.

Guertal, E. A., Sikora, E. J., Hagan, A. K., and Rodriguez-Kabana, R. (1997), 'Effect of winter cover crops on populations of southern root-knot and reniform nematodes', Agriculture, Ecosystems and Environment 70, 1–6.

Guo, B., Sleper, D. A., Nguyen, H. T., Arelli, P. R., and Shannon, J. G. (2006), 'Quantitative trait loci underlying resistance to three soybean cyst nematode populations in soybean PI 404198A', *Crop Science* 46(1), 224–233.

Guy, D. W. Jr., and Lewis, S. A. (1987), 'Interaction between *Meloidogyne incognita* and *Hoplolaimus columbus* on Davis soybean', *Journal of Nematology* 19, 346–51.

Haegeman, A., Mantelin, S., Jones, J. T., and Gheysen, G. (2012), 'Functional roles of effectors of plant- parasitic nematodes', *Gene* 492(1), 19–31.

Hallman, J. and Viaene, N. (2013) 'PM 7/119 (1) Nematode extraction', OEPP/EPPO Bulletin 43(3), 471–95, DOI:10.1111/epp.12077.

Handoo, Z.A., Mulrooney, R.P., and Skantar, A.M. (2010), 'First report of the sting nematode *Belonolaimus longicaudatus* on soybean in Delaware', *Plant Disease* 94, 133.

Hartman, G. L., Noel, G. R., and Gray, L. E. (1995), 'Occurrence of soybean sudden death syndrome in east-central Illinois and associated yield losses', *Plant Disease* 79, 314–418.

Hartman, G., Rupe, J., Sikora, E., Domier, L, Davis, J., and Steffey, K., eds., (2015), *Compendium of Soybean Diseases and Pests 5th Edition*, APS Press.

Hauge, B. M., Wang, M. L., Parsons, J. D., and Parnell, L. D. (2001), 'Nucleic acid molecules and other molecules associated with soybean cyst nematode resistance', U.S. Patent 20 030 005 491.

Heeren, J. R., Steffey, K. L., Tinsley, N. A., Estes, R. E., Niblack, T. L., and Gray, M. E. (2012), 'The interaction of soybean aphids and soybean cyst nematodes on selected resistant and susceptible soybean lines', *Journal of Applied Entomology* 136(9), 646–55.

Herman, M., Hussey, R. S., and Boerma, H. R. (1988), 'Interaction between *Meloidogyne incognita* and *Pratylenchus brachyurus* on soybean', *Journal of Nematology* 20(1), 79–84.

Hershman, D. E., and Bachi, P. R. (1995), 'Effect of wheat residue and tillage on *Heterodera glycines* and yield of doublecrop soybean in Kentucky', *Plant Disease* 79(6), 631–33.

Hershman, D. E., Heinz, R. D., and Kennedy, B. S. (2008), 'Soybean cyst nematode, *Heterodera glycines*, populations adapting to resistant soybean cultivars in Kentucky', *Plant Disease* 92(10), 1475–6.

Holguin, C. M., Baeza, J. A., Mueller, J. D., and Agudelo, P. (2015), 'High genetic diversity and geographic subdivision of three lance nematode species (*Hoplolaimus* spp.) in the United States', *Ecology and Evolution* 5, 2929–44.

Hughes, G. (1988), 'Spatial heterogeneity in crop loss assessment models', *Phytopathology* 78(7), 883–4.

Hunt, D., and Handoo, Z. (2009), 'Taxonomy, identification, and principal species', in Perry, R. N., Moens, M., and Starr, J. L. (eds.), *Root-knot Nematodes*, pp. 55–92. CABI.

Hymowitz, T. (1970), 'On the domestication of the soybean', *Economic Botany* 24(4), 408–21.

Hymowitz, T. (2004), 'Speciation and cytogenetics', in Boerma, H. R., and Specht, J. E. (eds.), *Soybeans: Improvement, Production, and Uses,* 3rd edn, Agronomy Monograph no. 16, pp. 97–136. American Society of Agronomy, Crop Science Society of America, Soil Science Society of America.

Hymowitz, T. (1990), 'Soybeans: the success story', in *Advances in New Crops*, Janick, J. and Simon, J. E. (eds.), pp. 159–63. Timber Press, Portland, OR.

Ichinohe, M. (1952), 'On the soybean cyst nematode, *Heterodera glycines* n. sp.', *Japanese Magazine of Applied Zoology* 17, 1–4.

Jackson, T. A., Smith, G. S., and Niblack, T. L. (2005), '*Heterodera glycines* infectivity and egg viability following nonhost crops and during overwintering', *Journal of Nematology* 37(3), 259–64.

Jaffe, H., Huettel, R. N., Demilo, A. B., Hayes, D. K., and Rebois, R. V. (1989), 'Isolation and identification of a compound from soybean cyst nematode, *Heterodera glycines*, with sex pheromone activity', *Journal of Chemical Ecology* 15(7), 2031–43.

Johnson, A. W. Handoo, Z. A., Baker, S. H, and Dowler, C. C. (1998), 'Crop yields and nematode population densities in triticale-cotton and triticale-soybean rotations', *Journal of Nematology* 30(3), 353–61.

Johnson, A. B., Kim, K. S., Riggs, R. D., and Scott, H. D. (1993), 'Location of *Heterodera glycines*-induced syncytia in soybean as affected by soil water regimes', *Journal of Nematology* 25(3), 422–6.

Kaitany, R., Melakeberhan, H., Bird, G. W., and Safir, G. (2000), 'Association of *Phytophthora sojae* with *Heterodera glycines* and nutrient stressed soybeans', *Nematropica* 30(2), 193–9.

Khan, M. W. (1993), *Nematode Interactions*, Chapman & Hall.
Kim, M., Hyten, D. L., Bent, A. F., and Diers, B. W. (2010), 'Fine mapping of the SCN resistance locus *rhg1- b* from PI 88788', *Plant Genome* 3(2), 81–9.
Kim, K. S., Vuon, T. D., Qiu, D., Robbins, R. T., Shannon, J. G., Li, Z., and Nguyen, H. T. (2016), 'Advancements in breeding, genetics, and genomics for resistance to three nematode species in soybean', *Theoretical and Applied Genetics* 129(12), 2295–2311. doi:10.1007/s00122-016-2816-x
Kimpinski, J., Sanderson, J.B., Gallant, C.E., and Arsenault, W.J. (2000), 'The effect of marigolds (*Tagetes* spp.) and other cover crops on *Pratylenchus penetrans* and on following potato crops', *Journal of Nematology* 32(4S), 531–6.
Koenning, S. R. (2002), 'Tolerance to *Hoplolaimus columbus* in glyphosate-resistant, transgenic soybean cultivars', *Journal of Nematology* 34, 370–3.
Koenning, S. R. (2004), 'Population biology', in *Biology and Management of Soybean Cyst Nematode, Second Edition*, D. P. Schmitt, J. A. Wrather, and R. D. Riggs (eds.), pp. 73–88. Schmitt and Associates, Marceline, MO.
Koenning, S. R. (2015), 'Root-knot nematodes', In Hartman, G., Rupe, J., Sikora, E., Domier, L, Davis, J., and Steffey, K. (eds.), *Compendium of Soybean Diseases and Pests* 5th edn, APS Press. pp. 98–100.
Koenning, S. R., and Anand, S. C. (1991), 'Effects of wheat and soybean planting date on *Heterodera glycines* population dynamics and soybean yield with conventional tillage', *Plant Disease*, 75, 301–3.
Koenning, S. R., Barker, K. R., and Schmitt, D. P. (1985), 'Influence of planting date on population dynamics and damage potential of *Pratylenchus brachyurus* on soybean', *Journal of Nematology* 17, 428–34.
Koenning, S. R., Schmitt, D. P, and Barker, K. R. (1985), 'Influence of selected cultural practices on winter survival of *Pratylenchus brachyurus* and subsequent effects on soybean yield', *Journal of Nematology* 17, 464–9.
Koenning, S. R., Schmitt, D. P., and Barker, K. R. (1993), 'Effects of cropping systems on population density of *Heterodera glycines* and soybean yield', *Plant Disease* 77(8), 780–6.
Koenning, S. R., and Wrather, J. A. (2010), 'Suppression of soybean yield potential in the United States by plant diseases from 2006 to 2009', *Plant Health Progress* doi:10. 1094/PHP-2010-1122-01-RS.
Kratochvil, R. J., Gallagher, E., Everts, K., and Sardanelli, S. (2004), 'Evaluation of crop rotation and other cultural practices for management of root-knot and lesion nematodes', *Agronomy Journal* 96, 1419–28.
Kruger, G., Westphal, A., LeRoy, A., and Xing, L. (2008), '*Meloidogyne incognita* resistance in soybean under midwest conditions', *Crop Science* 48, 716–26.
Kutner, M. H., Nchtsheim, C. J., and Neter, J. (2004), *Applied linear regression models, 4th ed.*, McGraw- Hill.
Lauritis, J. A., Rebois, R. V., and Graney, L. S. (1983), 'Development of *Heterodera glycines* Ichinohe on soybean, *Glycine max* (L.) Merr., under gnotobiotic conditions', *Journal of Nematology* 15(2), 272–80.
Lawn, D. A., and Noel, G. R. (1986), 'Field interrelationships among *Heterodera glycines, Pratylenchus scribneri*, and three other nematode species associated with soybean', *Journal of Nematology* 18, 98–106.
Lawn, D. A., and Noel, G. R. (1990), 'Effects of temperature on competition between *Heterodera glycines*, and *Pratylenchus scribneri* on soybean', *Nematropica* 20, 57–69.
Leach, M., Agudelo, P., and Lawton-Raub, A. (2012), 'Effect of crop rotations on *Rotylenchulus reniformis* population structure', *Plant Disease* 96:24–29.
Lee, T. G., Kumar, I., Diers, B. W. and Hudson, M. E. (2015), 'Evolution and selection of *Rhg1*, a copy-number variant nematode-resistance locus', *Molecular Ecology* 24(8), 1774–91.
Lightfoot, D., and K. Meksem (2002), 'Isolated polynucleotides and polypeptides relating to loci underlying resistance to soybean cyst nematode and soybean sudden death syndrome and methods employing same', U.S. Pat. Application Publ. No. 2002144310.

Lim, B., and Castillo, M. (1978), 'Interactions of *Meloidogyne incognita* and *Rotylenchulus reniformis* with selected soybean varieties', *Kalikasan* 7, 165–76.

Lim, S. M., and Chamberlain, D. W. (2001), 'Diseases of soybeans (Glycine max (L.) Merrill),' http://www.apsnet.org/publications/commonnames/Pages/Soybeans.aspx (accessed December 12, 2016).

Liu, X. Z., Li, J. Q., and Zhang, D. S. (1997), 'History and status of soybean cyst nematode in China', *International Journal of Nematology* 7(1), 18–25.

Liu, S., Kandoth, P. K., Warren, S. D., Yeckel, G., Heinz, R., Alden, J., Yang, C., Jamai, A., El-Mellouki, T., Juvale, P. S., Hill, J., Baum, T. J., Cianzio, S., Whitham, S. A., Korkin, D., Mitchum, M. G., and Meksem, K. (2012), 'A soybean cyst nematode resistance gene points to a new mechanism of plant resistance to pathogens', *Nature* 492(7428), 256–60.

Liu, Z., Griffin, T., and Kirkpatrick, T. L. (2014), 'Statistical and economic techniques for site-specific nematode management', *Journal of Nematology* 46(1), 12–17.

Liu, Z., Griffin, T. W., Kirkpatrick, T. L., and Monfort, W. S. (2015), 'Spatial econometric approaches to developing site-specific nematode management strategies in cotton production', *Precision Agriculture*, 16(5), 587–600.

Long, H. B., Sun, Y. F., Feng, T. Z., and Pei, Y. L. (2017), 'First Report of *Meloidogyne graminicola* on Soybean (*Glycine max*) in China', http://dx.doi.org/10.1094/PDIS-03-17-0334-PDN (posted 4/24/17).

Long, J. H., and Todd, T. C., (2001), 'Effect of crop rotation and cultivar resistance on seed yield and the soybean cyst nematode in full-season and double-cropped soybean', *Crop Science* 41(4), 1137–43.

Loof, P. A. A. (1991), 'The family Pratylenchidae', In Nickle, W. R., (ed.), *Manual of Agricultural Nematology*, Marcel Dekker. pp. 363–422.

Lopez-Nicora, H. D., and Niblack, T. L. (2012), 'A preliminary study of the composition of the gelatinous matrix of *Heterodera glycines*', *Journal of Nematology* 44(4), 474.

MacGuidwin, A. E. and Bender, B. E. (2012) 'Estimating population densities of root lesion nematodes, *Pratylenchus* spp., from soil samples using dual active and passive assays', *Plant Health Progress* doi:10.1094/PHP-2012-1120-01-RS.

Machado, A. C. Z., Inomoto, M. M., and Ferraz, L. C. C. B. (2007), 'Response of cotton cultivars to two Brazilian populations of *Pratylenchus brachyurus* (Godfrey) Filipjev & Sch. Stekh', *Journal of Cotton Science* 11, 288–92.

Mantelin, S., Thorpe, P., and Jones, J.T. (2015), 'Chapter Thirteen-Suppression of plant defences by plant-parasitic nematodes', *Advances in Botanical Research* 73. Elsevier.

Marburger, D., Conley, S., Esker, P., MacGuidwin, A., and Smith, D. (2014), 'Relationship between *Fusarium virguliforme* and *Heterodera glycines* in commercial soybean fields in Wisconsin', *Plant Health Progress* doi:10.1094/PHP-RS-13-0107.

Masler, E., Rogers, S., Halbrendt, J., Zasada, I., and Sardanelli, S. (2010), 'Effects of benzyl isothiocyanate on the reproduction of *Meloidogyne incognita* on *Glycine max* and *Capsicum annuum*', *Nematology* 12, 693–9.

Matson, A. L., and Williams, L. F. (1965), 'Evidence of a fourth gene for resistance to the soybean cyst nematode', *Crop Science* 5(5), 477.

McCarville, M. T., O'Neal, M., Tylka, G. L., Kanobe, C., and Macintosh, G. C. (2012), 'A nematode, fungus, and aphid interact via a shared host plant: implications for soybean management', *Entomologia Experimentalis et Applicata* 143(1), 55–66.

McCarville, M. T., Soh, D. H., Tylka, G. L., and O'Neal, M. E. (2014), 'Aboveground feeding by soybean aphid, *Aphis glycines*, affects soybean cyst nematode, *Heterodera glycines*, reproduction belowground', *PLoS ONE* 9(1), e86415. doi:10.1371/journal.pone.0086415.

McGawley, E. C., and Chapman, R.A. (1983), 'Reproduction of *Criconemoides simile, Helicotylenchus pseudorobustus*, and *Paratylenchus projectus* on soybean', *Journal of Nematology* 14, 572–5.

McGawley, E. C., and Overstreet, C. (1995), 'Reniform nematode', in Hartman, G. L., et al. (eds.), *Compendium of Soybean Diseases* 4th edn, pp. 54–55. APS Press.

McGawley, E. C., and Overstreet, C. (2015), 'Reniform nematode', in Hartman, G. L., et al. (eds.) *Compendium of Soybean Diseases* 5th edn, pp. 96–8. APS Press.

McLean, K. S., and Lawrence, G. W. (1993), 'Interrelationship of *Heterodera glycines* and *Fusarium solani* in sudden death syndrome of soybean', *Journal of Nematology* 25(3), 434–9.
McLean, K. S., and Lawrence, G. W. (1993), 'Localized Influence of *Heterodera glycines* on sudden death syndrome of soybean', *Journal of Nematology* 25(4), 674–8.
McSorley, R., and Dickson, D. W. (1989), 'Effects and dynamics of a nematode community on soybean', *Journal of Nematology* 21, 490–9.
McSorley, R. and Dickson, D. W. (1990), 'Vertical distribution of plant-parasitic nematodes in sandy soil under soybean', *Journal of Nematology* 22, 90–6.
McSorley, R., and Gallaher, R. N. (1993), 'Population dynamics of plant-parasitic nematodes on cover crops of corn and sorghum', *Journal of Nematology* 25, 446–53.
McSorley, R., and Gallaher, R. N. (1994), 'Effect of liming and tillage on soil nematode populations under soybean', *Soil and Crop Science Society of Florida Proceedings* 53, 31–35.
Melakeberhan, H. (1998), 'Pathogenicity of *Pratylenchus penetrans*, *Heterodera glycines*, and *Meloidogyne incognita* on soybean genotypes', *Journal of Nematology* 30, 93–9.
Melakeberhan, H., (2004), 'Physiological interaction between nematodes and their host plants', *Nematology, Advances and Perspectives*, pp.771–94. Tsinghua University Press, Beijing.
Melakeberhan, H., and Dey, J. (2003), 'Competition between *Heterodera glycines* and *Meloidogyne incognita* or *Pratylenchus penetrans*: independent infection rate measurements', Journal of Nematology 35, 1–6.
Melgar, J., Roy, K. W., and Abney, T. S. (1994), 'Sudden death syndrome of soybean: etiology, symptomatology, and effects of irrigation and *Heterodera glycines* on incidence and severity under field conditions', *Canadian Journal of Botany* 72(11), 1647–53.
Melito, S., Heuberger, A. L., Cook, D., Diers, B. W., MacGuidwin, A. E., and Bent, A. F. (2010), 'A nematode demographics assay in transgenic roots reveals no significant impacts of the *Rhg1* locus LRR- kinase on soybean cyst nematode resistance', *BMC Plant Biology* 10(1), 104.
Meyer, S., Juba, J., Nitao, J., Humber, R., Huettel, R., and Liu, X. (2004), 'Activity of fungal culture filtrates against soybean cyst nematode and root-knot nematode egg hatch and juvenile motility', *Nematology* 6, 23–32.
Miller, D. R., Chen, S. Y., Porter, P. M., Johnson, G. A., Wyse, D. L., Stetina, S. R., Klossner, L. D., and Nelson, G. A. (2006), 'Rotation crop evaluation for management of the soybean cyst nematode in Minnesota', *Agronomy Journal* 98(3), 569–78.
Miller, L. I. (1988), 'Morphological comparisons of cyst cone structures of one isolate each of *Heterodera schachtii, H. glycines* and one of their hybrids', *Journal of Nematology* 20(4), 648–9.
Miller, L. I. (1989), 'Morphological comparisons of second-stage juveniles of one isolate each of *Heterodera glycines, H. cruciferae* and one of their hybrids', *Journal of Nematology* 21, 574.
Mitchum, M. G., Wrather, J. A., Heinz, R. D., Shannon, J. G., and Danekas, G. (2007), 'Variability in distribution and virulence phenotypes of *Heterodera glycines* in Missouri during 2005', *Plant Disease* 91(11), 1473–6.
Mitchum, M. G. (2016), 'Soybean resistance to the soybean cyst nematode *Heterodera glycines*: An update', *Phytopathology* 106(12), 1444–50.
Mueller, J. D., and Noling, J. W. (1996), 'Standardization of reporting procedures for nematicide efficacy testing: a research and extension perspective', *Journal of Nematology* 28(4S), 575–85.
Mueller, J. D., Schmitt, D. P., Weiser, G. C., Shipe, E. R., and Musen, H. L. (1988), 'Performance of soybean cultivars in *Hoplolaimus columbus*-infested fields', *Journal of Nematology* 20(S), 65–9.
Musil, Kelsie M. (2016), 'Evaluations of biological control agents for the management of soybean cyst nematode (*Heterodera glycines*) in soybean (*Glycine max* L. Merr.)' *Theses, Dissertations, and Student Research in Agronomy and Horticulture. 118*. University of Nebraska-Lincoln http://digitalcommons.unl.edu/agronhortdiss/118 (accessed 18 April 2017)
Nakasono, K. (2004), 'Studies on morphological and physio-ecological variations of the reniform nematode, *Rotylenchulus reniformis* Linford and Oliveira, 1904 with an emphasis on differential geographical SSR markers for *Rotylechulus reniformis*: Distribution of amphimictic and parthenogenic populations in Japan', *Journal of Nematology* 36, 356–420.

Neher, D. A., and Olson, R. K. (1999), 'Nematode communities in soils of four farm cropping systems', *Pedobiologia* 43, 430–8.

Niblack, T. L. (2005), 'Soybean cyst nematode management reconsidered', *Plant Disease* 89(10), 1020–26.

Niblack, T. L., Arelli, P. R., Noel, G. R., Opperman, C. H., Orf, J. H., Schmitt, D. P., Shannon, J. G., and Tylka, G. L. (2002), 'A revised classification scheme for genetically diverse populations of *Heterodera glycines*', *Journal of Nematology* 34(4), 279–88.

Niblack, T. L., Baker, N. K., and Norton, D. C. (1992), 'Soybean yield losses due to *Heterodera glycines* in Iowa', *Plant Disease* 76(9), 943–8.

Niblack, T. L., and Chen, S. Y. (2004), 'Population density based management', in *Biology and management of soybean cyst nematode*, 2nd edn, D. P. Schmitt, J. A. Wrather, and R. D. Riggs (eds.), pp. 181–206. Schmitt & Associates of Marceline, Marceline, MO.

Niblack, T. L., Heinz, R. D., Smith, G. S., and Donald, P. A. (1993), 'Distribution, density, and diversity of *Heterodera glycines* in Missouri', *Journal of Nematology* 25(4S), 880–6.

Niblack, T. L., Hussey, R. S. and Boerma, H. R. (1986), 'Effects of interactions among *Heterodera glycines, Meloidogyne incognita*, and host genotype on soybean yield and nematode population densities', *Journal of Nematology* 18(4), 436–43.

Niblack, T. L., Lambert, K. N., and Tylka, G. L. (2006), 'A model plant pathogen from the kingdom Animalia: *Heterodera glycines*, the soybean cyst nematode', *Annu. Rev. Phytopathol*, 44, 283–303.

Niblack, T. L., Colgrove, A. L., Colgrove, K., and Bond, J. P. (2008), 'Shift in virulence of soybean cyst nematode is associated with use of resistance from PI 88788', *Plant Health Prog*, http://dx.doi.org/10.1094/PHP-2008-0118-01-RS.

Niblack, T. L., and Riggs, R. D. (2004), 'Variation in virulence phenotypes', in *Biology and Management of Soybean Cyst Nematode*, 2nd edn, D. P. Schmitt, J. A. Wrather, and R. D. Riggs (eds.), pp. 57–71. Schmitt & Associates of Marceline, Marceline, MO.

Niblack, T. L., and Tylka, G. L. (eds.) (2008), *Soybean Cyst Nematode Management Guide*, 5th edn North Central Soybean Research Program http://www.ncsrp.com/pdf_doc/SCN_Management.pdf.

Niblack, T. L., Tylka, G. L., Arelli, P., Bond, J., Diers, B., Donald, P., Faghihi, J., Ferris, V. R., Gallo, K., Heinz, R. D., Lopez-Nicora, H., Von Qualen, R., Welacky, T., and Wilcox, J. (2009) 'A standard greenhouse method for assessing soybean cyst nematode resistance in soybean: SCE08 (standardized cyst evaluation 2008)', *Plant Health Progress* doi:10.1094/PHP-2009-0513-01-RV.

Niblack, T. L., Tylka, G. L., and Riggs, R. D. (2004), 'Nematode pathogens of soybean', in Boerma, H. R., and Specht, J. E. (eds.), *Soybeans: Improvement, Production, and Uses*, 3rd edn, Agronomy Monograph no. 16, pp. 821–51. American Society of Agronomy, Crop Science Society of America, Soil Science Society of America.

Nitao, J., Fettinger, J., Chitwood, D., Meyer, S., and Schmidt, W. (2001), 'Nematode-antagonistic trichothecenes from *Fusarium equiseti*', *Journal of Chemical Ecology* 27, 859–69.

Nitao, J., Schmidt, W., Chitwood, D., Meyer, S., and Oliver, J. (2002), 'Isolation of flavipin, a fungus compound antagonistic to plant-parasitic nematodes', *Nematology* 4, 55–63.

Noe, J. P. (1993), 'Damage functions and population changes of *Hoplolaimus columbus* on cotton and soybean', *Journal of Nematology* 25, 440–5.

Noe, J. P., Imbriani, J. L., and Sasser, J. N. (1991), 'Maximizing the potential of cropping systems for nematode management', *Journal of Nematology* 23, 353–61.

Noel, G. R., (1992), 'History, Distribution, and Economics', in *Biology and Management of the Soybean Cyst Nematode*, R. D. Riggs and J. A. Wrather (eds.), pp. 1–14. APS Press.

Noel, G. R. (2004), 'Soybean response to infection', in *Biology and Management of Soybean Cyst Nematode*, 2nd *ed.*, D. P. Schmitt, J. A. Wrather, and R. D. Riggs (eds.), pp. 131–54. Schmitt & Associates of Marceline, Marceline, MO.

Nyczepir, A. P., and Lewis, S. A. (1979), 'Relative tolerance of selected soybean cultivars to *Hoplolaimus columbus* and possible effects of soil temperature', *Journal of Nematology* 11, 27–31.

Orbin, D. P. (1973), 'Histopathology of soybean roots infected with *Helicotylenchus dihystera*', *Journal of Nematology* 5, 37–40.

Overstreet, C., and McGawley, E. C. (1988), 'Influence of *Calonectria crotalariae* on reproduction of *Heterodera glycines* on soybean', *Journal of Nematology* 20(3), 457–67.
Overstreet, C., McGawley, E. C., and Russin, J. S. (1990), 'Interactions between *Calonectria* crotolariae and *Heterodera* gycines on soybean', *Journal of Nematology* 22(4), 496–505.
Owen, K. J., Thompson, J. P., Bell, K. L., and Clewett, T. G. (2014), 'Wheat biomass and yield increased when populations of the root-lesion nematode (*Pratylenchus thornei*) were reduced through sequential rotation of partially resistant winter and summer crops', *Crop and Pasture Science* 65, 227–41.
Pedersen, J. F., and Rodriguez-Kabana, R. (1991), 'Winter grass cover crop effects on nematodes and yields of double cropped soybean', *Plant and Soil* 131, 287–91.
Pedersen, P., Tylka, G. L., Mallarino, A., Macguidwin, A. E., Koval, N. C., and Grau, C. R. (2010), 'Correlation between soil pH, *Heterodera glycines* population densities, and soybean yield', *Crop Science* 50(4), 1458–64.
Pérez-Hernández, O., and Giesler, L. J. (2014), 'Quantitative relationship of soil texture with the observed population density reduction of *Heterodera glycines* after annual corn rotation in Nebraska', *Journal of Nematology* 46(2), 90.
Perry, R. N. (1996), 'Chemoreception in plant parasitic nematodes', *Annual Review of Phytopathology* 34(1), 181-199.
Perry, R. N., Moens, M., and Starr, J. L., eds. (2009), *Root-knot Nematodes*, CABI.
Plant Disease Management Reports, http://www.plantmanagementnetwork.org/pub/trial/search.aspx
Poromarto, S. H., and Nelson, B. D. (2010), 'Evaluation of northern-grown crops as hosts of soybean cyst nematode', *Plant Health Progress* http://dx.doi.org/10.1094/PHP-2010-0315-02-RS.
Poromarto, S. H., Gramig, G. G., Nelson Jr., B. D., and Jain, S. (2015), 'Evaluation of Weed Species from the Northern Great Plains as Hosts of Soybean Cyst Nematode', *Plant Health Progress* 16(1), 23.
Potter, J. W., and Fox, J. A. (1965), 'Hybridization of *Heterodera schachtii* and *H. glycines*', *Phytopathology* 55, 800–1.
Pudasaini, M. P., Schomaker, C. H., Been, T. H. and Moens, M. (2006), 'Vertical distribution of the plant- parasitic nematode, *Pratylenchus penetrans*, under four fi eld crops', *Phytopathology* 96, 226–33.
Rao-Arelli, A. P. (1994), 'Inheritance of resistance to *Heterodera glycines* race 3 in soybean accessions', *Plant Disease* 78(9), 898–900.
Rao-Arelli, A. P., Wrather, J. A., and Anand, S. C. (1992), 'Genetic diversity among isolates of *Heterodera glycines* and sources of resistance in soybean', *Plant Disease* 76(9), 894–6.
Raski, D. J. (1950), 'The life history and morphology of the sugar-beet nematodes, *Heterodera schachtii* Shmidt', *Phytopathology* 40(2), 135–52.
Rebois, R. V. (1973), 'Effect of soil temperature on infectivity and development of *Rotylenchulus reniformis* on resistant and susceptible soybeans, *Glycine max*', *Journal of Nematology* 5, 10–13.
Rebois, R. V., and Golden, A.M. (1985), 'Pathogenicity and reproduction of *Pratylenchus agilis* in field microplots of soybeans, corn, tomato, or corn-soybean cropping systems', *Plant Disease* 69, 927–9.
Rebois, R.V., and Huettel, R.N. (1986), 'Population dynamics, root penetration, and feeding behavior of *Pratylenchus agilis* in monoxenic root cultures of corn, tomato, and soybean', *Journal of Nematology* 18, 392–7.
Rebois, R. V., Madden, P. A., and Eldridge, B. J. (1975), 'Some ultrastructural changes induced in resistant and susceptible soybean roots following infection by *Rotylenchulus reniformis*', *Journal of Nematology* 7, 122–39.
Riggs, R. D. (2004), 'History and distribution', in *Biology and Management of Soybean Cyst Nematode*, 2nd edn, D. P. Schmitt, J. A. Wrather, and R. D. Riggs (eds.), pp. 9–40. Schmitt & Associates of Marceline, Marceline, MO.
Riggs, R. D., and Niblack, T. L. (1993) 'Nematode pests of oilseed crops and grain legumes', In K. Evans, D. Trudgill, and J. M. Webster (eds), *Nematodes in Temperate Agriculture*, pp. 209–58. Commonwealth Agricultural Bureaux.

Riggs, R. D., and Schmitt, D. P. (1988), 'Complete characterization of the race scheme for *Heterodera glycines*', *Journal of Nematology* 20(3), 392–95.

Robinson, A. F. (2007), 'Reniform in U. S. cotton: When, where, why, and some remedies', *Annual Review of Phytopathology* 45, 263–88.

Robinson, A. F., Inserra, R. N., Caswell-Chen, E. P., Vovlas, N., and Troccoli, A. (1997), '*Rotylenchulus* species: Identification, distribution, host ranges, and crop plant resistance', *Nematropica* 27, 127–80.

Rodriguez-Kabana, R., and Thurlow, D.L. (1980), 'Effect of *Hoplolaimus galeatus* and other nematodes on yield of selected soybean cultivars', *Nematropica* 10, 130–8.

Rogovska, N. P., Blackmer, A. M., and Tylka, G. L. (2009), 'Soybean yield and soybean cyst nematode densities related to soil pH, soil carbonate concentrations, and alkalinity stress index', *Agronomy Journal* 101(5), 1019–26.

Ross J. P. (1962), 'Physiological strains of *Heterodera glycines*', *Plant Disease Reporter* 46(11), 766–69.

Roy, K. W., Lawrence, G. W., Hodges, H. H., McLean, K. S., and Killebrew, J. F. (1989), 'Sudden death syndrome of soybean: *Fusarium solani* as incitant and relation of *Heterodera glycines* to disease severity', *Phytopathology* 79(2), 191–7.

Ruben, E., Jamai, A., Afzal, J., Njiti, V. N., Triwitayakorn, K., Iqbal, M. J., Yaegashi, S., Bashir, R., Kazi, S., Arelli, P., Town, C. D., Ishihara, H., Meksem, K., and Lightfoot, D. A. (2006), · 'Genomic analysis of the *rhg1* locus: Candidate genes that underlie soybean resistance to the cyst nematode', *Molecular Genetics and Genomics* 276(6), 503–516.

Rupe, J. C., Robbins, R. T., and Gbur Jr., E. E. (1997), 'Effect of crop rotation on soil population densities of *Fusarium solani* and *Heterodera glycines* and on the development of sudden death syndrome of soybean', *Crop Protection* 16(6), 575–80.

Rupe, J. C., Robbins, R. T., Becton, C. M., Sabbe, W. A., and Gbur Jr, E. E. (1999), 'Vertical and temporal distribution of *Fusarium solani* and *Heterodera glycines* in fields with sudden death syndrome of soybean', *Soil Biology and Biochemistry* 31(2), 245–51.

Russin, J. S., Layton, M. B., Boethel, D. J., McGawley, E. C., Snow, J. P., and Berggren, G. T. (1989), 'Development of *Heterodera glycines* on soybean damaged by soybean looper and stem canker', *Journal of Nematology* 21(1), 108–14.

Schmidt, L. M., Kelley, K., Doroh, M., Stetina, S. R., Hewlett, T. E., Green, A., and Simmons, L. J. (2010), 'Molecular and morphological characterization and biological control capabilities of a *Pasteuria* ssp. Parasitizing *Rotylenchulus reniformis*, the reniform nematode', *Journal of Nematology* 42, 207–17.

Schmitt, D. P. and Bailey, J.E. (1990), 'Chemical control of *Hoplolaimus columbus* on cotton and soybean', *Journal of Nematology* 22(4S), 689–94.

Schmitt, D. P., and Barker, K.R. (1981), 'Damage and reproductive potentials of *Pratylenchus brachyurus* and *Pratylenchus penetrans* on soybean', *Journal of Nematology* 13, 327–32.

Schmitt, D. P., and Ferris, H. (1998), 'Pathogenicity and damage levels', in *The cyst nematodes*, S.B. Sharma (ed.), pp. 239–65. Kluwer Academic Publishers.

Schmitt, D. P., and Imbriani, J. L. (1987), 'Management of *Hoplolaimus columbus* with tolerant soybean cultivars and nematicides', *Journal of Nematology* 22, 689–94.

Schmitt, D. P., and Noel, G. R. (1984), 'Nematode parasites of soybean', in Plant and Insect Nematodes, W. R. Nickle (ed.), pp. 13–59. Marcel Dekker.

Schmitt, D. P., and Shannon, G. (1992), 'Differentiating soybean responses to *Heterodera glycines* races', *Crop Science* 32(1), 275–77.

Schmutz, J., Cannon, S. B., Schlueter, J., Ma, J., Mitros, T., Nelson, W., Hyten, D. L., Song, Q., Thelen, J. J., Cheng, J., Xu, D., Hellsten, U., May, G. D., Yu, Y., Sakurai, T., Umezawa, T., Bhattacharyya, M. K., Sandhu, D., Valliyodan, B., Lindquist, E., Peto, M., Grant, D., Shu, S., Goodstein, D., Barry, K., Futrell-Griggs, M., Abernathy, B., Du, J., Tian, Z., Zhu, L., Gill, N., Joshi, T., Libault, M., Sethuraman, A., Zhang, X.-C., Shinozaki, K., Nguyen, H. T., Wing, R. A., Cregan, P., Specht, J.,

Grimwood, J., Rokhsar, D., Stacey, G., Shoemaker, R. C., and Jackson, S. A. (2010), 'Genome sequence of the palaeopolyploid soybean', *Nature*, 463(7278), 178–83.

Shannon, J. G., Arelli, P. R., and Young, L. D. (2004), 'Breeding for resistance and tolerance', in *Biology and Management of Soybean Cyst Nematode, 2nd ed.*, D. P. Schmitt, J. A. Wrather, and R. D. Riggs (eds.), Schmitt & Associates of Marceline, Marceline, MO. pp. 155–75.

Shannon, J. G., Nguyen, H., Anand, S., Robbins, R., Wrather, J., and Sleper, D. (2007a), 'Registration of 'Jake' soybean', *Journal of Plant Registrations* 1, 29–30.

Shannon, J. G., Nguyen, H., Anand, S., Wrather, J., and Sleper, D. (2007b), 'Registration of 'Stoddard' soybean', *Journal of Plant Registrations* 1, 28–9.

Shearin, Z., Hussey, R., Boerma, H., Finnerty, S., and Wood, E. (2009), 'A southern root-knot nematode resistance QTL linked to the T-locus in soybean', *Crop Science* 49, 467–72.

Silva, R. A., and Inomoto, M. M. (2002), 'Host-range characterization of two *Pratylenchus coffeae* isolates from Brazil', *Journal of Nematology* 34, 135–9.

Siddiqui, I., Heeb, S., and Haas, D. (2005), 'Extracellular protease of *Pseudomonas fluorescens* CHA0, a biocontrol factor with activity against the root-knot nematode *Meloidogyne incognita*', *Applied and Environmental Microbiology* 71, 5646–9.

Siddiqi, M. R. (1972), '*Rotylenchulus reniformis*', in Willmott, S., Gooch, P. S., Siddiqi, M. R., and Franklin, M. (eds.), *C. I. H. Descriptions of Plant Parasitic Nematodes*, Set 1, No. 5. CABI.

Sikora, E. J., and Noel, G. R. (1991), 'Distribution of *Heterodera glycines* races in Illinois', *Journal of Nematology* 23(4S), 624–8.

Sipes, B. S., Schmitt, D. P., and Barker, K. R. (1992), 'Fertility of three biotypes of *Heterodera glycines*', *Phytopathology* 82, 999–1001.

Smith, G. J., Wiebold, W. J., Niblack, T. L., Scharf, P. C., and Blevins, D. G. (2001), 'Macronutrient concentrations of soybean infected with soybean cyst nematode', *Plant and Soil* 235(1), 21–26.

Smith, G. S., Niblack, T. L., and Minor, H. C. (1991), 'Response of soybean cultivars to aldicarb in *Heterodera glycines*-infested soils in Missouri', *Journal of Nematology* 23(4S), 693–8.

Smith, G. S., Niblack, T. L., and Heinz, R. D. (1992), 'Recovery of *Heterodera glycines* from field-collected droppings of lesser snow geese (*Chen caerulescens*) in Missouri', *Journal of Nematology*, 24, 619 (Abstr.).

Sobczak, M., and Golinowski, W. (2009), 'Structure of cyst nematode feeding sites', in *Cell Biology of Plant Parasitism*, Berg, R. H., and Taylor, C.G. (eds.), pp. 153–187. Springer.

Sobczak M., Golinowski W. (2011). 'Cyst nematodes and syncytia', in *Genomics and Molecular Genetics of Plant–Nematode Interactions*, Jones J., Gheysen G., Fenoll C. (eds.), pp. 83–102. Dordrecht: Springer.

Stetina, S. R., Smith, J. R., and Ray, J. D. (2014), 'Identification of *Rotylenchulus reniformis* resistant *Glycine* lines', *Journal of Nematology* 46, 1–7.

Stirling, G. R. (2008), 'The impact of farming systems on soil biology and soilborne diseases: examples from the Australian sugar and vegetable industries - the case for better integration of sugarcane and vegetable production and implications for future research', *Australasian Plant Pathology* 37, 1–18.

Tabor, G. M., Tylka, G. L., Behm, J. E., and Bronson, C. R. (2003), '*Heterodera glycines* infection increases incidence and severity of brown stem rot in both resistant and susceptible soybean', *Plant Disease* 87(6), 655–61.

Tabor, G. M., Tylka, G. L., and Bronson, C. R. (2006), 'Soybean stem colonization by genotypes A and B of *Cadophora gregata* increases with increasing population densities of *Heterodera glycines*', *Plant Disease* 90(10), 1297–1301.

Todd, T. C. (1991), 'Effect of cropping regime on populations of *Belonolaimus* sp. and *Pratylenchus scribneri* in sandy soil', *Journal of Nematology* 23 (4S), 646–51.

Todd, T. C. (1993), 'Soybean planting date and maturity effects on *Heterodera glycines* and *Macrophomina phaseolina* in southeastern Kansas', *Journal of Nematology* 25(4S), 731–7.

Todd, T. C., Pearson, C. A. S., and Schwenk, F. W. (1987), 'Effect of *Heterodera glycines* on charcoal rot severity in soybean cultivars resistant and susceptible to soybean cyst nematode', *Journal of Nematology* 21, 35–40.

Triantaphyllou, A. C., and Hirschmann, H. (1962), 'Oogenesis and mode of reproduction in the soybean cyst nematode, *Heterodera glycines*', *Nematologica* 7(3), 235–41.

Triantaphyllou, A. C., and Hirschmann, H. (1964), 'Reproduction in plant and soil nematodes', *Annual Review of Phytopathology* 2(1), 57–80.

Tubajika, K. M., Tylka, G. L., Tachibana, H, and Yang, X. B. (1994), 'Incidence of brown stem rot as influenced by soybean cyst nematode (Abstr.)', *Phytopathology* 84, 1101.

Tylka, G. L. (2016), 'Understanding soybean cyst nematode HG types and races', *Plant Health Progress* 17, 149–51.

Tylka, G. L., and Marett, C. C. (2014), 'Distribution of the soybean cyst nematode, *Heterodera glycines*, in the United States and Canada: 1954 to 2014', *Plant Health Progress* doi:10.1094/PHP-BR-14-0006.

Tylka, G. L., Sanogo, C., Souhrada, S. K. (1998), 'Relationships among *Heterodera glycines* population densities, soybean yields, and soil pH', *Journal of Nematology* 30, 519–20 (Abstr.).

van Bezooijen, J. (2006), Methods And Techniques For Nematology, https://www.wur.nl/upload_mm/4/e/3/f9618ac5-ac20-41e6-9cf1-c556b15b9fa7_MethodsandTechniquesforNematology.pdf

Venkatesh, R., Harrison, S. K., and Riedel, R. M. (2000), 'Weed hosts of soybean cyst nematode (*Heterodera glycines*) in Ohio', *Weed Technology* 14(1), 156–60.

von Mende, N., Gravato Nobre, M. J., and Perry, R. N. (1998), 'Host finding, invasion and feeding', in *The Cyst Nematodes*, S. B. Sharma (ed.), pp. 217–38. Kluwer Academic Publishers.

Wang, J., Niblack, T. L., Tremain, J. A., Wiebold, W. J., Tylka, G. L., Marrett, C. C., Noel, G. R., Myers, O., and Schmidt, M. E. (2003), 'Soybean cyst nematode reduces soybean yield without causing obvious aboveground symptoms', *Plant Disease* 87(6), 623–8.

Wang, K. -H., and McSorley, R. (2005), 'Effects of soil ecosystem management on nematode pests, nutrient cycling, and plant health', APSnet Feature doi:10.1094/APSnetFeature/2001-0105.

Warnke, S. A., Chen, S., Wyse, D. L., Johnson, G. A., and Porter, P. M. (2008), 'Effect of rotation crops on hatch, viability and development of *Heterodera glycines*', *Nematology* 10(6), 869–82.

Wartman, F. S., and Bernard, E. C. (1985), 'Reproduction of *Pratylenchus alleni* (Nematoda) on soybean and other field and vegetable crops', *Tennessee Farm And Home Science* – Tennessee Agricultural Experiment Station 135, 3–5.

Weaver, D., Garden, E., and Rodriguez-Kabana, R. (1998), 'Velvetbean and bahiagrass as rotation crops for management of *Meloidogyne* spp. and *Heterodera glycines* in soybean', *Journal of Nematology* 30(4S), 563–8.

Webb, D. M., Baltazar, B. M., Rao-Arelli, A. P., Schupp, J., Clayton, K., Keim, P., and Beavis, W. D. (1995), 'Genetic mapping of soybean cyst nematode Race-3 resistance loci in the soybean PI 437.654', *Theoretical and Applied Genetics* 91(4), 574–581.

Westphal, A., Li, C., Xing, L., McKay, A., and Malvick, D. (2014), 'Contributions of *Fusarium virguliforme* and *Heterodera glycines* to the disease complex of sudden death syndrome of soybean', *PLoS ONE* 9(6), e99529. doi:10.1371/journal.pone.0099529.

Westphal, A., Robinson, A. F., Scott, A. W., and Santini, J. B. (2004), 'Depth distribution of *Rotylenchulus reniformis* under crops of different host status and after fumigation', *Nematology* 6, 97–107.

Westphal, A., Xing, L. J., Pillsbury, R., and Vyn, T. J. (2009), 'Effect of tillage intensity on population densities of *Heterodera glycines* in intensive soybean production systems', *Field Crops Research* 113(3), 218–26.

Winkler, H. E., Hetrick, B. A. D., and Todd, T. C. (1994), 'Interactions of *Heterodera glycines*, *Macrophomina phaseolina*, and Mycorrhizal Fungi on soybean in Kansas', *Journal of Nematology* 26(4S), 675–82.

Winstead, N. N., Skotland, C. B., and Sasser, J. N. (1955), 'Soybean cyst nematode in North Carolina', *Plant Disease Reporter* 39(1), 9–11.

Wrather, J. A., and Anand, S. C. (1988), 'Relationship between the time of infection with *Heterodera glycines* and soybean yield', *Journal of Nematology* 20(3), 439–42.

Wrather, J. A., Anderson, T. R., Arsyad, D. M., Tan, Y., Ploper, L. D., Porta-Puglia, A., Ram, H. H., and Yorinori, J. T. (2001), 'Soybean disease loss estimates for the top ten soybean-producing countries in 1998', *Canadian Journal of Plant Pathology* 23(2), 115–21.

Wrather, A., Shannon, G., Balardin, R., Carregal, L., Escobar, R., Gupta, G. K., Ma, Z., Morel, W., Ploper, D., and Tenuta, A. (2010), 'Effect of diseases on soybean yield in the top eight producing countries in 2006', *Plant Health Progress* doi:10.1094/PHP-2010-0125-01-RS.

Xia, Y., Xie, S., Ma, X., Wu, H., Wang, X., and Gao, X. (2011), 'The *purL* gene of *Bacillus subtilis* is associated with nematicidal activity', *Federation of European Microbiological Societies Microbiology Letters* 322, 99–107.

Xing, L., and Westphal, A. (2006), 'Interaction of *Fusarium solani* f. sp. *glycines* and *Heterodera glycines* in sudden death syndrome of soybean', *Phytopathology* 96(7), 763–70.

Yates, J., Boerma, H., Finnerty, S., and Hussey, R. (2010), 'Three soybean plant introductions possess unique resistance to peanut root-knot nematode', *Crop Science* 50, 118–22.

Ye, W., and Robbins, R. T. (2004), '*Longidorus biformis* n. sp. and *L. glycines* n. sp. (Nematoda: Longidoridae): two amphimictic species from Arkansas', *Journal of Nematology* 36, 1–13.

Yen, J.-H., Niblack, T. L., and Wiebold, W. J. (1995), 'Dormancy of *Heterodera glycines* in Missouri', *Journal of Nematology*, 27(2), 153–63.

Young, L. D. (1998), 'Managing soybean resistance to *Heterodera glycines*', *Journal of Nematology* 30(4S), 525–9.

Yu, N., Lee, T. G., Rosa, D. P., Hudson, M., and Diers, B. W. (2016), 'Impact of *Rhg1* copy number, type, and interaction with *Rhg4* on resistance to *Heterodera glycines* in soybean', *Theoretical and Applied Genetics* 129(12), 2403–12.

Yu, Q., and Potter, J. (2008), 'Selective nematicidal activity of nicotine', *Journal of Food, Agriculture & Environment* 6, 428–32.

Yue, P., Sleper, D. A., and Arelli, P.R. (2001), 'Mapping resistance to multiple races of *Heterodera glycines* in soybean PI 89772', *Crop Science* 41(5), 1589–1595.

Zasada, I., Ferris, H., Elmore, C. L., Roncoroni, J. A., MacDonald, J. D., Bolkan, L. R., and Yakbe, L. E. (2003), 'Field application of brassicaceous amendments for control of soilborne pests and pathogens', *Plant Health Progress* DOI:10.1094/PHP-2003-1120-01-RS.

Zasada, I., Halbrendt, J., Rogers, S., and Masler, E. (2009), 'Behavioural response of *Meloidogyne incognita* to benzyl isothiocyanate', *Nematology* 11, 603–10.

Zasada, I., Melakeberhan, H., Koenning, S., Tylka, G., Avendano, F., Li, Y., and Logan, T. (2008), 'Potential of an alkaline-stabilized biosolid to manage nematodes: case studies on soybean cyst and root-knot nematodes', *Plant Disease* 92, 4–13.

Zaworski, E. R. (2014). 'Effects of ILeVO® on soybean sudden death syndrome and soybean cyst nematode', University of Iowa MS Thesis http://gradworks.umi.com/15/84/1584688.html (accessed 18 April 2017)

Zhang, L., Dean, R. A., Knap, H. T., and Lewis, S. A. (1998), 'Diversity among a *Heterodera glycines* field isolate and derived inbreds based on RAPD analysis and reproduction on soybean genotypes', *Journal of Nematology* 30(4), 477–84.

Zheng, J., Li, Y., and Chen, S. (2006), 'Characterization of the virulence phenotypes of *Heterodera glycines* in Minnesota', *Journal of Nematology* 38(3), 383–90.

Zirakparvar, M. E. (1980), 'Host range of *Pratylenchus hexincisus* and its pathogenicity on corn, soybean, and tomato', *Phytopathology* 70, 749–53.

Zirakparvar, M. E. (1982), 'Susceptibility of soybean cultivars and lines to *Pratylenchus hexincisus*', *Journal of Nematology* 14, 217–20.

Chapter 4

Key factors limiting sustainable insect pest management in soybeans

M. E. O'Neal, Iowa State University, USA; and R. Cox, EarthEmpower Consulting and Investment, Mexico

1 Introduction
2 Invasive species
3 Insect resistance management (IRM), especially for Bt-soybeans
4 Conclusion
5 Future trends
6 Where to look for further information
7 References

1 Introduction

1.1 Sustainable crop production and pest management

If sustainability implies the preservation of resources for future generations, then pest management tools, like insecticides and insect-resistant varieties, can be considered resources that should be preserved for future generations. Overuse of these tools can result in resistance such that their ability to prevent yield loss is lost. Replacements can be found in the form of new active ingredients or novel plant traits. However there is growing evidence that as resistance to a single toxin increases, the rate of resistance developing for other toxins increases due to cross-resistance (Whalon et al. 2008). Furthermore, the use of insecticides within areas of intense agricultural activity affects the delivery of ecosystem services like pollination (Kremen et al. 2002) when non-target insects like bees are also killed. In an effort to reduce such non-target impacts, entomologists have developed an approach referred to as integrated pest management (IPM) to minimize insecticide use (Stern et al. 1959). Much of the theory for applying IPM principles was developed with regard to soybean production (e.g. Stone and Pedigo 1972). In this chapter, we will explore whether IPM can lead to sustainable pest management for soybeans.

Stern et al. (1959) described many negative consequences from the preventative use of insecticides and outlined an alternative that many consider the first articulation of IPM. Central to this approach is restricting the use of pesticides to when the abundance of the target pest is sufficient to reduce yield. IPM focuses on the concepts of the gain threshold, economic injury level (EIL) and economic threshold (ET). These values can be calculated for specific pests and

http://dx.doi.org/10.19103/AS.2017.0034.19

crop combinations. In general, the gain threshold is reached when the cost of suppressing yield loss due to an insect pest equals the value of the crop. EIL is the lowest number of insects that will cause yield loss equal to the gain threshold. Once a pest population reaches an EIL, a farmer has lost value, so entomologists calculate an ET, often in terms of a pest density, at which point a management action should be taken to prevent a pest population from reaching the EIL (Higley and Boethel 1994). To apply these principles, farmers are required to scout their fields to determine what pests are present and if their density is approaching an ET. By limiting the use of insecticides to only those situations in which they are needed to prevent yield loss, the risk of insecticide resistance and environmental degradation could be reduced.

1.2 Integrated pest management practices for soybean

Not all pests have an EIL or ET calculated for determining the need for an insecticide and the timing of its application. However, IPM is not limited to insecticides, and a management programme should consider all possible approaches to disrupt and limit the impact of a target pest (Pedigo and Rice 2014). Preventing yield loss from insect pests can also include the use of biological control and insect-resistant soybeans. Each of these strategies has advantages and disadvantages, with the target pest often dictating which form is the primary tool used by farmers. Below, we discuss these general practices, highlighting a few examples of where they have been used for soybean production.

In general, natural enemies can cause significant mortality to insect herbivores, limiting their capacity to reach populations that exceed an ET. These natural enemies include predators, parasitoids and pathogens, which may attack various stages of an insect pest's growth (i.e. egg, larvae, adult). When these natural enemies are managed, either through their rearing and release or through their conservation in and around the target crop, a famer is practising biological control (Hayek 2004). In general, there are three approaches to using natural enemies to manage pests: Importation (or classical), Augmentation and Conservation (for examples, see Landis and Orr 2017). The utility of each approach varies by pest. For example, invasive species that have escaped the natural enemies that limit their growth within their native range are ideal candidates for a biological control programme that imports these natural enemies into the exotic range of the pest. Such a practice will be discussed in the case study below. Augmentation differs from importation because the natural enemy reared and released can already be found in the field, but its population is too low to suppress a target pest. Augmentation is typically not used for field crops like soybeans due to the large area that needs to be covered. In regions where there are sufficient resources to rear and release larger amounts of natural enemies, an augmentation plan can work. For example, Brazilian farmer cooperatives developed a programme to rear *Beauveria bassiana*, a bacterial pathogen that was sprayed on soybeans to kill lepidopteran pests (Li et al. 2010). Conservation biological control seeks to improve the habitat within and around a targeted crop field to increase the abundance and impact of natural enemies on pests. The impact of conservation biological control has been explored for pest of soybeans, but with limited success (Cox et al. 2014).

Insect-resistant soybeans are commercially available, often in the form of genetically modified (GM) crops with genes that are toxic to specific insect herbivores. Currently, these are available in China (Yu et al. 2014), Argentina, Brazil, Paraguay and Uruguay (Anon. 2016). To date, insect-resistant GM soybeans are not commercially available in the United States and European Union. This limited distribution of GM-based traits is due in part to government regulation and trait protection (i.e. establishing intellectual property).

Resistance to insects can be found in the germplasm of soybean (e.g. Hesler et al. 2013), but the commercial availability of this form of resistance is limited, as will be discussed below in the case study.

Worldwide, farmers have access to insecticides that can be applied to both soybean seed and foliage. The application of insecticides to soybean seeds began with the discovery and registration of neonicotinoids, a class of systemic insecticides that can move internally through a plant. Their initial use in treating soybean crops within the United States was for the management of the bean leaf beetle (*Cerotoma trifurcata*, Bradshaw and Rice 2003), which had caused significant yield loss in Minnesota, Iowa and parts of Wisconsin during a ten-year period beginning in the early 1990s. Beginning with emergency exemptions and eventually ending with a complete federal label from the EPA, neonicotinoids were approved for use on soybeans. Within a decade, the use of neonicotinoids as a seed treatment on soybeans became very common, accounting for 30–60% of commercial soybeans sold in the United States (Douglas and Tooker 2015). This almost exponential increase in use in the United States mirrors the global use of neonicotinoids on both seed and foliage in other field crops as well (Simon-Delso et al. 2015). Estimates of the economic value for seed-applied insecticides in the United States varies by region (Johnson et al. 2009; North et al. 2016), in part due to regional differences in pest complexes. In a review of research on the efficacy of neonicotinoid seed treatments for soybeans, the United States Environmental Protection Agency questioned their value for most of the soybean growing regions of the United States (Myers and Hill 2014). Furthermore, there is growing evidence that the use of neonicotinoids is responsible for environmental damage (Sanchez-Bayo 2014; Krupke et al. 2012; Mogren and Lundgren 2016), with several NGOs calling for their restricted use or outright ban. To what extent a cost-benefit analysis will result in restrictions on neonicotinoid use in the United States is not clear, though such regulatory action has been taken in the European Union. The widespread, preventative approach to insecticide use could accelerate the occurrence of resistance in target pests. Recently, resistance has been documented for the soybean aphid (details below). Replacing these active ingredients can be challenging as the time and effort to discover and bring new chemistries to market can require ~10 years and $10–30 million. This example highlights how bringing insecticides to market is challenging, with multiple issues influencing the registration process.

1.3 Use of pest management approaches globally and in the United States

Pest management practices for soybean production – biological control, host plant resistance and insecticides in an IPM context – vary widely by geography. In countries with substantial infrastructure for research and extension, a well-implemented IPM programme is more likely as it requires resources for crop scouting, measurement of growing degree days, access to market values of the crop and ongoing research to maintain and share information about ETs and injury levels (Parsa et al. 2014). Such resources are often not available for soybean production in developing countries, possibly making IPM inappropriate for resource-poor farmers (Morse and Buhler 1997). As noted by Parsa et al. (2014) the potential barriers to the adoption of IPM vary based on the resources available to farmers. So even though the value of restricting the use of insecticide could make IPM a valuable approach in developing countries where the cost of an insecticide can be a significant input, a lack of collective action and availability of IPM experts can limit its adoption. Exploring the sociological and economic factors that restrict IPM is beyond the scope of this review. For an example of a more IPM-based approach that has replaced a preventative approach to insecticide use in a resource-poor farming system, consider the management of leaf hoppers in rice fields within Indonesia (Matteson 2000).

In the discussion of key issues for sustainable pest management, we use two case studies to highlight the biology of the target pest and components of the economic returns that affect the use of pest management in soybeans. These are taken from two of the leading soybean-producing regions of the world, the United States and Brazil. These examples focus primarily on large-scale producers that employ conventional (i.e. not organic) production practices. However, some key issues and some of the tactics can be appropriate for organic production.

2 Invasive species

In the United States, soybeans are an exotic plant introduced from Asia. As a result of this introduction, soybeans escaped many of the insect pests that feed on the plant within its native range (Kogan and Turnipseed 1987; Kogan 1998). Historically, soybeans have required little need for insecticides in the states that produce most of the crop. Despite this escape, several insect species that are either native or endemic to the United States use soybeans as a host, often inflicting significant damage. Outbreaks had been uncommon, limited to the more southern states of the continental United States where several stink bug and lepidopteran species populations can frequently build and thus reduce soybean yields (Musser et al. 2015).

Despite this extended honeymoon period, insect pests that feed on soybeans in the United States have spread due to invasive species that have recently established themselves in the major soybean-growing regions of the United States. This includes the invasion by the soybean aphid (*Aphis glycines*), the kudzu bug (*Megacopta cribraria*) and the brown marmorated stink bug (*Halyomorpha halys*). Each of these species is capable of establishing large populations on soybeans in the United States, resulting in economic damage. Invasive species are a threat to sustainable production as they require the use of insecticides when previously this was not required. This may disrupt existing pest management programmes by reducing the population of natural enemies within a crop field, allowing for secondary pests to recolonize treated fields. A major driver for the emergence of invasive pests is global trade (Work et al. 2005). Soybean-producing regions that are closest to highly trafficked areas, like airports or harbours, are at risk for exposure to invasive species. For example, both the soybean aphid and kudzu bug are thought to have been introduced to the United States in cities with international airports, like Chicago, Illinois and Atlanta, Georgia, respectively (Ragsdale et al. 2011; Hosokawa et al. 2014; Ruberson et al. 2013).

In general, invasive species are not often a significant pest in their native range but produce significantly larger populations in their introduced range. Ecologists explain this by invoking the Enemy Release hypothesis, noting that when a species is brought into a new place, it often arrives without the natural enemies that feed on it within its native range (Crawley 1989). Due to this ecology of invasive species in general, a common approach to their management is an importation biological control programme (Hayek

2004). Such a programme seeks to reunite the natural enemies that prevent these pests from developing large populations in their home range. Currently, efforts to establish a population of natural enemies that attack invasive pests of soybeans in the United States are underway. This approach is not without problems as the introduced natural enemies may attack non-target organisms, including native insects, or they may have unexpected negative impacts on the environment and other cropping systems. For an example of these negative consequences, consider the review of the invasion of *Harmonia axyridis* (Koch 2003; Koch and Galvan 2008), an invasive species and predator of the soybean aphid in North America that is the dominant lady beetle species found in soybean fields throughout the Midwest soybean-growing region (Gardiner et al. 2009a; Schmidt et al. 2008).

2.1 Case study: developing an IPM programme for an invasive soybean pest: Soybean aphid in the US Midwest

An example of an invasive pest that has had a large impact on soybean production in the United States is the soybean aphid (*Aphis glycines*). The impact of soybean aphids in North America has been reviewed in several places, with a focus on its ecology and management (Ragsdale et al. 2011; Hodgson et al. 2012b; Tilmon et al. 2011). This aphid feeds almost exclusively on soybeans during the summer and overwinters on the woody perennial, common buckthorn (*Rhamnus cathartica*), also invasive in North America. Despite *Rhamnus* often being several kilometres away from soybean fields, outbreaks of soybean aphids occur across the Midwest region of the United States. This is due to the ability of the soybean aphid (and many other aphid species) to disperse over great distances during the spring and fall when they move between their two host plants in anticipation of seasonal changes. Although local dispersal from *Rhamnus* to soybeans can help explain within-field abundance of soybean aphids early in the growing season, soybean aphids migrate between these two plants at the landscape scale (Bahlai et al. 2015). Therefore, even though a farmer in the Midwest may practise crop rotation, say for the management of other insect pests like corn rootworms (Gray et al. 2009), migrating soybean aphids can discover soybean fields, even though location of soybeans within the landscape changes yearly. In the United States, several endemic predatory insects are found in soybean fields and feed on the soybean aphid (Schmidt et al. 2008; Rutledge et al. 2004), capable of suppressing populations below an ET (Gardiner et al. 2009b). However, outbreaks continue, though less frequently and intensively than in the initial five years of the aphid's occurrence in the United States (Bahlai et al. 2015).

With the arrival of the soybean aphid to North America, applied entomologists at both University and Federal research programmes began a series of experiments to determine methods to prevent yield loss from the aphid's feeding. A multi-state, multi-year study revealed the relationship between aphid density (i.e. number of aphids per plant) and yield loss (Ragsdale et al. 2007). From this relationship an EIL (~675 aphids/plant) and an ET (250 aphids/plant) were calculated based on a range of crop values and input costs. This data set was developed from naturally occurring infestations during the reproductive stages of the soybean plant (R1–R4). Initial recommendations encouraged farmers to scout fields during these reproductive stages to determine if populations reached the ET and were increasing. This is time-consuming, especially when a tall (~1 m) plant has many side branches. To reduce the time needed to assess if an aphid population requires a foliar application of insecticide, Hodgson et al. (2004) developed a sampling plan ('speed scouting') that can allow a field to be evaluated within 8.6 minutes compared to 57 minutes when aphids are counted on each plant. This speed scouting approach was tested in multiple, commercial fields (Hodgson et al. 2007), and was found to protect yield compared to untreated plots. Decisions based on speed scouting were consistent with whole plant estimates 79% of the time, with differences occurring in speed scouting suggesting an insecticide application was needed when it was not. Based on these findings, speed scouting is considered a conservative method to make estimates, and farmers are encouraged to take multiple samples to more accurately assess populations. The development of EIL and ET, with the guidelines for scouting and assessing the population density of soybean aphids, is the basis for the IPM programme recommended to soybean farmers in the US Midwest region (Hodgson et al. 2012b).

Beyond determining the need for and timing of a foliar application of insecticide, there is little else that a farmer in the United States can do to manage soybean aphid outbreaks. Cultural practices like crop rotation (noted above) have no or limited impact on the aphid. The sequence of crops within a rotation can increase the risk of a soybean aphid outbreak, if it produces physiological stress on the soybean plant such that it is a better host for the aphid. When soybeans are potassium deficient, soybean aphid populations grow faster and are more likely to reach ET and EILs than in soils grown in fields with sufficient potassium (Bruulseman et al. 2010). A soybean plant that suffers from potassium deficiency is a better host of the aphid (i.e. more nutritious) than a healthy soybean plant. Farmers are recommended to manage potassium levels in the soil through fertilization, especially in regions where alfalfa is part of the crop rotation.

Pest managers using an IPM approach are encouraged to include multiple tactics (Pedigo and Rice 2014; Stern et al. 1959), beyond insecticides. Both biological control and aphid-resistant soybeans have been explored as management options. As noted earlier, several predatory insects were noted as a source of mortality for soybean aphids in North America, but are insufficient to prevent outbreaks. Efforts to improve upon the biological control of soybean aphids in the United States have taken several forms, including using volatiles to attract natural enemies into a soybean field (Mallinger et al. 2011), adding habitat for conserving natural enemies (Cox et al. 2014) and importation of natural enemies from Asia to the United States (Heimpel et al. 2004).

Beyond insecticides, yield loss from soybean aphids can be prevented with aphid-resistant soybean varieties. Soon after the discovery of the soybean aphid (*Aphis glycines*) in the United States, breeders discovered genes that confer resistance to *Aphis glycines* (or *Rag*-genes) in soybean germplasm. Commercial cultivars containing *Rag*1 were first sold in 2010, yet their use is limited, in part because of survival by soybean aphids on resistant cultivars occasionally at economically damaging levels (Hesler et al. 2013). Unlike a cultivar with only one *Rag* gene, a *Rag*1 + *Rag*2 pyramid protects soybeans from economically damaging levels of soybean aphids without the use of insecticides (McCarville et al. 2014). However, even this pyramid is not aphid-free. Several virulent biotypes capable of surviving on *Rag*-containing soybean lines including one that survives on a *Rag*1 + *Rag*2 pyramid (i.e. Biotype 4; Alt and Ryan-Mahmutagic 2013), have been discovered in the United States. Unique pyramids have been developed with the capacity to suppress aphid populations containing biotype 4 (Ajayi-Oyetunde et al. 2016).

Beyond the use of *Rag* genes found within the soybean germplasm, there is the potential to add novel aphid resistance to soybeans through genetic modification. Soybean aphid mortality was produced through Bt-proteins modified to increase exposure to a phloem-feeding insect (Chougule et al. 2013), resulting in a novel source of resistance. Because this is a novel trait with no evolutionary history with the soybean aphid, the occurrence of biotypes virulent to a Bt-protein is expected to be lower than that of biotypes virulent to *Rag* genes. Regardless of the source of resistance incorporated into a soybean plant to protect it from the soybean aphid, long-term use of an insect-resistant variety will require a resistance management plan. This topic will be discussed in the second case study.

Despite efforts to advance both biological control and host plant resistance for soybean aphids, insecticides are the main management tactic available to US farmers to manage the spread of soybean aphid. How sustainable a soybean aphid management is can be a function of how frequently farmers apply IPM recommendations and how much insecticide is used within the region affected by insect pests. Before soybean aphids arrived in the United States, few, if any, insecticides were used on soybeans in the states that grow the majority of soybeans (Illinois, Iowa and Minnesota). Since its establishment in 2001, insecticide use has increased 130%, mostly in the form of foliar-applied insecticides (Ragsdale et al. 2011), though seed-applied insecticide use has also increased (Hodgson et al. 2012a). As noted above, all of the elements of an IPM-based approach for using insecticides were developed for the soybean aphid, including an EIL and ET, and a scouting method that allows for rapid scouting of an aphid population (Ragsdale et al. 2007; Hodgson et al. 2012b). While these methods were being developed and shared, surveys of farmers in the Midwest revealed that insecticide use varied yearly (13–89%), with the majority of farmers (84–94%) identifying scouting reports as 'very important'. Curiously, over the course of the survey period (2004–07) the percentage of farmers that identified the ET as the lowest aphid density for profitable insecticide spraying decreased over time, reaching a high of 74% but reducing to the lowest percentages in the last year of the survey for all states (MN = 50%, MI = 47%, IA = 51%). This is especially interesting as, compared to a preventative approach, the economic value of this IPM-based recommendation for the soybean aphid has been shown to be more profitable for farmers (Johnson et al. 2009; Krupke et al. 2017). For example, the likelihood of exceeding the gain threshold was greater when insecticides were applied based on need as opposed to a calendar basis (i.e. preventatively), 84–70%, respectively (Johnson et al. 2009).

So, if an IPM approach to using insecticide is profitable, why do these soybean farmers not adopt IPM-based strategies? In general, there are two possible mistakes farmers can make when employing an ET to apply an insecticide; (I) an insecticide can be applied when it is not needed or (II) not applied when needed. Johnson et al. (2009) estimated that the cost of an insecticide and its application was approximately \$35.82/ha (\$14.5/acre), resulting in a gain threshold for applying a foliar insecticide of 116 kg/ha, which is approximately 3.6% of the average yield of an Iowan soybean field. An untreated outbreak can reduce soybean yield by as much as 40% (Ragsdale et al. 2011), suggesting the consequences of the first mistake (I) are minor compared to the second mistake (II). Without insurance to compensate farmers in case this latter mistake is made, there is more incentive for farmers to apply insecticide when not needed for this pest. Furthermore, Orlowski et al. (2016) noted that a single application of an insecticide exceeded a gain threshold from 37% to 93%, across a range of increasing crop values. This insecticide was applied independent of the density of soybean aphids or any other pests. Because the insecticides used by Orlowski et al. (2016) are toxic to a wide variety of insect pests, it is unclear if the yield protection was due to the suppression of soybean aphids or combinations of other insect pests. What is clear is that from the perspective of a soybean farmer, there can be great, immediate value in using insecticides regardless of pest pressure. Unfortunately, this can lead to the soybean aphid experiencing selection pressure that could produce a population resistant to insecticides, which has occurred with other aphids (Hemingway 2000). Thus, the preventative use of insecticides can erode their value for crop production, making this approach unsustainable.

Resistance to foliar-applied insecticides would be very damaging to soybean production where the soybean aphid is common. Currently, a limited variety of active ingredients are available to soybean farmers, with the most commonly used ingredients being pyrethroids, primarily lambda-cyhalothrin. During the initial decade of the aphid's occurrence in the United States, lambda-cyhalothrin was used in many studies and efficacy trials of various other insecticides (Hodgson et al. 2016). Repeatedly, these trials demonstrated remarkable mortality almost immediately, with populations reduced by as much as 99%. Recently, failures to prevent soybean aphid outbreaks with pyrethroids have been attributed to populations resistant to these insecticides (Hanson et al. 2017). The full range of these resistant populations is not fully known. Currently, it appears to be limited to West-Central Minnesota, with reports of failures in the bordering states (Iowa, South Dakota, North Dakota). If this resistance spreads, it is likely to cause yield losses, especially for farmers that apply insecticides in a preventative manner without scouting or assessing the efficacy of the insecticide after application. In this scenario, continued selection pressure for insecticide resistance will lead to greater frequency of resistant aphids within population.

Going forward, the potential for a reduction in the use of insecticides is possible if more pest management tactics are available to soybean farmers. Efforts to release parasitoid wasps from the soybean aphids' native range to attack the aphid in the United States continue (Hopper and Diers 2014; O'Neal unpublished data). Such importation biological control programmes can have a remarkably favourable cost-benefit ratio, with many examples of success (Bale et al. 2008; Kenis et al. 2017). However, most of the examples of successful establishment of an exotic natural enemy and subsequent suppression of the target pest are most associated with perennial crops. A recent review of importation biological control for insect pests of woody plants suggests that the success rate of such importation biological control programmes can be low (Kenis et al. 2017).

Based on comparisons with other field crops, the use of aphid-resistant cultivars may have a more immediate impact on the sustainable production of soybeans. Regardless of which form aphid-resistant soybeans take, the impact of this technology could significantly reduce the use of insecticides. In a recent lifecycle assessment of the environmental impacts of the major field crops in the United States (wheat, cotton, corn and soybean, see Yi and Suh 2015), only soybean production was found to have a negative impact on the health of freshwater ecosystems. This increase is due to an increase in insecticide use for soybean aphid management. The authors note that unlike corn and cotton, whose pests are managed by insect-resistant cultivars developed through genetic modification (i.e. GM), the availability of insect resistance for soybeans is commercially limited in the United States. By shifting pest management from insecticides to insect-resistant varieties, GM crops could reduce the environmental impact of agriculture. However, in regions where GM soybeans with resistance to an insect pest is common (i.e. South America), their sustainable use is not guaranteed as noted below. For the soybean aphid, regardless of the source of the resistance, there is a need to develop insect resistance management (IRM) that limits the occurrence of virulent aphid biotypes that can survive on aphid-resistant plants. As noted in the next case study, an IRM plan is required to prolong the life of insect-resistant varieties.

3 Insect resistance management (IRM), especially for Bt-soybeans

For soybean to reap the benefits of insect-resistant varieties enjoyed by these other crops, an IRM plan may be required. In the United States, an IRM plan is required for registration of a GM variety. Such plans are not required for the registration of insecticides, but the Insecticide Resistance Action Committee (www.irac-online.org) provides guidelines for selecting and using insecticides to prevent resistance and cross resistance.

In anticipation of the widespread use of insect-resistant crops, manufacturers, federal regulators and farmers were involved in developing and implementing plans to limit the occurrence of resistance to these crops. Several strategies are available and have been reviewed (Tabashnik et al. 2008). In general, the two main strategies include a high dose and refuge. The high dose is defined by the insect-resistant variety producing a high dose of a toxin that kills enough of the pest population to make inheritance of resistance functionally recessive (i.e. only homozygous resistant individuals survive, assuming resistance is an autosomal trait controlled by a single locus with two alleles). A refuge is planted to produce susceptible adults that mate with the resistant adults, reducing the likelihood that homozygous resistant progeny are produced. Depending upon the crop and target insect, the susceptible plants of the refuge may be planted next to the main crop (i.e. a block refuge) or interspersed within the main crop (i.e. a refuge in a bag). However, if the target pest is polyphagous, such that it can survive on multiple plants, then a natural refuge may be possible. Such a refuge does not require the planting of susceptible plants alongside or within the resistant crop.

When these strategies are made available by the manufacturer and adopted by farmers, the sustainable use of insect-resistant crops has been observed (Tabashnik et al. 2008). However, where the recommended IRM plan was not fully implemented, failures have been observed (Gassmann et al. 2011). In the case study below, we explore how the use of GM soybeans in South America has fared, noting the extent to which IRM was practised.

3.1 Case study 2: managing multiple pests with Bt-soybean in South America

Bt-soybean has become a common part of the soybean production landscape in South America, especially in Brazil and potentially Argentina. Bt-soybean, specifically the expression of the Cry1Ac protein from *Bacillus thuringiensis*, is a trait owned by Monsanto Company and made available through a number of different seed companies depending on the country, including both national and international (i.e. Syngenta) agricultural companies (Silva et al. 2016). It is commonly known commercially as INTACTA RR2 PRO, seed with a suite of technologies combining insect resistance and herbicide resistance. This product can be a source of mortality and is advertised as a management tool for common lepidopteran pests of soybean, including *Anticarsia gemmatalis*, *Chrysodeixis includens* and *Helicoverpa armigera* (Bernardi et al. 2012; Yu et al. 2013). This Bt-protein does not provide protection from lepidopteran pests within the genus *Spodoptera* (Silva et al. 2016; Bernardi et al. 2014).

When Bt-soybeans were used in Brazil, insecticide usage was reduced (Kaskey 2016). Such a change suggests that this technology may lead to more sustainably produced soybeans, by limiting the environmental impacts of soybean production. For example, many of the insecticides used on soybeans are broad spectrum, referring to their toxicity to all insects and not just a target pest. Replacing broad-spectrum insecticides that kill both the target insect and non-target insects with a more specific pest management approach, like Bt, can help conserve beneficial insects. This benefit was confirmed in a study conducted in Brazil that showed a significant decrease in pest populations and no effect on beneficial insect populations when INTACTA soybeans were used without an insecticide (Justiniano et al. 2014). Switching to a more selective insecticide can allow non-target pests to re-emerge as pests, resulting in secondary pest outbreaks. Such outbreaks have occurred in other crops when Bt-traited cultivars are used (Catarino et al. 2015). There is potential for this to occur with Bt-soybean. For example, *Spodoptera* spp. are a potential secondary pest for INTACTA soybeans, though this may be limited if natural enemies of these pests are conserved (Bernardi et al. 2014).

The sustainable use of the Cry1Ac protein in soybeans may be limited by the use of this toxin in other crops, like corn and cotton. The occurrence of resistance may be exacerbated when such crops are grown in a tropical climate like that found in Brazil. Crop production can occur throughout a calendar year without a fallow period (i.e. winter in temperate regions). Irrigation can allow for production in a dry season, allowing for continuous crop production. In these conditions, an insect that feeds on a crop can experience multiple generations within a year. When Bt-crops are planted in this continuous manner, the risk for resistance can be increased. For example, field resistance to Bt-corn by *Spodoptera frugiperda* has occurred in Brazil in regions where corn is produced continuously (Farias et al. 2014). Although a refuge may have been recommended, it is unclear how well it was adopted in this region of Brazil.

Unlike the United States, where the use of Bt crops and the required refuge is regulated at a national level, the refuge regulation is less clear in Argentina and Brazil. In Argentina, private companies are leading the efforts to plan against resistance. A committee has been formed to manage resistance issues resulting from the use of Bt crops; however, it is industry led, not a government initiative or law. The Insecticide Resistance Action Committee (IRAC) is a specialist technical group of the industry association CropLife, which is an international trade association representing agricultural businesses. In 2015, IRAC provided recommendations for managing resistance for Bt in soybean, cotton and corn in Brazil. With regard to soybean, this recommendation includes several practices including a 20% refuge of a non-Bt cultivar and the selection and timing of insecticides in the entire pest complex that attacks soybean. Although these recommendations may be effective in delaying the occurrence of resistance, to what extent they are being followed by farmers will determine if they truly work. In our review of this topic, we found outreach material from both research organizations and government agencies promoting the use of refuges but no mention to the legal requirement of a refuge.

Finally, managing resistance in soybeans will be additionally challenged by invasive species. Recently, *Helicoverpa armigera*, a polyphagous pest that feeds on several field crops, has been found in Brazil (Tay et al. 2013). Agriculture in Brazil allows for multiple seasons of crop production with overlapping cropping seasons. If farmers in a region where *Helicoverpa armigera* is established use the same Bt-protein in cultivars of maize, soybean and cotton, then providing a refuge may be challenging. To date, field-derived resistance to Cry1Ac by *Helicoverpa armigera* has not been reported in Brazil (Dourado et al. 2016). However, resistance to Cry1Ac for *Helicoverpa armigera* has been reported in the lab and field (Akhurst et al. 2003; Zhang et al. 2012; Liu et al. 2010), suggesting that *Helicoverpa armigera* is capable of overcoming this toxin. As noted by Dourado et al. (2016), planting a refuge of *Helicoverpa armigera*-susceptible crops is not mandatory in Brazil, thus making the sustainable use of this trait, either alone or in a pyramid with other Bt traits, a challenge.

4 Conclusion

The goal of this chapter was to describe factors that are leading to increased adoption of preventative/prophylactic use of insect pest management tools. When a preventative approach is adopted, unnecessary use of insecticides and pest-resistant cultivars will continue, which in turn can erode the value of these tools for crop production. Although such an approach may be economically beneficial to farmers, we have noted that there are biological consequences that can negatively impact soybean production as well as the surrounding environment when a preventative approach is adopted. We have suggested that using the principles of IPM can be an improvement to a preventative approach, by limiting the use of pest management tools to when they are needed. By reducing the use of insecticides, the selection pressure that insects experience may be reduced, leading to a more sustainable use of these tools.

As noted above, there are barriers that prevent the IPM approach from being adopted by farmers. The risk (both perceived and real) of not applying an insecticide may prevent farmers from switching from a preventative to IPM-based approach. Furthermore, farmers may not be aware of EIL or ET as a tool for determining the need for and timing of pest management. A lack of information may be most prevalent in regions where substantial research and extension infrastructure do not exist. But even within regions of soybean production with such an infrastructure, limited information can slow the adoption of IPM. For example, invasive species continue to arrive and establish within the United States. Recently, the kudzu bug (*Megacopta cribraria*) and brown marmorated stink bug (*Halyomorpha halys*) have invaded the United States. Each insect is capable of establishing large populations on soybeans in the United States, resulting in economic damage to soybeans. Applied entomologists have developed preliminary thresholds for each of these pests within regions of the United States where they have established themselves (Owens et al. 2013; Seiter et al. 2015, 2016). To what extent these ETs will be employed by farmers as these pests expand their host range is not clear. Based on surveys of soybean farmers in the North Central region of the United States, the occurrence of an invasive pest like the soybean aphid may have briefly increased scouting before a preventative approach to insecticide application became common. However, surveys of farmers in the Southern United States suggest that a culture of scouting fields is much more common (Musser et al. 2015) and predates the occurrence of these invasive pests. Efforts to identify factors that contribute to the adoption of IPM for soybean production will require addressing regional differences in farming practices as well as addressing the perceived and real risks associated with IPM.

For pest management tools like seed-applied insecticides and insect-resistant varieties, farmers may not know if a pest outbreak is likely to occur at the time of planting. In anticipation of widespread adoption, insect-resistant crops, agribusiness, scientists and regulatory agencies have developed and implemented IRM plans to limit the occurrence of resistance to at least the insect-resistant varieties. To date, we are not aware of similar plans for seed-applied insecticides. To what extent these plans are adopted fully by farmers will determine how sustainably these tools can be used.

In the end, sustainability not only refers to the environmental soundness of the practice but also to the economic and social viability. Though IPM and IRM can provide environmental and production benefits, the question remains of their real and perceived economic benefit for soybean farmers, especially in South America. The added cost of an input may not always result in additional returns (see Johnson et al. 2009). As with any tool, it is important to evaluate the benefits and costs to determine in each situation what combination of tools can provide the most effective, sustainable and cost-effective practice.

5 Future trends

A major assumption in our discussion is that IPM can lead to more sustainable pest management through the reduction in the use of a given pest management tool. *A priori*, IPM can lead to reduced insecticide use compared to a prophylactic or calendar-based approach. However, IPM does not imply no insecticide use. To what extent even a limited amount of use can result in resistance is not clear, and will clearly depend upon the pest and crop production system. Future research will need to address if integration of IPM within an IRM programme can help achieve both the farmer's short-term goals of maximizing production, while working towards the long-term goals of preventing or, at least, delaying resistance.

Research studies on South American soybean production need to strongly focus on pest and beneficial insect community dynamics among the multiple cropping seasons and cropping systems that include soybean. Special attention is required to understand population and community dynamics extending beyond insects into pathogen management, especially given the increased soybean disease pressure resulting in fungicide use and declining entomopathogen (disease of insects) organisms.

Worldwide, investments are needed in the basic infrastructure for IPM programmes, so that sustainable pest management in soybean can be accessible to all farmers. This investment may also require training scientists from around the world in developing economic thresholds and economic injury levels with regard to key pests of soybean in their respective regions. If reducing insecticide use while maintaining high production levels is the goal, it will require different approaches in different cultural, political and agroecological contexts. The IPM tool box is designed to be adapted across differing circumstances but may require the guidance of soybean IPM experts to achieve high-quality results.

Going forward, addressing pest management within the context of how soybeans are grown may best help farmers understand the value of not only insecticides but other pesticide inputs as well. As noted by Davis et al. (2012), when soybeans were grown in a three- or four-year rotation, the amount of inputs and cost of production were lower than those in a two-year rotation with corn. A substantial reduction in herbicide use was attributed to the longer rotations. In the pests reviewed above, crop rotation alone likely has limited value in preventing outbreaks. Both the soybean aphid and *Helicoverpa armigera* migrate into fields after the plants have emerged. However, crop rotation may alter the physical and biochemical nature of the soil, which in turn could affect the plant's physiology. As noted above, soybean following alfalfa is more at risk for soybean aphid outbreaks due to depletion of potassium in the soil (Bruulesma et al. 2010). Other preceding crops can affect soybean aphid populations as well (Lundgren et al. 2017), though the mechanisms are not yet determined. A systems-level approach used by Davis et al. (2012) could help to determine which agronomic practices (e.g. length of a crop rotation and the types of plants within the rotation) result in more sustainable soybean production. Hopefully, a synthesis of the ideas presented in this box can lead to such results.

6 Where to look for further information

There are many textbooks that can be a source of information on integrated pest management, and we have cited them within the text (Pedigo and Rice 2014). For more information, especially where textbooks are hard to find, we recommend visiting 'Radcliffe's IPM World Textbook'. This is an online resource for all crops; https://ipmworld.umn.edu/.

Information sources are most abundant for soybean production systems in the United States, where many universities have been conducting investigations.

For soybean pest management guidance in South America, most references are available in Spanish and Portuguese. National research organizations and universities in Argentina, Brazil, and other countries in South America have published extensive scientific findings related to soybean production and the challenges being faced in the region.

Specifically we recommend EMBRAPA from Brazil and INTA from Argentina as sources of significant bodies of research and published findings:

EMBRAPA (Brazilian Agriculture Research Corporation): https://www.embrapa.br/soja.

INTA (National Institute for Agricultural Technology): https://inta.gob.ar.

Many stakeholders are currently involved in promoting insect resistance management plans, and information about their recommendations for refuges can be found here:

Insect Resistance Action Committee: http://www.irac-online.org/.

Insect Resistance Management Program (Programa de Manejo de Resistencia de Insectos) http://www.programamri.com/.

From Monsanto Company's South America Website – http://www.intactarr2pro.com.ar/refugio.

Other regions of the world with soybean production, such as sub-Saharan Africa offer less information regarding pest management challenges and practices. However, we recommend looking for a contact point in the ministry of agriculture or the associated national agriculture research institute. Though there is not always information available online, national scientists may be able to guide you for more information.

7 References

Ajayi-Oyetunde, O. O., B.W. Diers, D. Lagos-Kutz, C. B. Hill, G. L. Hartman, U. Reuter-Carlson and C. A. Bradley. 2016. Differential reactions of soybean isolines with combinations of aphid resistance genes *Rag*1, *Rag*2, and *Rag*3 to four soybean aphid biotypes. *Journal of Economic Entomology*, 109: 1431–7.

Akhurst, R. J., W. James, L. J. Bird and C. Beard. 2003. Resistance to the Cry1Ac delta-endotoxin of *Bacillus thuringiensis* in the cotton bollworm, *Helicoverpa armigera* (Lepidoptera: Noctuidae). *Journal of Economic Entomology*, 96: 1290–9.

Alt, J. and M. Ryan-Mahmutagic. 2013. Soybean aphid biotype 4 identified. *Crop Science*, 53: 1491–5.

Anon. 2016. Semilleros, Variedades, Comercios Y Multiplicadores Autorizados. The Monsanto Company. http://www.intactarr2pro.com.py/semilleros (accessed June 2017).

Bahlai, C. A., W. vander Werf, M. O'Neal, L. Hemerik and D. A. Landis. 2015. Shifts in dynamic regime of an invasive lady beetle are linked to the invasion and insecticidal management of its prey. *Ecological Applications*, 25: 1807–18. http://dx.doi.org/10.1890/14-2022.1.

Bale, J. S., J. C. van Lenteren and F. Bigler. 2008. Biological control and sustainable food production. *Philosophical Transactions of the Royal Society B*, 363: 761–76.

Bernardi, O., G. S. Malvestiti, P. M. Dourado, W. S. Oliveira, S. Martinelli, G. U. Berger, G. P. Head and C. Omoto. 2012. Assessment of the high-dose concept and level of control provided by MON 87701Å~MON 89788 soybean against Anticarsia gemmatalis and Pseudoplusia includens (Lepidoptera: Noctuidae) in Brazil. *Pest Management Science*, 68: 1083–91.

Bernardi, O., R. J. Sorgatto, A. D. Barbosa, F. A. Domingues, P. M. Dourado, R. A. Carvalho, S. Martinelli, G. P. Head and C. Omoto. 2014. Low susceptibility of *Spodoptera cosmioides, Spodoptera eridania* and *Spodoptera frugiperda* (Lepidoptera: Noctuidae) to genetically-modified soybean expressingCry1Ac protein. *Crop Protection*, 58: 33–40.

Bradshaw, J. D. and M. E. Rice. 2003. Bean leaf beetles: A current and historical perspective. *Integrated Crop Management Newsletter*, 3: 20–1. IC-490. http://www.ipm.iastate.edu/ipm/icm/2003/3-17-2003/beanleafbeetles.html.

Bruulseman, T., C. Difonzo and C. Gratton. 2010. How potassium nutrition can suppress soybean aphids. *Better Crops*, 94: 10–13, https://www.ipni.net/ppiweb/bcrops.nsf/$webindex/9DF22A43E763A8C4852577230049DC39/$file/BC22010-Page11-13.pdf (accessed June 2017).

Catarino, R., G. Ceddia, F. J. Areal and J. Park. 2015. The impact of secondary pests on *Bacillus thuringiensis* (Bt) crops. *Plant Biotechnology Journal*, 13: 601–12.

Chougule, N., H. Li, S. Liu, L. B. Linz, K. E. Narva, T. Meade and B. C. Bonning. 2013. Retargeting of the *Bacillus thuringiensis* toxin Cyt2Aa against hemipteran insect pests. *Proceedings of the National Academy of Sciences*, 21: 8465–70.

Crawley, M. J. 1989. Insect herbivores and plant population dynamics. *Annual Reviews of Entomology*, 34(1): 531–64.

Cox, R., M. E. O'Neal, R. Hessel, L. A. Schulte and M. Helmers. 2014. The impact of prairie strips on aphidophagous predator abundance and soybean aphid predation in agricultural catchments. *Environmental Entomology*, 43: 1185–97.

Davis, A. S., J. D. Hill, C. A. Chase, A. M. Johanns and M. Liebman. 2012. Increasing cropping system diversity balances productivity, profitability and environmental health. *PLoS ONE*, 7(10): e47149.

Douglas, M. R. and J. F. Tooker. 2015. Large-scale deployment of seed treatments has driven rapid increase in use of neonicotinoid insecticides and preemptive pest management in US field crops. *Environmental Science & Technology*, 49: 5088–97

Dourado, P. M., F. B. Bacalhau, D. Amado, R. A. Carvalho, S. Martinelli, G. P. Head and C. Omoto 2016. High susceptibility to Cry1Ac and low resistance allele frequency reduce the risk of resistance of *Helicoverpa armigera* to Bt soybean in Brazil. *PLoS ONE*, 11(8): e0161388. https://doi.org/10.1371/journal.pone.0161388.

Farias, J. R., D. A. Andow, R. J. Horikoshi, R. J. Sorgatto, P. Fresia, A. C. dos Santos and C. Omoto. 2014. Field-evolved resistance to Cry1F maize by Spodoptera frugiperda (Lepidoptera: Noctuidae) in Brazil. *Crop Protection*, 64: 150–8.

Gardiner, M. M., D. A. Landis, C. Gratton, N. Schmidt, M. E. O'Neal, E. Mueller, J. Chacon, G. E. Heimpel and C. D. DiFonzo. 2009a. Landscape composition influences patterns of native and exotic lady beetle abundance. *Diversity and Distributions*, 15: 554–64.

Gardiner, M. M., D. A. Landis, C. Gratton, C. D. DiFonzo, M. E. O'Neal, J. M. Chacon,, T. M. Wayo, N. P. Schmidt, E. E. Mueller and G. E. Heimpel. 2009b. Landscape diversity enhances biological control of an introduced crop pest in the north-central USA. *Ecological Applications*, 19: 143–54.

Gassmann, A. J., J. L. Petxold-Maxwell, R. S. Keweshan and M. W. Dunbar. 2011. Field-evolved resistance to Bt maize by western corn rootworm. *PLoS ONE*, 6(7): e22629.

Gray, M. E., T. W. Sappington, N. J. Miller, Moeser, J. and M. O. Bohn. 2009. Adaptation and Invasiveness of western corn rootworm: Intensifying research on a worsening pest. *Annual Review of Entomology*, 54: 303–21.

Hanson, A. A., J. Menger-Anderson, C. Silverstein, B. D. Potter, I. V. MacRae, E. W. Hodgson and R. L. Koch. 2017. Evidence for soybean aphid (Hemiptera: Aphididae) resistance to pyrethroid insecticides in the upper Midwestern United States. *Journal of Economic Entomology* 110: 2235–46.

Hayek, A. 2004. Natural Enemies: An Introduction to Biological Control. Cambridge Univesity Press, Cambridge, UK. ISBN: 0521653385.

Heimpel, G. E., D. W. Ragsdale, R. C. Venette, K. R. Hopper, R. J. O'Neil, C. E. Rutledge and Z. Wu. 2004. Prospects for importation biological control of the soybean aphid: Anticipating potential costs and benefits. *Annals of the Entomological Society of America*, 97: 249–58.

Hemingway, J. 2000. The molecular basis of two contrasting metabolic mechanisms of insecticide resistance. *Insect Biochemistry and Molecular Biology*, 30: 1009–15.

Hesler, L. S., M. V. Chiozza, M. E. O'Neal, G. C. MacIntosh, K. J. Tilmon, D. I. Chandrasena, N. A. Tinsley, S. R. Cianzio, A. C. Costamagna, E. Cullen, C. D. DiFonzo, B. D. Potter, K. Steffey and K. J. Koehler. 2013. Performance and prospects of *Rag* genes for management of soybean aphid. *Entomologia Experimentalis et Applicata*, 147: 201–16.

Higley, L. G. and D. J. Boethel. 1994. *Handbook of Soybean Insect Pests*. ESA Handbook Series 1. Entomological Society of America, Lanham, MD.

Hodgson, E. W., E. C. Burkness, W. D. Hutchison and D. W. Ragsdale. 2004. Enumerative and binomial sequential sampling plans for soybean aphid (Homoptera: aphididae) in soybean. *Journal of Economic Entomology*, 97: 2127–36.

Hodgson, E. W., B. P. McCornack, K. A. Koch, D. W. Ragsdale, Johnson, K. D, M. E. O'Neal, E. Cullen, H. Kraiss, C. D. DiFonzo and L. M. Behnken. 2007. Field validation of Speed Scouting for soybean aphid. Online. *Crop Management*. doi:10.1094/CM-2007-0511-01-RS.

Hodgson, E. W., M. R. Kemis and B. N. Geisinger. 2012a. Assessment of Iowa soybean growers for insect pest management practices. *Journal of Extension*, 50(4): 4RIB6.

Hodgson, E. W., B. P. McCornack, K. Tilmon and J. J. Knodel. 2012b. Management recommendations for soybean aphid (Hemiptera: Aphididae) in the United States. *Journal of Integrated Pest Management*, 3(1): E1–E10.

Hopper, K. R. and B. W. Diers. 2014. Parasitism of soybean aphid by Aphelinus species on soybean susceptible versus resistant to the aphid. *Biological Control*, 76: 101–6.

Hosokawa, T., N. Nikoh and T. Fukatsu. 2014. Fine-scale geographical origin of an insect pest invading North America. *PLoS ONE* 9(2): e89107. doi:10.1371/journal.pone.0089107

Johnson, K. D., M. E. O'Neal, D. W. Ragsdale, C. D. DiFonzo, S. M. Swinton, P. M. Dixon, B. D. Potter, E. W. Hodgson and A. C. Costamagna. 2009. Probability of economical management of soybean aphid (Hemiptera: Aphididae) in North America. *Journal of Economic Entomology* 102: 2101–8.

Justiniano, W., M. Fernandes and C. Viana. 2014. Diversity, composition and population dynamics of arthropods in the genetically modified soybeans roundup ready® RR1 (GT 40-3-2) and intacta RR2 PRO® (MON87701 × MON89788). *Journal of Agricultural Science*, 6: 33–44.

Kaskey, J. 2016. Monsanto Killing DuPont Insecticide Sales With New Soybeans. *Bloomberg*. https://www.agweb.com/article/sales-surge-for-monsantos-new-intacta-soybeans-blmg/ (accessed June 2017).

Kenis, M., B. P. Hurley, A. E. Hajek and M. J. W. Cock. 2017. Classical biological control of insect pests of trees: Facts and figures. *Biological Invasions*, 19: 3401–17. doi:10.1007/s10530-017-1414-4.

Koch, R. L. 2003. The multicolored Asian lady beetle, Harmonia axyridis: A review of its biology, uses in biological control and non-target impacts. *Journal of Insect Science*, 3: 1–16.

Koch, R. L. and T. L. Galvan. 2008. Bad side of a good beetle: The North American experience with Harmonia axyridis. *BioControl*, 53: 23–35.

Kogan, M. 1998. Integrated pest management: Historical perspectives and contemporary developments. *Annual Review of Entomology*, 43: 243–70.

Kogan, M. and S. G. Turnipseed. 1987. Ecology and management of soybean arthropods. *Annual Review of Entomology*, 32:507–38.

Kremen, C., N. M. Williams and R. W. Thorp. 2002. Crop pollination from native bees at risk from agricultural intensification. *Proceedings of the National Academy of Science*, 99: 16812–16.

Krupke, C. H., G. J. Hunt, B. D. Eitzer, G. Andino and K. Given. 2012. Multiple routes of pesticide exposure for honey bees living near agricultural fields. *PLoS ONE*, 7: e29268. doi:10.1371/journal.pone.0029268.

Krupke, C. H., A. M. Alford, E. M. Cullen, E. W. Hodgson, J. J. Knodel, B. McCornack, B. D. Potter, M. I. Spigler, K. Tilmon and K. Welch. 2017. Assessing the value and pest management window provided by neonicotinoid seed treatments for management of soybean aphid (aphis glycines Matsumura) in the upper Midwestern United States. *Pest Management Science*, 73: 2184–93.

Landis, D. A. and D.B. Orr. 2017. Biological control: Approaches and applications. Radcliffe's IPM World Textbook. https://ipmworld.umn.edu/landis. (accessed 18 December 2017).

Li, Z. Z., S. B. Alves, D. W. Roberts, M. Z. Fan, I. Delalibera, J. Tang, R. B. Lopes, M. Faria and D. E. N. Rangel. 2010. Biological control of insects in Brazil and China: history, current programs and reasons for their successes using entomopathogenic fungi. *Biocontrol Science and Technology*, 20: 117–36.

Liu, F., Z. Xu, Y. C. Zhu, F. Huang, Y. Wang, H. Li, C. Gao, W. Zhou and J. Shen. 2010. Evidence of field-evolved resistance to Cry1Ac-expressing Bt cotton in *Helicoverpa armigera* (Lepidoptera: Noctuidae) in northern China. *Pest Management Science*, 66: 155–61.

Lundgren, J. G., L. S. Hesler and R. L. Anderson. 2017. Preceding crop affects soybean aphid abundance and predator-prey dynamics in soybean. *Journal of Applied Entomology*, 141(9): 669–76.

Mallinger, R. E., D. B. Hogg and C. Gratton. 2011. Methyl salicylate attracts natural enemies and reduces populations of soybean aphids (Hemiptera: Aphididae) in soybean agroecosystems. *Journal of Economic Entomology*, 104, 115–24.

Matteson, P. C. 2000. Insect pest management in tropical Asian irrigated rice. *Annual Review of Entomology* 45: 549–74.

McCarville, M. T., M. E. O'Neal, B. D. Potter, K. J. Tilmon, E. M.Cullen, B. P. McCornack, J. F. Tooker and D. A. Prischmann-Voldseth. 2014. One gene versus two: a regional study on the efficacy of single gene versus pyramided resistance for soybean aphid management. *Journal of Economic Entomology*, 107: 1680–7.

Mogren, C. L. and J. G. Lundgren. 2016. Neonicotinoid-contaminated pollinator strips adjacent to cropland reduce honey bee nutritional status. *Scientifc Reports* 6: 29609. Doi: 10.1038/srep29608.

Morse, S. and W. Buhler. 1997. IPM in developing countries: The danger of an ideal. *Integrated Pest Management Reviews*, 2: 175–85.

Musser, F. R., A. L. Catchot, J. A. Davis, D. A. Herbert, G. M. Lorenz, T. Reed, D. D. Reisig and S. D. Stewart. 2015. 2014 soybean insect losses in the Southern US. *Midsouth Entomologist*, 8: 35–48.

Myers, C. and E. Hill. 2014. Benefits of neonicotinoid seed treatments to soybean production. *Environmental Protection Agency*. https://www.epa.gov/pollinator-protection/benefits-neonicotinoid-seed-treatments-soybean-production (accessed June 2017).

North, J. H., J. Gore, A. L. Catchot, S. D. Stewart, G. M. Lorenz, F. R. Musser, D. R. Cook, D. L. Kerns and D. M. Dodds. 2016. Value of neonicotinoid insecticide seed treatments in mid-south soybean (Glycine max) production systems. *Journal of Economic Entomology*, 109: 1156–60.

Orlowski, J. M., B. Y. Haverkamp, R. G. Laurenz, D. A. Marburger, E. W. Wilson, S. N. Casteel, S. P. Conley, S. L. Naeve, E. D. Nafziger, K. L. Roozeboom, W. J. Ross, K. D. Thelen and C. D. Lee. 2016. High-input management systems effect on soybean seed yield, yield components, and economic break-even probabilities. *Crop Science*, 56: 1988–2004.

Owens, D. R., D. A. Herbert, G. P. Dively, D. D. Reisig and T. P. Kuhar. 2013. Does Feeding by *Halyomorpha halys* (Hemiptera: Pentatomidae) reduce soybean seed quality and yield? *Journal of Economic Entomology*, 106: 1317–23.

Parsa, S., S. Morse, A. Bonifacio, T. C. B. Chancellor, B. Condori, V. Crespo-Pérez, S. L. A. Hobbs, J. Kroschel, M. N. Ba, F. Rebaudo, S. G. Sherwood, S. J. Vanek, E. Faye, M. A. Herrera and O. Dangles. 2014. Obstacles to integrated pest management adoption in developing countries. *Proceedings of the National Academy of Sciences*, 111(10): 3889–94.

Pedigo, L. P. and M. Rice. 2014. *Entomology and Pest Management*. 6th Edition. Waveland Press, Inc., Long Grove, IL. ISBN: 1-4786-2285-7.

Ragsdale, D., B. P. McCornack, R. C. Venette, B. D. Potter, I. V. Macrae, E. W. Hodgson, M. E. O'Neal, K. D. Johnson, R. J. O'Neil, C. D. DiFonzo, T. E. Hunt, P. Glogoza and E. M. Cullen. 2007. Economic threshold for soybean aphid (Hemiptera: Aphididae). *Journal of Economic Entomology*, 100: 1258–67.

Ragsdale, D. W., D. A. Landis, J. Brodeur, G. E. Heimpel and N. Desneux. 2011. Ecology and management of the soybean aphid in North America. *Annual Review of Entomology*, 56: 375–99.

Ruberson, J. R., K. Takasu, G. D. Buntin, J. E. Eger Jr., W. A. Gardner, J. K. Greene, T. M. Jenkins, W. A. Jones, D. M. Olson, P. M. Roberts, D. R. Suiter and M. D. Toews. 2013. From Asian curiosity to eruptive American pest: *Megacopta cribraria* (Hemiptera: Plataspidae) and prospects for its biological control. *Applied Entomology and Zoology*, 48: 3–13.

Rutledge, C., R. O'Neal, T. Fox and D. Landis. 2004. Soybean aphid predators and their use in integrated pest management. *Annals of the Entomological Society of America*, 97: 240–8.

Sanchez-Bayo, F. 2014. The trouble with neonicotinoids. *Science*, 346: 806. doi:10.1126/science.1259159.

Schmidt, N. P., M. E. O'Neal and P. M. Dixon. 2008. Aphidophagous predators in Iowa soybean: A community comparison across multiple years and sampling methods. *Annals of the Entomological Society of America*, 101: 341–50.

Seiter, N. J., D. Pozo-Valdivia, A. I. Greene, J. K. Reay-Jones, P. M. Roberts and D. D. Reisig. 2015. Action thresholds for managing *Megacopta cribraria* (Hemiptera: Plataspidae) in soybean base on sweep-net sampling. *Journal of Economic Entomology*, 108: 1818–29.

Seiter, N. J., D. Pozo-Valdivia, A. I. Greene, J. K. Reay-Jones, P. M. Roberts and D. D. Reisig. 2016. Management of *Megacopta cribraria* (Hemiptera: Plataspidae) at different stages of soybean (Fabales: Fabaceae) development. *Journal of Economic Entomology*, 109: 1167–76.

Silva, G., B. Bueno, B. Bortolotto, O. Santos and A. Pomari-Fernandes. 2016. Biological characteristics of black armyworm *Spodoptera cosmioides* on genetically modified soybean and corn crops that express insecticide Cry proteins. *Revista Brasileira de Entomologia*, 60: 255–9.

Simon-Delso, N., V. Amaral-Rogers, L. P. Belzunces, J. M. Bonmatin, M. Chagnon, C. Downs, L. Furlan, D. W. Gibbons, C. Ciorio, V. Girolami, D. Goulson, D. P. Kruetzweiser, C. H. Krupke, M. Liess, E. Long, M. McField, P. Mineau, E. A. D. Mitchell, C. A. Morrissey, D. A. Noome, L. Pisa, J. Settele, J. D. Stark, A. Tapparo, H. van Dyck, J. Van Praagh, J. P. Van der Sluijs, P. R. Whitehorn and M. Wiemers. 2015. Systemic insecticides (neonicotinoids and fipronil): Trends, uses, mode of action and metabolites. *Environmental Science and Pollution Research*, 22: 5–34.

Stern, V. M., R. F. Smith, R. van den Bosch and K. S. Hagen. 1959. The integrated control concept. *Hilgardia*, 29: 81–101.

Stone, J. D. and L. P. Pedigo. 1972. Development and economic injury level of green cloverworm (Lepidoptera: Noctuidae) on soybean in Iowa. *Journal of Economic Entomology*, 65: 197–201.

Tabashnik, B. E., A. J. Gassmann, D. W. Crowder and Y. Carriere. 2008. Insect resistance to Bt crops: evidence versus theory. *Nature Biotechnology*, 26: 199.

Tay, W. T., M. F. Soria, T. Walsh, D. Thomazoni, P. Silvie, G. T. Behere, C. Anderson and S. Downes. 2013. A Brave New World for an Old World Pest: *Helicoverpa armigera* (Lepidoptera: Noctuidae) in Brazil. *PLoS ONE*, 8(11): e80134.

Tilmon, K. J., E. W. Hodgson, M. E. O'Neal and D. W. Ragsdale. 2011. Biology of the soybean aphid, *Aphis glycines* (Hemiptera: Aphididae) in the United States. *Journal of Integrated Pest Management*, 2(2): A1–A7

Turnipseed, S. G. and M. Kogan. 1976. Soybean entomology. *Annual Review of Entomology*, 21: 247–82.

Whalon, M. E., D. Mota-Sanchez and R. M. Hollingworth. 2008. Analysis of global pesticide resistance in arthropods, pp. 5–31. In: M. E. Whalon, D. Mota-Sanchez and R. M. Hollingworth (Eds), *Global Pesticide Resistance in Arthropods*. CABI Publishing, Oxon, England.

Work, T. T., D. G. McCullough, J. F. Cavey and R. Komsa. 2005. Arrival rate of nonindigenous insect species into the United States through foreign trade. *Biological Invasions*, 2: 323–32.

Yi, Y. and S. Suh. 2015. Changes in environmental impacts of major crops in the US. *Environmental Research Letters*, 10: 094016.

Yu, H., Y. Li, X. Li, J. Romeis and K. Wu. 2013. Expression of Cry1Ac in transgenic Bt soybean lines and their efficiency in controlling lepidopteran pests. *Pest Management Science*, 69: 1326–33.

Yu, H., Y. Li, X. Li and K. Wu. 2014. Arthropod abundance and diversity in transgenic Bt soybean. *Environmental Entomology*, 43: 1124–34.

Zhang, H., W. Tian, J. Zhao, L. Jin, J. Yang, C. Liu, Y. Yang, S. Wu, K. Wu, J. Cuj, B. E. Tabashnik and Y. Wu. 2012. Diverse genetic basis of field-evolved resistance to Bt cotton in cotton bollworm from China. *Proceedings of the National Academy of Sciences*, 109: 10275–80.

Chapter 5

Advances in disease-resistant varieties of soybean

David R. Walker, USDA-ARS, USA

1 Introduction

2 Resistance to major soybean diseases and nematode pests

3 Selection for disease resistance

4 Contribution of soybean genomics research to breeding strategies

5 Pathogen genomics

6 Disease resistance from *G. soja* and perennial *Glycine* species

7 Transgene-mediated disease resistance

8 Summary and future trends

9 Where to look for further information

10 References

1 Introduction

Soybean [*Glycine max* (L.) Merr.] plants are potential hosts to at least 200 pathogens, of which about 35 are considered economically important (Hartman 2015). Sustainable management of these diseases to reduce yield losses requires combining host plant resistance with cultural practices and prudent use of chemical applications if a disease appears likely to exceed an economic injury threshold before or during seed development (Hartman and Hill 2010).

Diseases that can detrimentally affect soybean yield and seed quality range from seed rot and 'damping-off' of seedlings to foliar blights and seed infections (Hartman et al. 2015c). Susceptible soybean plants are vulnerable to infection by viruses, bacteria, mollicutes (bacteria that lack cell walls), oomycetes and fungi, and can be parasitized by nematodes from several genera. Some of the most damaging diseases of soybean grown in North America are listed in Table 1. Wrather and Koenning (2009) estimated that the four most economically important soybean diseases and pests in the United States between 1996 and 2007, in order of importance, were soybean cyst nematode (SCN; *Heterodera glycines*); *Phytophthora* root and stem rot (PRR; *Phytophthora sojae*); seedling diseases caused by *Fusarium* spp., *Pythium* spp. and *Rhizoctonia solani*; and charcoal rot (*Macrophomina phaseolina*). SCN and PRR are routinely listed as the first and second most economically important diseases of soybean grown in North America (Koenning and Wrather 2010). This is partly because they are soil pathogens, so once established in a

http://dx.doi.org/10.19103/AS.2017.0034.20

Table 1 Some of the major diseases of soybean in the United States and Canada and the relative importance of resistance in northern and southern production regions

Disease (acronyms used in text are in parentheses)	Pathogen/pest	Class	Relative importance of resistance in northern United States production regions	Relative importance of resistance in southern United States production regions
Bacterial blight	*Pseudomonas savastanoi* pv. *glycinea*	Bacteria	High	Moderate-Low
Bacterial pustule	*Xanthomonas axonopodis* pv. *glycines*	Bacteria	Moderate-High	High
Phytophthora root and stem rot (PRR)	*Phytophthora sojae* (primarily)	Oomycetes	High	High in Mid-South; moderate elsewhere
Pythium seed rot, damping-off and root rot	*Pythium* spp. (taxonomic revision of the genus *Pythium* has been proposed)	Oomycetes	High	High in Mid-South; moderate elsewhere
Anthracnose	*Colletotrichum truncatum* and other *Colletotrichum/Glomerella* spp.	Fungi	Moderate	Moderate
Brown stem rot	*Cadophora gregata*	Fungi	High	Moderate-Low
Charcoal rot	*Macrophomina phaseolina*	Fungi	Moderate	High
Frogeye leaf spot (FLS)	*Cercospora sojina*	Fungi	Low-Moderate	High
Phomopsis seed decay (PSD)	*Diaporthe longicolla* (primarily)	Fungi	Moderate	High
Rhizoctonia damping-off and root rot	*Rhizoctonia solani*	Fungi	High	Moderate
Sclerotinia stem rot	*Sclerotinia sclerotiorum*	Fungi	High	Low
Soybean rust (SBR)	*Phakopsora pachyrhizi*	Fungi	Low	Moderate-High
Stem canker (northern)	*Diaporthe caulivora*	Fungi	High	Moderate
Stem canker (southern)	*Diaporthe aspalathi* (syn. *D. phaseolorum* var. *meridionalis*)	Fungi	Low	High
Sudden death syndrome (SDS)	*Fusarium virguliforme*	Fungi	High	Moderate
Soybean cyst nematode (SCN)	*Heterodera glycines*	Nematodes	High	High
Root-knot nematodes	*Meloidogyne incognita* (southern); *M. arenaria* (peanut); *M. javanica* (Java)	Nematodes	Low	High

field, they will remain a threat in subsequent years and cannot be eradicated. In addition, PRR is primarily a problem in the major production regions of the northern United States and adjacent regions in Canada. Because of the destructive potential of these diseases and their widespread distribution in both northern and southern soybean-producing areas in the United States, few modern cultivars lack resistance to SCN or PRR. In contrast, losses attributed to some other diseases can vary considerably from one growing season to another, largely due to differences in precipitation and temperatures. For example, white mould, or *Sclerotinia* stem rot (*Sclerotinia sclerotiorum*) (SSR), was ranked as the second most destructive disease in the northern United States in 2004, but caused only negligible losses in the southern states that year, and had not even been among the top ten disease problems in the northern states in 2003 (Wrather and Koenning 2006).

Different environmental conditions in northern and southern soybean production regions of the United States account in part for some differences in the most economically important pathogens and diseases (Table 1). In the northern states, sudden death syndrome [SDS; *Fusarium viruliforme* (previously called *F. solani* f. sp. *glycines*) in North America], brown stem rot [BSR, caused by the fungus *Cadophora gregata* (syn. *Phialophora gregata*)] and seedling diseases caused by various oomycete and fungal pathogens are generally a larger and more frequent problem than they are in most southern states (Wrather and Koenning 2006). In southern production areas, anthracnose (*Colletotrichum* spp.), frogeye leaf spot (FLS; *Cercospora sojina*) and southern stem canker [SSC, *Diaporthe aspalathi* (syn. *D. phaseolorum* var. *meridonalis*)] tend to be more of a problem than in states farther north. In certain growing seasons, however, diseases that are generally more of a problem in southern states can cause devastating losses in the north-central states, as charcoal rot (*Macrophomina phaseolina*) did in 2003 (Wrather and Koenning 2006).

In addition to pathogens that cause diseases of high economic importance, some other pathogens might seldom cause major yield losses, but nevertheless produce highly visible symptoms that would be unacceptable to many producers. Examples include *Soybean mosaic virus* (SMV) and *Bean pod mottle virus*, which are widespread in North America, but are generally not considered to be major threats to yield, though they can be a problem in certain year-location environments (Wrather and Koenning 2006). Similarly, bacterial blight (caused by *Pseudomonas savastanoi* pv. *glycinea*) and bacterial pustule (BP, caused by *Xanthomonas axonopodis* pv. *glycines*) are thought to rarely cause major yield losses, but infections can produce an abundance of foliar lesions that are likely to cause concern among producers. Thus there is an incentive for seed companies to avoid commercializing lines that can develop severe symptoms. Finally, there are certain diseases like northern stem canker and SSC that are now relatively uncommon because they were so devastating in the past that it is likely that most modern cultivars will have been derived from crosses in which both parents were resistant.

Soybean breeders in North America have been highly successful in developing cultivars with effective levels of resistance to many of the major soybean pathogens. A group of soybean cultivars that had been released between 1923 and 2008 had a 100% incidence of bacterial diseases (bacterial blight and bacterial pustule), brown spot, anthracnose, *Cercospora* blight, charcoal rot, and pod and stem blight at a location in east-central Illinois in 2011, but the disease severity ratings for bacterial disease and brown spot were significantly higher for cultivars released prior to 1970 than on ones released after 1989 (Hartman et al. 2015a). The same trend was also seen in ratings for root and stem diseases, and the majority of the foliar and stem/root diseases observed on plants were negatively correlated with year of release. The correlation was strongest for the bacterial diseases, illustrating the success that breeders and pathologists have achieved in developing and selecting lines with resistance to oomycete, fungal and bacterial diseases (Hartman et al. 2015a).

The development of disease-resistant soybean cultivars for the United States and Canada has been successful despite the narrow genetic base of North American cultivars (Gizlice et al. 1994) and the comparatively low genetic diversity within cultivated germplasm (Hyten et al. 2006). The combination of the small number of ancestral Asian landraces that most modern North American cultivars are descended from and intensive selection for improved agronomic traits in lines derived from them resulted in the loss of approximately 78% of the rare alleles (freq ≤ 0.01) that were present in the Asian landraces (Hyten et al. 2006). As soybean seed production became more important, breeding efforts concentrated on crosses with lines that possessed traits essential for high seed yield and an increasingly mechanized production system, and disease resistance was rarely a high priority.

The need to breed for resistance to diseases and pests that were prevalent in North America has had a mixed effect on the genetic diversity of cultivars developed for the United States and Canada, increasing the number of germplasm accessions used as sources of resistance genes in recent decades, while also limiting diversity because a small number of early cultivars with resistance to major diseases were used very extensively in breeding programmes during the second half of the Twentieth Century. Although a few of the most important ancestors of North American cultivars had some resistance to major soybean diseases and nematode pests, the Chinese accession CNS (Clemson Non-Shatter), which is resistant to BP, was the only major ancestor that was used primarily because of its resistance to a disease (Narvel et al. 2001). Carter et al. (2004) attributed phenotypic distinction as being an important factor in the success that North American breeders have had in introgressing exotic resistance alleles into adapted material, at least when a single major gene was involved. The USDA Soybean Germplasm Collection has been a valuable source of resistance genes that have been backcrossed from exotic plant introductions (PIs) into the genetic backgrounds of breeding lines and cultivars that yield well in the United States and other major producer countries. Efforts to develop cultivars with resistance to different diseases have been influenced by pathogenic variation in pathogen populations, the potential for a pathogen to evolve new races or strains, and the level of disease resistance initially present in existing cultivars and elite lines (Carter et al. 2004).

Advances in biotechnology and bioinformatics have had a major impact on soybean breeding methods and our understanding of genes that condition disease and pest resistance. While DNA markers have been used in marker-assisted selection (MAS) for disease resistance for several decades, often in the absence of knowledge about the actual gene(s) that they were used to tag, the development of much denser linkage maps and completion of a draft assembly of the *G. max* genome (Schmutz et al. 2010) opened up new possibilities. This resource dramatically increased the efficiency of discovering and confirming the locations and identities of genes involved in resistance and a wide variety of other traits important to maximizing seed yield and composition in soybean.

2 Resistance to major soybean diseases and nematode pests

Examples of soybean diseases that cause major economic losses in the United States and resistance to them are briefly reviewed below, but the list is by no means comprehensive.

Numerous viruses can cause disease and can be a particular problem when virus-susceptible exotic PIs are used in breeding programmes as sources of resistance genes for other pathogens, but virus diseases are not included in this discussion because they are estimated to cause only about 1.2% yield losses worldwide, compared to 8.9% losses caused by pathogenic bacteria and fungi (Oerke 2006).

2.1 Major diseases caused by bacteria

Bacterial blight [*Pseudomonas savastanoi* pv. *glycinea* (syn. *P. syringae* pv. *glycinea*)]

Bacterial blight is one of the most common diseases of soybean worldwide. It occurs primarily when weather conditions are cool and wet, especially early in the growing season (Tolin and Lacy 2004; Zhao 2015). In addition to causing small, angular lesions on leaves, petioles and pods, the pathogen can also infect seeds. Severe infections can cause premature defoliation and yield loss, though most infections do not result in serious losses (Zhao 2015). Even when bacterial blight does not pose a serious risk to yield though producers are likely to find the appearance of lesions on cultivars objectionable.

Four resistance genes (*Rpg1-b*, *Rpg2*, *Rpg3* and *Rpg4*) have been reported in soybean, and each interacts with a different avirulence gene in the pathogen (Zhao 2015). The *Rpg1-b* gene was mapped to a region on chromosome 13 (Linkage Group F) that contains a cluster of resistance genes (Ashfield et al. 1998, 2003). These genes include *Rps3* (for resistance to *Phytophthora sojae*; Diers et al. 1992), *Rsv1* for resistance against SMV (Gore et al. 2002) and *Rpv1* for resistance to peanut mottle virus (Gore et al. 2002), as well as the *Rpg1-r* gene (Ashfield et al. 1998). The *Rpg1-b* gene is involved in recognition of *P. savastanoi* pv. *glycinea* strains with the *avrB* avirulence gene, whereas the *Rpg1-r* gene provides resistance to strains with the *avrRpm1* avirulence gene. Nine physiological races of *P. savastanoi* pv. *glycinea* have been characterized using the differentials 'Cheppewa', 'Harosoy', 'Lindarin', 'Norchief', 'Peking', 'Merit', 'Flambeau', 'Acme' and 'Centennial' (Siegel et al. 2008).

'Williams 82' (Bernard and Cremeens 1988) has the *Rpg2* gene and is resistant to bacterial strains with the *AvrA* gene (Ludwig and Tenhaken 2000). Since this cultivar was used in many breeding programmes in the United States, it may have been a source of bacterial blight resistance for many current cultivars. The line LL89-605 has bacterial blight resistance derived from PI 437821 (Nickell et al. 1994). Since bacterial blight symptoms are often very evident on infected plants and seeds, it is likely that highly susceptible lines are often discarded during selection, but there is also a high probability that the elite parents used in many contemporary crosses will both be resistant to bacterial blight.

Bacterial pustule

BP is caused by *Xanthomonas axonopodis* pv. *glycines* (syn. *X. campestris* pv. *glycines*). Lesions with raised centres, which form primarily on the abaxial sides of infected leaves, distinguish BP from bacterial blight. It is frequently seen in Brazil, China, India, Korea and the United States (Goradia et al. 2009; Kim et al. 2010; Narvel et al. 2001). While bacterial blight is more common in cooler weather, BP occurs when the weather is warm and moist, so it tends to be a problem primarily in the southern United States, though it can also occur in the Midwest during hot summers with abundant precipitation. Although severe infections can cause defoliation and decreased seed weights, BP has not been implicated in serious yield losses. Like bacterial blight, however, BP is of some concern as a 'cosmetic' disease that would make susceptible cultivars unacceptable to some producers. Selection for BP resistance was therefore a high priority in breeding programmes for the southern United States during the early decades of cultivar development in that region (Narvel et al. 2001).

A single recessive gene, *rxp*, conditions immunity or high levels of resistance to North American isolates of *X. axonopodis* pv. *glycines* (Narvel et al. 2001). The *rxp* gene was identified in CNS (PI 548445), which was one of the major ancestral lines of North American soybean cultivars released between 1947 and 1988 (Gizlice et al. 1994), and is also in the pedigree of many modern cultivars adapted to the southern United States. The *Rxp* locus was mapped to chromosome 17 (linkage group D2), and a molecular pedigree analysis confirmed that 12 resistant cultivars with CNS in their pedigree had marker genotypes identical to CNS in that genomic region (Narvel et al. 2001; Kim et al. 2010). The continued durability of *rxp*-mediated resistance over several decades, despite the widespread deployment of the gene in many of the soybean cultivars grown in the United States, is notable. In an experiment that screened 525 glyphosate-tolerant soybean cultivars for resistance to a moderately aggressive Illinois isolate of *X. axonopodis* pv. *glycines*, Goradia et al. (2009) observed BP symptoms on 152 (29%) of them. Within this group, a number of the cultivars appeared to be moderately susceptible or moderately resistant based on the number of pustules per plant relative to the highly susceptible control line. Williams 82, which also inherited resistance from CNS, was classified as moderately resistant, presumably because its resistance was partially overcome by the high inoculum dosage applied in the greenhouse assay used (Goradia et al. 2009).

Diseases caused by oomycetes

Soil-borne diseases are considered one of the two major causes of yield loss in soybean grown in North America, and a concern in all soybean production areas (Wrather and Koenning 2009; Wrather et al. 2010). Two of the most important and widespread genera of pathogens responsible for seed rot and damping-off of seedlings, as well as for reduced vigour of surviving plants, are *Phytophthora* and *Pythium*. These are actually oomycetes, and although they bear some similarities to soil-borne fungal pathogens, they are taxonomically unrelated. While an increase in seed treatment with fungicides in recent decades has reduced problems with seed and seedling disease, in addition to early infection of surviving plants, the seed treatments provide little if any protection as the roots of plants grow beyond the zone in which the fungicides were effective in suppressing the pathogens and infection. Resistance to common soil-borne oomycetes is therefore highly desirable, if not essential, in cultivars bred for North America.

Phytophthora root and stem rot

Phytophthora sojae, the primary oomycete that causes PRR, is one of the most economically important soybean pathogens in several major soybean-producing countries (Dorrance and Grünwald 2009). Another species, *P. sansomeana*, also causes PRR in China and in some fields in the United States, and has a broad host range that includes Douglas fir, maize and weed species (Hansen et al. 2009; Schmitthenner and Dorrance 2015). PRR was estimated to be the second most important disease that contributed to reduced soybean yields in the United States between 1996 and 2009 (Koenning and Wrather 2010; Wrather and Koenning 2009). The disease is most often a problem on soybean planted in soils that experience prolonged saturation with water, which facilitates the movement of motile, asexual zoospores towards soybean roots. *P. sansomeana*, which was first described in 2009, can also cause PRR, and may be the more abundant species in some fields (Hansen et al. 2009). PRR management involves a combination of cultural practices like crop rotation, host resistance and seed treatments with fungicides that are active against oomycete pathogens. Despite the considerable increase in the use of fungicide seed treatments in North America, most of the commercial soybean cultivars available in North America in 2016 carry at least one gene for resistance to *P. sojae*, and many bred for northern production areas also have quantitative partial resistance. Fungicide treatments presumably help to reduce or delay infections of seeds and seedlings, but are not likely to provide long-term protection. Therefore, resistance is likely to benefit seedling survival and plant vigour, particularly since the fungicide-impregnated seed coats get pushed out of the soil following germination.

In North America yield losses to PRR have been mitigated for more than 50 years by planting cultivars carrying single dominant *Rps* genes. Thus far, 21 *Rps* genes or alleles have been identified and mapped to eight loci on five chromosomes (Grau et al. 2004; Li et al. 2016; Schneider et al. 2016). Cultivars carrying the *Rps*1a allele were widely grown in the north-central United States in the 1960s, but within a few years several races of SCN had appeared that were virulent on plants with this gene (Grau et al. 2004). Most of the cultivars currently on the market have either the *Rps*1c or *Rps*1k allele, despite the fact that populations of *P. sojae* able to overcome resistance from these genes had already been reported in the 1990s (Grau et al. 2004). The heavy reliance on a few alleles has resulted in intense selection pressure on the pathogen and the emergence of new races and populations of *P. sojae* with broader virulence (Schneider et al. 2016; Stewart et al. 2016). More than 200 physiological races of *P. sojae* have been reported in North America, and 55 of them have been described (Grau et al. 2004; Schneider et al. 2016). At least ten different virulence pathotypes exist just on the island of Hokkaido in northern Japan (Sugimoto et al. 2012). The reaction patterns for 45 out of 49 Hokkaido isolates on eight differentials did not correspond with the reactions to isolates representing the 55 races previously described in the USDA collection, illustrating the considerable pathotype diversity in this species internationally.

Ten *Rps* genes and alleles have been mapped to chromosome 3, including six alleles at the *Rps*1 locus, *Rps*7 and *Rps*9 (Demirbas et al. 2001; Li et al. 2016; Sugimoto et al. 2011; Weng et al. 2001). *Rps*8 and five alleles of *Rps*3 have been mapped on chromosome 13 (Dermirbas et al. 2001; Gordon et al. 2006; Sandhu et al. 2005). *Rps*2 and *Rps*UN2 are on chromosome 16, and *Rps*4, *Rps*5, *Rps*6 and *Rps*JS are located on chromosome 18 (Demirbas et al. 2001; Lee et al. 2014; Sandhu et al. 2005). For a more comprehensive list of *Rps* mapping studies, see Li et al. (2016).

To prolong PRR resistance and the useful life of *Rps* genes, Buzzell and Anderson (1982) proposed combining the race-specific genes with other genes that condition non-specific partial resistance (sometimes referred to as field resistance or tolerance). Partial resistance appears to be rate-limiting, possibly restricting colonization of the host tissue (Tooley and Grau 1984). Private sector breeders clearly believe in the wisdom of combining partial resistance quantitative trait loci (QTL) with an *Rps* gene, since they have largely focused on this tactic instead of pyramiding multiple *Rps* genes. Most of the cultivars currently marketed in the US Midwest are described as having field tolerance/partial resistance in addition to a major gene (usually *Rps*1c or *Rps*1k), though a few cultivars do carry a second *Rps* gene like *Rps*3a.

Much work has been done to identify sources of partial resistance, or tolerance to PRR. To identify germplasm with partial resistance it is first necessary to ascertain that it does not have a major resistance gene. This is done by conducting a preliminary hypocotyl inoculation assay with complex pathotypes (Dorrance et al. 2008; Schmitthenner 1985). The hypocotyl inoculation assay is typically fatal to lines with partial resistance and no *Rps* gene. The lines that succumb are then screened again using a moderately aggressive *P. sojae* isolate in a greenhouse layer test that allows identification of partial resistance based on differences in root growth and appearance, with non-inoculated plants as controls (Dorrance et al. 2008; Schneider et al. 2016). QTL associated with partial resistance to PRR in lines like 'Conrad' have been mapped by linkage analysis using an interval mapping in biparental populations (Tucker et al. 2010; Weng et al. 2007). Li et al. (2010) identified three QTL associated with tolerance to PRR in Conrad and five additional QTL with resistance alleles from 'Hefeng 25', a tolerant cultivar from north-eastern China. They demonstrated that a decrease in disease was positively correlated with an accumulation of tolerance QTL in lines derived from a cross between these two partially resistant parents. Association mapping has also been used successfully to identify previously reported and newly discovered QTL associated with partial resistance (Huang et al. 2016; Li et al. 2016; Schneider et al. 2016; Sun et al. 2014).

Pythium rot and damping-off (*Pythium* spp.)

At least 30 species of *Pythium* and related oomycetes are potential pathogens of soybean and are a major component of the soybean seedling disease complex (Rothrock et al. 2015). *Pythium* spp. infections result in reduced emergence, damping-off and sometimes stunting and wilting later in the season as a result of reduced root development. Relationships among *Pythium* spp. have been reassessed in the last decade on the basis of molecular analyses and morphological differences, and a taxonomic revision that would separate the species into four new genera has been proposed (Uzuhashi et al. 2010). Since the species pathogenic on soybean are similar in many regards, they will be treated here as members of the same genus to minimize confusion. Commonly encountered species that are virulent on soybean include *P. aphanidermatum*, *P. dissotocum*, *P. inflatum*, *P. myriotylum*, *P. torulosum*, *P. irregulare* (syn. *Globisporangium irregulare*), *P. sylvaticum* (syn. *G. sylvaticum*) and *P. ultimum* (syn. *G. ultiumum*). Accurate estimations of the effects of *Pythium* spp. infections on soybean yields are difficult, since they are often confounded with losses caused by a suite of soil-borne pathogens that may include *Phytophthora sojae*, *Rhizoctonia solani* and *Fusarium* spp. (Rothrock et al. 2015).

No cultivars with high levels of resistance to *Pythium* spp. pathogens seem to be available at present, but some cultivars and lines with moderate resistance to certain species or isolates have been reported. Compared to a more susceptible cultivar, 'Archer' has had fewer disease symptoms, denser stands, higher root weights and less root discolouration after inoculation with *P. ultimum* and other species of *Pythium* (Bates et al. 2008; Kirkpatrick et al. 2006). Based on the reactions of Archer-derived lines to *P. aphanidermatum*, Rosso et al. (2008) discovered a resistance gene locus on chromosome 13 (formerly LG F). Ellis et al. (2013) screened a set of soybean accessions that have partial resistance to *Phytophthora sojae* for resistance to *P. irregulare*, and found that PI 424354 had a high level of partial resistance. This Korean land race is also resistant to *Fusarium graminearum*, and to at least 12 races of *Phytophthora sojae* (Dorrance and Schmitthenner, 2000; Ellis et al. 2013).

Diseases caused by fungal pathogens

The diseases discussed below are among the most important that reduce soybean yields or increase production costs in North America. Some cause yield losses annually, while the incidence and severity of others can vary considerably in different regions or growing seasons.

Anthracnose (*Colletotrichum* spp.)

Anthracnose is widespread in soybean growing areas of the world, where it can reduce stands, seed quality and yields up to 26% in the United States, and by considerably more in some tropical areas (Yang and Hartman 2015a). Infection can occur at any stage of plant development. Screening assays have been developed for anthracnose resistance, and several germplasm accessions, including eight that were important gene donors to modern North American soybean cultivars, have incomplete resistance (Manandhar et al. 1988; Yang and Hartman 2015b). These included 'Mandarin', 'Mandarin 507' and 'Boone'. Although commercial cultivars are not advertised as being resistant to anthracnose and directed screening for resistance is probably not being conducted, it is likely that highly susceptible lines tested in locations with warm and humid weather conditions are often eliminated due to poor yields or obvious disease symptoms.

Brown stem rot [*Cadophora gregata* (syn. *Phialophora gregata*)]

BSR causes necrosis of the pith and vascular tissues on the interior of soybean stems, and infections sometimes cause interveinal chlorosis and necrosis on leaves. BSR is common in the northern states of the US Midwest and in Canada, and has also been found in Mexico, South America, Egypt, the former Yugoslavia and Japan (Malvick et al. 2015). Symptoms are worst when hot, dry weather follows cool, wet conditions during the pod filling stages of development, and yield losses exceeding 30% have been reported. Two types of the pathogen have been described; type A causes foliar symptoms in addition to internal browning of the stem, whereas the symptoms caused by type B are generally limited to internal stem browning. Differential interactions between the two pathogen types with resistant and susceptible soybean cultivars may indicate that they infect different parts of the vascular system (Hughes et al. 2002; Impullitti and Malvick 2014).

Attempts to map BSR resistance genes using biparental mapping populations and relatively low marker densities initially led researchers to conclude that there were three linked but independent resistance genes located in the same region of chromosome 16 (Linkage Group J): *Rbs1* from germplasm line L78-4094, *Rbs2* from PI 437833 and *Rbs3* from PI 437970 (Bachman et al. 2001). Resistance at all three loci was conditioned by a dominant allele, and a resistance allele at any one of the loci is sufficient to provide a high level of BSR resistance, but not immunity (Bachman et al. 2001; Willmot and Nickell 1989). The cultivar Bell and many other lines that inherited resistance to SCN from PI 88788 also have BSR resistance, and the gene responsible for that was also mapped to the same region on chromosome 16 (Patzoldt et al. 2005). BSR resistance genes in three PIs from south-central China (PI 594638B, PI 594650A and PI 594858B) were also mapped to that region (Perez et al. 2010). Despite the previous evidence for independent loci, however, a recent genome-wide association study (GWAS) involving 4735 germplasm accessions indicated that there is most likely only one *Rbs* locus in the implicated region of chromosome 16 (Rincker et al. 2016). Earlier evidence from Perez et al. (2010) that

PI 594637, another accession from south-central China, might have a BSR resistance gene elsewhere in the genome was not supported by the results of the GWAS (Rincker et al. 2016). Nearly 400 germplasm accessions with BSR resistance have been identified, but only two (PI 84946-2 and PI 88788) are known to have been used in breeding programmes to develop BSR-resistant cultivars (Rincker et al. 2016). This, together with the fact that resistance genes from other regions of the genome have yet to be found, is a cause of some concern, even though the resistance trait has been durable thus far.

Charcoal rot (*Macrophomina phaseolina*)

Macrophomina phaseolina can infect more than 500 plant species, including soybean, common bean, peanut and maize (Pawlowski et al. 2015). Charcoal rot is one of the most important soybean diseases in the United States, reducing yields by up to 15% (Mengistu et al. 2011a), and it was estimated to have been the sixth most economically important soybean disease internationally in 2006 (Wrather et al. 2010). Symptoms on adult plants are most often observed when the plants experience heat and drought stress. Although charcoal rot is considered to be primarily a problem in the tropics and in the southern United States, it sometimes occurs as far north as the northern Midwest (Mengistu et al. 2013). The fungus survives as microsclerotia in soybean debris and in the soil, or in the seed coats of seeds from infected plants. It can infect plants at any time during the growing season, and infected seeds often rot after being planted. Soybean plants that become infected soon after emergence often do not develop visible symptoms in the upper taproot and lower stem until the R7 to R8 developmental stages (i.e. after completion of seed development). Cultural management practices that minimize drought and other stresses reduce disease, but irrigation is often unavailable or unfeasible for soybean fields in many areas where charcoal rot is likely to occur.

A few lines have been reported to be partially resistant to charcoal rot, but all or most commercial cultivars are considered to be susceptible, and no immune germplasm has been identified (Mengistu et al. 2011a, 2013). Six moderately resistant lines were identified by Mengistu et al. (2011a) that had levels of resistance at least equal to the moderately resistant check line DT97-4290 (Paris et al. 2006). Mengistu et al. (2013) subsequently evaluated the reactions of 628 maturity group (MG) 00 to VII PIs from the USDA Soybean Germplasm Collection, and identified 45 with at least moderate resistance in the field. Among the most resistant were PI 594302 (MG VII), PI 567562A (MG IV), PI 506764 (VII) and PI 567334 (MG VI), all of which had less disease than DT97-4290. Obtaining consistent and reliable disease rating data on large numbers of plants or breeding lines has been a challenge. Until recently, no reliable greenhouse assay was available, so field evaluations remained the standard (Mengistu et al. 2013). A cut-stem inoculation technique showed promise, however, and has some advantages over field evaluations (Twizeyimana et al. 2012). Pawlowski et al. (2015) used this assay to screen 70 PIs that Gizlice et al. (1994) had identified as major contributors of genes to US cultivars released prior to 1989. PI 548414 (MG 000), PI 548302 (MG II) and PI 548178 (MG III) had resistance at least equal to that of DT97-4290. The authors suggested that these could be particularly useful as sources of resistance for cultivars developed for the northern United States, since previously reported sources were from later MGs.

Frogeye leaf spot (*Cercospora sojina*)

FLS is a common disease in many soybean-producing countries. Although it is a problem primarily in the southern United States, it also occurs in the Midwest, where the incidence of FLS might be increasing (Grau et al. 2004; Mian et al. 2008). *C. sojina* causes mostly foliar disease, though stems, pods and seeds can also become infected. Yield loss, which can exceed 30% on infected plants, results from a reduction in photosynthetic leaf area and premature defoliation. Planting resistant cultivars has been an important component of reducing losses to FLS in the United States and China (Mian et al. 2008). FLS is most effectively managed through a combination of host resistance, cultural practices that destroy crop residues and prudent fungicide applications (Wise and Newman 2015). Host resistance and the identification of germplasm with novel resistance genes have become more important as populations of the fungus in some production areas have evolved resistance to the quinone outside inhibitor fungicides (a.k.a. strobilurins) (Zeng et al. 2015).

Numerous physiological races of *C. sojina* have been reported by researchers in the United States, China and Brazil, but the lack of a standard set of differentials made the total number impossible to determine. Mian et al. (2008) identified 11 races after inoculating a set of 12 differentials with 93 isolates from the three countries. Because of resistance gene × fungal race interactions, the authors emphasized the importance of knowing both the soybean genotype and the pathogen race to fully understand the reaction of the host plant. At least three resistance genes have been named, and additional genes have been reported, but not yet confirmed to be located at different loci. *Rcs1* from 'Lincoln' confers resistance to *C. sojina* race 1 (Athow and Probst 1952), and the *Rcs2* gene from 'Kent' conditions resistance to race 2 (Athow et al. 1962). The *Rcs3* gene from 'Davis' provides a high level of resistance to race 5 and all other known US races of the fungus (Boerma and Phillips 1983; Phillips and Boerma 1982). Additional genes conferring resistance to race 5 were found in the cultivars Ransom, Stonewall and Lee (Pace et al. 1993), but those sources of resistance have not been widely exploited. 'Peking' also carries a single dominant gene that conditions resistance to a number of *C. sojina* isolates (Baker et al. 1999). In a screen of 522 MG 00 to VII accessions, Mengistu et al. (2011b) found 75 with resistance to race 11 of *C. sojina*, and only three of the 75 had marker haplotypes around the *Rcs3* locus that were identical to Davis. The accessions PI 594891 and PI 594774 were found to have a high level of resistance similar to that of lines with the *Rcs3* gene, but it is conditioned by single dominant genes on a different chromosome (Pham et al. 2015).

The *Rcs3* gene locus is located on chromosome 16 (LG J), approximately 10 cM from a cluster of resistance genes that includes the *Rbs* gene(s) for resistance to BSR, and 20 cM from a second cluster that contains the *Rps2*, *Rmd* and *Rj2* resistance genes (Mian et al. 1999; Polzin et al. 1994). The single dominant resistance genes *Rcs*(PI 594891) and *Rcs*(PI 594774) from those highly resistant PIs have been mapped to the same region of chromosome 13 (Pham et al. 2015). Three of five candidate genes had DNA sequence differences compared to the susceptible line 'Blackhawk', and combined evidence from genotype, phenotype and haplotype analyses indicates that the two PIs either carry different genes or different alleles of the same gene. The tagging of these genomic regions with single nucleotide polymorphisms (SNPs) that have been developed into robust and breeder-friendly Kompetitive Allele Specific PCR (KASP) markers will facilitate MAS of plants carrying the genes (Pham et al. 2015).

Phomopsis seed decay [caused primarily by *Diaporthe longicolla* (syn. *Phomopsis longicolla*)]

Phomopsis seed decay (PSD) can be severe when a growing season is hot and humid during seed development and maturation, so it tends to be a more frequent problem in the southern United States (Li et al. 2015). Although the primary cause of PSD is *D. longicolla*, other fungi in the *Diaporthe/Phomopsis* complex can also cause this disease. PSD became a more significant problem in the Mid-South production area following adoption of the early soybean production system (ESPS), which involves planting primarily indeterminate MG III and IV cultivars in early April and harvesting them in August (Mayhew and Caviness 1994). While the ESPS improves the likelihood of having adequate moisture during the pod filling period in July and early August, high humidity during that period also promotes PSD. Farther north, PSD tends to be more of a problem in areas near major waterways (due to heavier dews) or where rainfall has been abundant during the growing season (Sinclair 1993).

PI 80837, PI 82264, PI 181550 and 'Delmar' were reported to be resistant to PSD in the early 1970s, and PI 417479 was found to have similar resistance in the 1980s (Brown et al. 1987). PI 200510, PI 209908 and Arksoy were used as sources of PSD resistance in the North Carolina State University breeding programme in the 1980s (Ross 1986). More recently, eight additional germplasm accessions with resistance to PSD in Mississippi were identified (Li et al. 2011). These potential sources of novel resistance genes were PI 424324B, PI 458130, PI 506647, PI 567270C, PI 567381B, PI 567521, PI 567635 and PI 594858A, and several of these have been used in USDA-ARS breeding programmes in Stoneville, MS, and Urbana, IL. However, in a three-year evaluation of PSD resistance among 143 MG III through V commercial cultivars adapted to the Mid-South growing region, Mengistu et al. (2012) found only two resistant and one moderately resistant cultivars.

Inheritance of resistance to PSD has been investigated by several research groups and efforts have been made to map and tag PSD resistance genes. Analyses of breeding lines developed from PI 417479 indicated that PSD resistance in those was controlled by two complementary dominant nuclear genes (Minor et al. 1995). Segregation for resistance in populations derived from PI 360841 indicated that the accession has two complementary dominant resistance genes (Smith et al. 2008). Sun et al. (2013) identified two PSD resistance QTLs on chromosomes 6 and 10 of the Korean cultivar Taekwangkong, both of which were in genomic regions that also influenced days to maturity, suggesting a biological correlation between the traits.

Rhizoctonia damping-off and root rot [*Rhizoctonia solani* (syn. *Thanatephorus cucumeris*)]

Rhizoctonia root rot occurs worldwide and is common in North America (Yang and Hartman 2015). Seedling stand reductions result from pre- and post-emergence damping-off and seedling blight, and older plants are sometimes plagued with root rot and stunting. Severe epidemics can result in yield losses close to 50% (Yang and Hartman 2015). As a facultative saprophyte, *R. solani* can survive as sclerotia in the soil or as mycelia in plant debris from the previous season, but survival of the pathogen is reduced by high soil temperatures and saturated soil.

Soybean cultivars currently marketed in North America are not known to be resistant to *Rhizoctonia* root and hypocotyl rot (Bradley et al. 2005). Partial resistance has been found among some modern cultivars and older land races that were important ancestors of North American soybean varieties (Bradley et al. 2001). Muyolo et al. (1993) reported finding the cultivars Asgrow 7986, Centennial, Hardee, Pella, RA 606 and Victory to be partially resistant. Bradley et al. (2001) later found partial resistance to an Illinois isolate of the fungus in 21 of the North American ancestral lines and 20 commercial cultivars grown in the 1990s. Three ancestral lines that had apparently served as sources of resistance for cultivars found to be resistant were CNS, Mandarin (Ottowa) and Jackson. Although none of the resistant ancestral lines had lower disease severity ratings than the partially resistant check, Centennial, it is possible that some carry different genes. *R. solani* includes at least 14 different anastomosis groups, and little is known about interactions between these and different resistance genes, so more investigations are needed to determine the breadth of the partial resistance discovered thus far using limited numbers of isolates. Bradley et al. (2005) reported that partial resistance in the cultivar Savoy is quantitatively inherited and concluded that selection for resistance in the F_2 generation should be made on the basis of $F_{2:3}$ family means rather than on individual F_2 plants. The increased use of seed treatments with fungicides is likely to reduce stand losses to *R. solani* and other soil-borne pathogens, but it is unlikely to provide sufficient protection to older susceptible plants, so *Rhizoctonia* resistance remains an important trait (Bradley et al. 2001).

Sclerotinia stem rot (*Sclerotinia sclerotiorum*)

Sclerotinia sclerotiorum causes stem rot and produces white, fluffy mycelia on the surface of stems and pods, so the symptoms are often referred to as 'white mould' (Hoffman et al. 2002). SSR can be a problem on soybean grown in Canada, the United States, Argentina, Brazil, South Africa, India and Nepal (Hartman et al. 2015). Disease incidence and severity, as well as resulting yield losses, are generally greater when weather conditions are cool and moist, and infection occurs during the flowering period. In certain growing seasons, SSR has been estimated to be among the three most important biotic causes of yield loss in soybean (Wrather et al. 1997). Lesions develop around the stem and impede the transport of water, nutrients and photosynthates to pods and seeds (Grau and Hartman 2015). Seed from infected plants are often discoloured or covered with white mycelia. Due to the broad host range of this pathogen, crop rotation is of limited value for disease management.

Environmental and cultural conditions that favour high soybean yields also promote SSR (Grau and Hartman 2015). A dense canopy and a cool, moist soil provide ideal conditions for the fungus to infect and colonize stem tissue via flowers. A trend towards earlier planting dates promotes denser canopies and therefore conditions favourable for the pathogen. Late-maturing cultivars also tend to develop denser canopies that could increase disease problems (Grau and Hartman 2015).

No race variation in *S. sclerotiorum* has been reported, and the physiological resistance identified thus far appears to be partial and quantitatively inherited (Arahana et al. 2001; Kim and Diers 2000). Resistance derived from 'Williams' has been widely used in North American breeding programmes (Kim et al. 1999). Although this resistance has proven stable, it is insufficient to provide adequate protection when environmental conditions promote infection and growth of the pathogen. A screening of more than 6400 PIs from the USDA Soybean Germplasm Collection for *Sclerotinia* resistance in the field identified seven PIs with resistance at least as strong as that of the resistant check (Hoffman et al. 2002). Putatively resistant accessions were also evaluated using a petiole inoculation method that produced ratings that were correlated with field disease severity ratings of PIs from MGs I and II. Despite the fact that SSR is primarily a disease when and where cool and wet weather occurs, Hartman et al. (2000) identified resistant germplasm accessions in several perennial *Glycine* spp. from Australia which could eventually be used as unique sources of resistance via backcrossing or by gene cloning followed by transformation.

Part of the challenge in identifying and utilizing resistance to SSR has been the difficulty of obtaining consistent results in field and greenhouse evaluations, and in getting ratings from greenhouse assays that accurately reflect resistance in the field (Chun et al. 1987; Hoffman et al. 2002). Development of disease in the field is highly influenced by both macro- and microenvironmental differences, particularly in temperature, light intensity and humidity (Nelson et al. 1991; Wegulo et al. 1998). The reactions of soybean lines screened in the field are influenced by escape mechanisms as well as physiological resistance (Kim and Diers 2000; Kim et al. 2000). Even though infection in the field begins on floral tissue, a cut-stem inoculation method worked well for screening soybean, dry bean and sunflower plants for resistance (Vuong et al. 2004). Alternative inoculation methods for greenhouse assays like the 'spray-mycelium' and 'drop-mycelium' inoculation techniques tested by Chen and Wang (2005) might also be useful to reduce time and expenses required to screen germplasm for resistance.

Germplasm with very high levels of resistance to *S. sclerotiorum* has not been identified, but numerous cultivars and germplasm accessions with partial resistance have been reported (Hoffman et al. 2002). Grau et al. (1982) found that the cultivars Corsoy, Hodgson and Hodgson 78 were less susceptible than other cultivars grown in naturally infested fields in Wisconsin, and those were among a group of cultivars that developed less disease in Michigan as well (Chun et al. 1987). Boland and Hall (1987) identified the cultivars Maple Arrow, Ace, Maple Presto and McCall as being resistant to *S. sclerotiorum* in field trials conducted in Ontario. Both Wegulo et al. (1998) and Kim et al. (2000) identified the cultivars Corsoy 79 and Novartis S19-90 as having consistently lower disease severities in various assays. In a broad study to screen 6520 MG 0 to IV germplasm accessions for resistance in Ontario and several north-central states, 68 soybean PIs with partial resistance were found (Hoffman et al. 2002). Of these, the seven with the highest levels of resistance were PI 153282, PI 189931, PI 196157, PI 398637, PI 417818 and PI 561331. Evidence of resistance was observed in excised leaf inoculation and petiole inoculation assays, as well as in the field.

Kim and Diers (2000) identified three QTL associated with lower disease severity, but concluded that two of the loci were likely to have been involved with disease escape rather than physiological resistance. Using data from a detached leaf assay, Arahana et al. (2001) found 28 putative QTL for resistance to SSR on 15 different linkage groups. The amount of phenotypic variation explained by each ranged from 4 to 10%, and seven QTL, each on a different linkage group, were detected in multiple populations. Four regions on soybean chromosomes 7, 10, 13 and 15 of PI 391589A and PI 391589B were found to be associated with partial resistance to *S. sclerotiorum* (Guo et al. 2008). The researchers also found evidence for additional putative QTL associated with resistance. The application of association mapping combined with a genotyping by sequencing approach might be useful for discovering additional genomic regions associated with SSR resistance, or for better characterizing regions that have already been reported (Iquira et al. 2015). Since only partial resistance has been found in soybean, some researchers have investigated the potential of a transgenic approach to obtain high levels of resistance. Cunha et al. (2010) observed a considerable reduction of disease on some plants that had been transformed with a decarboxylase gene (*oxdc*) for an enzyme that degrades oxalic acid produced by the pathogen. At least one breeding line with partial resistance to SSR has been released (Diers et al. 2006), but the extent to which it or other partially resistant germplasm has been used in commercial breeding programmes is unclear.

SSR (white mould) resistance/susceptibility ratings are frequently listed for soybean cultivars adapted to the north-central United States, with values that indicate poor to moderate resistance. This indicates that SSR resistance is important to producers and that the private seed companies are evaluating the relative resistance of new cultivars that are being developed for the northern production regions in North America.

Soybean rust (*Phakopsora pachyrhizi*)

Although soybean rust (SBR) is an occasional threat to soybean production in the mid-southern and south-eastern states of the United States, it frequently causes devastating yield losses in tropical and subtropical production regions worldwide, and was estimated to have caused the highest disease losses by far in the top eight soybean-producing countries in 2006 (Wrather et al. 2010). SBR has been managed effectively with appropriate fungicides, but the cost of applying those, often several times during a growing season, substantially increases production costs. SBR was first identified in the continental US in November of 2004 (Schneider et al. 2005), and among the reasons why it has not caused major yield losses in North America was the well-coordinated and unprecedented response of pathologists in establishing a network of sentinel plots to monitor the spread of annual epidemics, together with a sharp increase in the use of fungicides on soybean grown in the southern United States (Kelly et al. 2015; Sikora et al. 2014).

Resistance to SBR is characterized by either a Type 0 (i.e. immune) reaction or the development of reddish-brown (RB) foliar lesions with few uredinia and little or no sporulation (Bromfield 1984; Bromfield et al. 1980). In contrast, susceptible soybean plants develop a so-called TAN reaction with foliar lesions that typically have multiple uredinia per lesion, from which profuse sporulation occurs. The immune type of reaction, characterized by the absence of macroscopically visible lesions, is relatively rare, and occurs only when plants with certain resistance (*Rpp*) genes are challenged with specific fungal isolates (Bromfield 1984). The actual range of reactions observed, however, often includes some reactions that are intermediate between the three simple classes (Bromfield 1984; Yamanaka et al. 2010). Four resistant germplasm accessions with four different dominant resistance genes were identified in the 1970s and 1980s: PI 200492 (*Rpp1*), PI 230970 (*Rpp2*), PI 462312 (*Rpp3*) and PI 459025B (*Rpp4*) (Bromfield and Hartwig 1980; Hartwig 1986; Hartwig and Bromfield 1983; McLean and Byth 1976). Two groups from Brazil independently and simultaneously discovered a fifth *Rpp* locus in several different germplasm accessions (Garcia et al. 2008; Pierozzi et al. 2008). A sixth resistance locus (*Rpp6*) was subsequently found by Li et al. (2012).

Following the appearance of SBR in Paraguay and Brazil in 2001 (Yorinori et al. 2005), efforts to find additional novel sources of *Rpp* genes were intensified. Miles et al. (2006) screened 16 595 PIs from the USDA Soybean Germplasm Collection using greenhouse assays in which seedlings were inoculated with a mixture of four foreign isolates of *P. pachyrhizi*. From two rounds of assays they selected 805 PIs that had low visual severity and/or RB lesions. Subsequent studies were conducted to evaluate the resistance of the selected accessions to field populations of the fungus in Paraguay (Miles et al. 2008) and in the southern United States (Walker et al. 2011, 2014a). There was surprisingly little overlap between the accessions that were most resistant in Paraguay and those with the most resistance to US *P. pachyrhizi* populations. PI 567102B and PI 567104B are among the few germplasm accessions that were highly resistant in the southern United States as well as in Paraguay (Miles et al. 2008; Walker et al. 2011, 2014a,b). The challenge of choosing sources of resistance for breeding programmes that are likely to be effective over time and across locations is significant due to the considerable pathotype diversity in and among *P. pachyrhizi* populations and isolates, despite the apparent absence of a sexual stage (Paul et al. 2013, 2015).

The SBR resistance genes *Rpp1*, *Rpp2*, *Rpp3* and *Rpp4* have been mapped to soybean chromosomes 18, 16, 6 and a second region on 18, respectively (Garcia et al. 2008; Hyten et al. 2007, 2009; Silva et al. 2008). Additional SBR resistance genes were identified and mapped to a fifth independent locus (*Rpp5*) on Gm03 (Linkage Group N) in PI 200526, PI 200487, PI 471904 and PI 200456. The *Rpp5* locus appears to be the location of at least three different resistance alleles, including a recessive resistance allele in PI 200456 that was the first reported for an SBR resistance gene (Garcia et al. 2008). The *Rpp6* gene was identified in PI 567102B, and was mapped to a different arm of chromosome 18 from the *Rpp1* and *Rpp4* loci (Li et al. 2012). An *Rpp* gene was also mapped to this locus in PI 567104B, another accession that has remained highly resistant to US populations of *P. pachyrhizi* (Liu et al. 2016; Walker et al. 2014a). Using bulked segregant analysis, Harris et al. (2016) discovered that 52 out of 75 PIs with resistance to US populations of the fungus had a resistance gene in the *Rpp3* region of chromosome 6. When challenged with eight domestic and foreign *P. pachyrhizi* isolates, 42 of the PIs had the same reaction profiles as either PI 462312 (source of the original *Rpp3* gene) or the Japanese cultivar Hyuuga, which has resistance alleles at the *Rpp3* and *Rpp5* loci (Kendrick et al. 2011).

A number of agronomically competitive breeding lines with SBR resistance have been released in the United States. Diers et al. (2014) backcrossed the *Rpp1*, *Rpp1-b* and *Rpp?*(Hyuuga) genes, in addition to one of the resistance alleles from the *Rpp5* locus, into a MG II breeding line and a MG IV breeding line adapted to the Midwest. At the University of Georgia, Boerma et al. (2011) developed a MG VI line carrying the *Rpp?*(Hyuuga) gene, which is currently suspected of being the same allele as the *Rpp3* gene from PI 462312 (Kendrick et al. 2011). King et al. (2016) developed four isolines of a MG VII line from Georgia that carry the *Rpp1*, *Rpp2*, *Rpp3* or *Rpp4* gene. The MG V, determinate, Asgrow cultivar AG5232, which had the *Rpp1* gene, was released in 2012. It was grown for a few years in the Mid-South, but was discontinued because of a nearly complete shift in demand among producers for indeterminate varieties. The Brazilian seed company Tropical Melhoramento e Genética Ltda (TMG) developed a series of rust-resistant cultivars with what they called Inox® technology. The first two cultivars from this series, TMG 801 and TMG 803, were released for the 2009/10 growing season, and each carried a single *Rpp* gene. Since then, additional improved cultivars with at least one *Rpp* gene have been released, and five TMG cultivars available for the 2016/2017 growing season had the Inox® technology (www.tmg.agr.br/cultivares/soja). Lemos et al. (2011) demonstrated the advantages of combining the *Rpp2* and *Rpp4* genes with one of the resistance alleles from the *Rpp5* locus to reduce disease symptoms.

Changes in gene expression levels of plants with the *Rpp2* gene showed a distinct biphasic response to infection that was similar to that seen in susceptible plants, but the second phase of differential expression was more intense and occurred 1–2 days earlier in the resistant plants (van de Mortel et al. 2007). Meyer et al. (2009) identified five candidate genes at the *Rpp4* locus of PI 459025B by using forward and reverse genetic approaches in combination with sequence data from 'Williams 82'. Fine-mapping, quantitative real-time reverse transcription (qRT)-PCR and virus-induced gene silencing were subsequently used to determine which of the candidate genes was most likely to be the *Rpp4* gene.

Stem canker (*Diaporthe caulivora*)

Stem canker was first reported on soybean in Iowa in the 1940s and subsequently became widespread in the upper Midwest until producers stopped planting popular, but highly susceptible cultivars like 'Hawkeye' and 'Blackhawk' (Backman et al. 1985). Resistance in cultivars developed for the north-central United States and for Ontario, Canada, henceforth became an important, if not essential, trait to prevent major yield losses. In the early 1980s stem canker symptoms started to appear on soybean in the southern United States production regions (Backman et al. 1985).

Both northern stem canker and SSC are caused by pathogens from the *Diaporthe/Phomopsis* complex. The taxonomy of members of this complex that are pathogenic on soybean is complicated, and has been revised based on the results of molecular analyses (Gomes et al. 2013). The pathogen that causes northern stem canker was long referred to as *Diaporthe phaseolorum* var. *caulivora*, but was elevated to a separate species, *D. caulivora*, in 2011. The pathogen responsible for SSC, previously referred to as *D. phaseolorum* var. *meridionalis*, was found to be identical to *D. aspalathi*, and is not closely related to *D. phaseolorum*, despite morphological similarities (Gomes et al. 2013). It should be noted that prior to 1989, most researchers considered both pathogens *D. phaseolorum* var. *caulivora*, and many researchers have continued to use the nomenclature that treats the pathogens as *caulivora* and *meridionalis* varieties of *D. phaseolorum*. Both of the stem canker pathogens have been found to co-exist in Illinois, Wisconsin and in Argentina (Lu et al. 2010). Early symptoms of both types of canker consist of reddish lesions that typically first appear near a node that is low on the stem. These lesions darken and elongate into a sunken canker that can eventually girdle the stem in northern stem canker, though this seldom happens from SSC. Leaf symptoms include interveinal chlorosis and necrosis, and severe infections can be fatal to the host plant. A study by Rupe et al. (1999) confirmed that leaf symptoms do not appear until at least 34–41 days after infection, and are seldom seen before the host reaches reproductive stages of development.

Northern stem canker is seldom seen in production fields, and only 14 out of 63 Iowa fungal isolates with *Diaporthe/Phomopsis* characteristics were confirmed to be *D. caulivora* (Lu et al. 2010). One possible reason for this is that breeding lines susceptible to *D. caulivora* might be eliminated from selection during development and yield testing. Furthermore, most regionally adapted elite parents used for crossing are likely to carry resistance genes and produce resistant progeny. Information about the status and level of northern stem canker resistance is seldom provided by private seed companies.

D. aspalathi (syn. *D. phaseolus* var. *meridionalis*) is considered more aggressive on soybean than *D. caulivora* (Backman et al. 1985). The *Rdm1*, *Rdm2*, *Rdm3* and *Rdm4* genes (originally called *Rdc1*, *Rdc2*, *Rdc3* and *Rdc4*) reside at different loci and condition high levels of resistance to SSC. A fifth resistance gene, *Rdm5*, is linked to the *Rpm4* locus in 'Hutcheson', and is generally co-inherited, but low levels of recombination sometimes occur (Chiesa et al. 2009, 2013). Results from a study by Chiesa et al. (2013) indicated that the *Rdm2* gene has incomplete penetrance when challenged with certain *D. aspalathi* isolates, and that percent penetrance is influenced by genetic background as well as interactions with the pathogen isolate.

Sudden death syndrome [*Fusarium virguliforme* (formerly called *F. solani* f. sp. *glycines*)]

SDS is caused by *F. virguliforme* in North America, and can be caused by additional *Fusarium* spp. in South America (Hartman et al. 2015). Under extreme conditions, yield losses can approach 100%. This disease is unusual in that the most obvious symptoms appear in the foliage even though the fungal infection is restricted to the roots. The above-ground symptoms, which typically appear as plants reach reproductive stages of development, are caused by the transport of a toxin produced by the fungus (Brar et al. 2011). Early foliar symptoms appear as interveinal chlorotic spots, which often become necrotic with time. Severe foliar symptoms early in the reproductive stages of development, premature defoliation and pod abortion can result in substantial yield losses during wet growing seasons, particularly in fields with heavy SCN populations or other factors that can exacerbate disease (Hartman et al. 2015b). The broad host range of *F. virguliforme*, which includes weed and crop species, limits the value of crop rotation for reducing yield losses (Kolander et al. 2012).

A number of sources of resistance to SDS have been reported (Hartman et al. 1997; Mueller et al. 2002, 2003; Rupe et al. 1991), and have been used in breeding programmes. Few cultivars are marketed as being highly resistant to SDS, but Mueller et al. (2003) identified nine SDS-resistant lines among 90 ancestral lines representing approximately 99% of the genes in US soybean cultivars released between 1947 and 1988. However, none of these ancestral lines were major contributors to the modern germplasm pool, based on pedigree analyses. Twelve SDS-resistant cultivars and elite lines from public breeding programmes were also reported in the same paper. An interesting observation made by the authors was that certain lines with moderate SDS resistance, were progeny of parents that were rated as susceptible to SDS, which suggests an additive or synergistic effect of combining genes with low individual contributions to resistance. However, none of the cultivars, breeding lines or ancestral lines was more resistant to SDS than PI 567374, which was used as a moderately resistant check. Mueller et al. (2002) found 57 PIs out of 6037 that were screened to have at least moderate SDS resistance, and recommended them as new and potentially novel sources of resistance genes.

At least 14 different confirmed QTL associated with SDS resistance or reduced symptom development have been reported (Luckew et al. 2013). Some of these are associated with true resistance and variation in the severity of root rot, while others are associated with variation in foliar disease (i.e. 'leaf scorch') severity, perhaps influencing the host's ability to degrade or tolerate toxins produced by the fungus (Triwitayakorn et al. 2005). Some QTL are associated with variation in both root and foliar disease severity, and QTL detected in some mapping populations are not detected in others. In any case, the genetic complexity of SDS and the fact that disease symptoms are influenced by environmental conditions increase the challenge of developing cultivars with effective resistance. Luckew et al. (2013) evaluated the usefulness of 10 of the known resistance QTL and found that five of them (*qRfs4*, *qRfs5*, *qRFS7*, *qRfs12* and *Rfs16*) were significantly associated with at least one of several disease measurement parameters across multiple populations. They concluded that breeders should focus on stacking combinations of these QTL to achieve useful levels of resistance. For a more comprehensive report on the history and current status of mapping QTL for SDS resistance, selecting for resistance and using novel approaches to develop resistant germplasm, see the review by Lightfoot (2015).

2.2 Nematodes

Resistance to SCN (*Heterodera glycines*) and some root-knot nematodes (*Meloidogyne* spp.) will be discussed briefly here because of the economic importance of these nematodes as major causes of yield loss. Soybean is parasitized by nematodes from the genera *Heterodera, Meloidogyne, Pratylenchus, Hoplolaimus* and *Belonolaimus*. All live in the soil, or are soil-borne, and live in or on roots (Niblack et al. 2004). Although they can cause stunting and chlorosis, these symptoms resemble those caused by some other pathogens, and in many cases, symptoms can be absent or very subtle, leaving producers unaware of the yield losses caused by nematode parasitism. Damage can occur directly or as the result of other pathogens associated with nematodes, notably the association of SCN with *Fusarium virguliforme*, the cause of SDS, and *Phialophora gregata*, which causes BSR (Niblack et al. 2004). The physiological stresses put on infested plants can also make them more vulnerable to nutrient deficiencies and other abiotic stresses.

Soybean cyst nematode

The SCN (*Heterodera glycines*) is estimated to be the most economically important parasite or pathogen of soybean in the United States, and is among the most important biotic threats to soybean yields internationally (Wrather et al. 2010; Wrather and Koenning 2006). SCN is native to East Asia and was first detected in the United States in 1954 (Niblack et al. 2004). It has now spread to nearly all of the soybean-producing regions of the United States except for the Northeast (Tylka and Marett 2014). The losses attributed to SCN in North America are due to its widespread distribution, ability to parasitize soybean plants over a range of weather conditions and soil types, and to the fact that once SCN becomes established in a field, it is virtually impossible to eradicate (Niblack et al. 2004). Another problem with SCN is that it can cause substantial yield losses without inducing obvious above-ground symptoms that would trigger concern in producers. Symptoms are sometimes influenced by the presence of other pathogens (Gao et al. 2006).

While rotating soybean with maize or another non-host crop can help in managing SCN populations and reducing yield losses, SCN eggs are able to remain viable for years in the soil, and planting other crops for more than a year results in diminished benefits to reducing SCN population sizes after the first year of rotation (Niblack et al. 2004). Since SCN cannot be managed exclusively with crop rotation and by treatment of seeds with nematicides and biological control agents, host resistance is a critical component of management. For several decades SCN resistance has been nearly essential for any cultivar bred for the US Midwest and north-central production regions, and it is important in most other production regions. Planting resistant cultivars in infested fields reduces nematode populations in subsequent growing seasons as well as increasing yields. The challenge of breeding for SCN resistance is considerable because of the great variation in the ability of *H. glycines* populations to parasitize hosts with specific resistance genes (Niblack and Riggs 2015). Virulence profiles on a set of seven indicator lines can be used to classify nematode populations or strains into 'HG types', which are more accurate descriptions than races for SCN (Niblack et al. 2002).

SCN resistance is quantified by the relative numbers of female nematodes or cysts that can develop on the roots of infested plants (Luedders and Dropkin 1983). Although more than 120 different sources of *Rhg* (resistance to *H. glycines*) genes have been identified (Arelli et al. 2000), breeders have exploited very limited numbers of sources and genes. The most important source of resistance by far has been PI 88788, despite years of evidence from different fields that many SCN populations were adapting to the resistance gene(s) inherited from this accession (Niblack et al. 2008). More than 94% of the soybean cultivars available in Illinois and 97% of the cultivars marketed in Iowa in 2015 had resistance derived from PI 88788 (Kim et al. 2011; Tylka and Mullaney 2015). The persistent popularity of PI 88788-derived resistance in soybean breeding programmes is due largely to its continued efficacy against many nematode populations and the lack of a deleterious effect on yield (Kim et al. 2011).

Due to the importance of SCN as a soybean pest, loci associated with SCN resistance were among the first disease resistance loci mapped in soybean using DNA markers (Concibido et al. 1994; Mahalingam and Skorupska 1995; Webb et al. 1995). Much of the resistance of PI 88788 is explained by the presence of resistance alleles at the *rhg1* complex locus, which was named *rhg1-b* in this accession (Concibido et al. 1997; Kim et al. 2010). The *rhg1-b* haplotype of PI 88788 has been shown to consist of ten tandem repeats of a sequence containing three tightly linked and apparently unrelated genes which collectively contribute to SCN resistance (Cook et al. 2012, 2014). The *rhg1* complex locus has also been detected in genetic mapping populations derived from a number of other SCN-resistant soybean accessions, including Peking, PI 437654, PI 209332, PI 90763, PI 89772, PI 88287 and PI 404198A (Kim et al. 2010). PI 88788 also carries a confirmed QTL (cqSCN-003) on chromosome 16 (formerly LG J) (Glover et al. 2004).

The second most widely exploited source of SCN resistance in North American breeding programmes has been Peking (PI 548402), which was the major source of resistance up until the 1990s (Diers et al. 1997). Peking also carries resistance alleles at the *rhg1* complex locus, but is estimated to have only three copies of the 31-kb region that contains the three resistance genes (Cook et al. 2014). Peking was found to carry a second resistance gene at the *Rhg4* locus in addition to resistance alleles at the *rhg1* complex locus (Concibido et al. 1997; Mahalingam and Skoprupska 1995). There is epistasis between the *rhg1* and *Rhg4* loci in Peking, such that the *rhg1* locus/haplotype by itself does not condition the level of resistance provided by the locus from PI 88788 (likely due to the lower copy number), but its efficacy is higher if the *Rhg4* gene from Peking is also present (Cook et al. 2014). In contrast, stacking the Peking *Rhg4* gene together with the *rhg1-b* gene does not enhance resistance. The gene at the *Rhg4* locus was found to encode a Ser hydroxymethyl transferase (Liu et al. 2012). The need to introgress two independent *Rhg* genes and the increased risk of linkage drag associated with that is probably one of the reasons why Peking has not been used as a resistance source to the same extent as PI 88788.

PI 88788 was first reported to be resistant to SCN in 1962, and Peking and PI 90763 were reported in 1970 (McCann et al. 1982). Other SCN-resistant germplasm accessions that have been used in a few breeding programmes include PI 90763, PI 209332 and PI 437654. PI 437654 was the source of 'Hartwig' and CystX resistance, which has had limited commercial use, and it also carries a resistance gene at the *Rhg4* locus (Cook et al. 2014). Diers et al. (1997) reported that PI 437654, along with PI 438489B, was resistant to five different races (now referred to as HG types) of *H. glycines*. The broad resistance of PI 437654 might be due to a large number of SCN resistance QTL. Wu et al. (2009) found putative QTL on eight different chromosomes, and many of those appeared to be involved in epistasis. Resistance QTL from PI 90763 and PI 404198A have also been mapped (Guo et al. 2005, 2006). The association between resistance to SDS and some sources of SCN resistance (Gelin et al. 2006) was mentioned previously. The *Rfs2/Rhg1* locus on soybean chromosome 18 was shown to be involved in co-inheritance of resistance to SDS in the roots and reduced root infestation by SCN (Srour et al. 2012). A resistance allele at *Rfs2* interferes with the spread of *Fusarium virguliforme* root infections, and in most PIs and cultivars, resistance alleles at the *Rhg1* locus are required for partial resistance to all SCN HG groups.

Nematode resistance from a number of germplasm accessions, including PI 88788 and PI 209332, has been associated with yield penalties in non-infested fields or environments with low SCN pressure (Kabelka et al. 2006; Kopisch-Obuch et al. 2005). In moderately or highly infested fields, however, near-isogenic lines with the *rhg1* resistance locus had significantly higher yields than lines with the susceptibility locus (Brucker et al. 2005). In any case, suspicions about yield penalties for incorporating SCN resistance genes or QTL that are not already in agronomically elite genetic backgrounds are probably one of the main reasons why private sector breeders have been reluctant to utilize more varied sources of SCN resistance.

Two SCN resistance genes have been identified in the *Glycine soja* accession PI 468916 (Kabelka et al. 2006). After being introgressed into two elite (i.e. predominantly *G. max*) genetic backgrounds, neither gene had a detrimental effect on yield (Kabelka et al. 2006). These genes were shown to enhance SCN resistance when combined with either the *rhg-1b* gene cluster from PI 88788 or the *rhg1* locus and *Rhg4* gene derived from PI 437654 (Kim et al. 2011). The perennial *G. tomentella* accession PI 483218 ($2n = 78$) was found to be immune to *H. glycines* HG Type 0 (Bauer et al. 2007).

Root-knot nematodes

The southern root-knot nematode (*Meloidogyne incognita*) and the peanut root-knot nematode (*M. arenaria*) are major pests of soybean grown in the south-eastern US, and the Javanese root-knot nematode (*M. javanica*) can also reduce yields (Boerma and Hussey 1992; Yates et al. 2010). Some *Meloidogyne* species have wide host ranges and cause considerable damage globally. Resistance has been a major breeding objective for several decades in south-eastern breeding programmes like the one at the University of Georgia (Boerma and Hussey 1992). Southern root-knot nematode resistance is particularly important for cultivars from later MGs; more than half of the MG V and VI cultivars that were recommended for Georgia in 2015 were susceptible to *M. incognita*, only one out of 18 recommended cultivars from MGs VII and VIII was susceptible. However, root-knot nematodes became a greater problem in the Mid-South production region as the early soybean production system (ESPS) became more popular in the region, since few cultivars from MG V and earlier had resistance (Koenning 2015). In addition to causing direct damage to soybean plants, root-knot nematode injury can reduce nodule formation by *Bradyrhizobium* spp. and increase vulnerability to infection by vascular fungal pathogens (Koenning 2015).

Resistant and tolerant cultivars are important for minimizing losses to root-knot nematodes. 'Avery', 'Forrest', 'Kirby', 'Palmetto', PI 417444 and PI 96354, and the breeding lines D3-3349, D86-3429 and G93-9009 have been used in breeding for resistance to *M. incognita* (Koenning 2015). Sources of resistance to *M. arenaria* that have been used in breeding programmes include 'Forrest', 'Jackson', PI 200538, PI 230977 and the breeding line G93-9106. *M. javanica* resistance derived from 'Gordon', 'Coodetec 201', PI 230977, PI 80466, PI 595099 and the breeding line G93-9223 has been used (Koenning 2015). In a screening of 2370 soybean PIs, PI 200538 and PI 230977 were identified as being highly resistant to both peanut and Javanese root-knot nematodes (Luzzi et al. 1987). Resistant breeding lines have been developed from crosses with these sources (Yates et al. 2010). PI 594403, PI 594427C and PI 594651B are other potentially unique sources of peanut root-knot nematode resistance (Harris et al. 2003). Although the accessions develop more galls than PI 200538, the female nematodes feeding on PI 594403 and PI 594427 produced the same or lower numbers of eggs. Yates et al. (2010) demonstrated that PI 494403, PI 594427C and PI 594651B likely carry different resistance genes from PI 200538. The development of soybean lines and cultivars with resistance to the various root-knot nematodes has been facilitated by the identification of markers for MAS to select plants with QTL from resistant germplasm accessions (Tamulonis et al. 1997a,b). A major QTL was identified on chromosome 13 (LG F) and a minor QTL on chromosome 15 (LG E) are associated with resistance to peanut root-knot nematodes (Tamulonis et al. 1997a). Ha et al. (2010) used SSR markers to show that a major QTL from the cultivar Palmetto that conditions resistance to the southern root-knot nematode (*M. incognita*) was inherited by resistant cultivars that were genotyped.

3 Selection for disease resistance

Besides yield potential, relative maturity and herbicide tolerance, disease resistance traits are among the main descriptors that seed companies use to characterize their cultivars. Based on the abundance of publications and patent claims dealing with gene/QTL mapping and MAS, and on discussions with soybean breeders, private and many public sector breeding programmes in the United States and Canada are increasingly using MAS to develop disease-resistant elite lines and new cultivars. MAS can substantially reduce the costs of breeding by eliminating plants and lines that lack the targeted genes in early generations and allowing breeders to focus resources on those lines that do carry the resistance genes or QTL. In addition, MAS allows breeders (or breeding teams) to pyramid multiple genes/QTL that condition resistance to the same pathogen in order to obtain lines with broader and more durable resistance. A good example of this would be the common objective of combining partial resistance QTL with at least one major R gene for *Phytophthora* resistance. In cases where linkage drag associated with an introgressed gene is a concern, MAS with a dense group of markers could allow plants with rare crossovers between the desirable resistance gene and loci with undesirable alleles to be identified. Although simple sequence repeat (SSR) markers are still used to a limited extent for MAS, single nucleotide polymorphism (SNP) markers have become the primary marker type used for MAS because of the ability to automate genotyping and to determine the genotype at thousands of SNP loci evenly distributed around the genome quickly (Song et al. 2013; Yuan et al. 2014).

Most, if not all, breeding programmes still rely on phenotypic assays to some extent, however. It is likely that high-yielding lines with putative resistance based on MAS are screened to confirm that they have the expected resistance phenotype before being considered for release. In other cases, it might be more cost-effective or reliable to screen for resistant plants or lines on a mass scale in breeding nurseries planted in fields that have a history of disease or which can be inoculated with one or more pathogens and/or managed in a way that promotes disease. Many of the references cited earlier for specific diseases provide descriptions of phenoptypic assays for the field or greenhouse.

It is quite likely that in the process of evaluating breeding lines, some degree of resistance or tolerance to common soybean pathogens is retained in some of the most promising lines as a result of natural selection and avoiding or discarding plants and seeds with obvious and severe disease symptoms. For example, susceptibility to seedling diseases that severely reduce stand densities is likely to reduce the yield from a plot if the density is below the threshold at which remaining plants are able to compensate through more vigorous growth and branching. Plants or lines with severe foliar diseases may be noted during the growing season and excluded from further agronomic evaluations. This is likely to be a common occurrence in the case of plants infected with a virus that causes visible symptoms and/or reduces pod set and yield, for example. Symptomatic plants are unlikely to be selected or to produce yields equivalent to those of virus-free lines.

4 Contribution of soybean genomics research to breeding strategies

A complete assembly of the *Glycine max* cv. Williams 83 genome sequence was published in 2010 after considerable struggles to deal with challenges posed by repetitive DNA sequences and duplicated genomic regions resulting from the ancient polyploid origin of cultivated soybean (Schmutz et al. 2010). This reference genome sequence had already become a valuable resource for genetic and genomic studies to learn more about the structure and content of the soybean genome, and it had facilitated the fine-mapping of genes and QTL for many traits, as well as the identification of candidate genes for those traits (Stupar and Specht 2013). It is beyond the scope of this chapter to provide an extensive discussion about the implications of genomics research on breeding soybean for disease resistance, but the contrast between the widely deployed *Rps1*-k and *rhg1* loci provides an interesting glimpse at the structural diversity that exists among disease resistance loci.

Five classes of effector-specific R proteins are known, and these interact with pathogen proteins (effectors) that might have evolved to promote infection and reproduction of the pathogen by subverting processes involved in host defence (Martin et al. 2003). The majority of proteins encoded by known resistance (R) genes contain nucleotide-binding (NB) and leucine-rich repeat (LRR) domains, and these can be divided into three classes (McHale et al. 2006). Four similar genes encoding two different classes of coiled coil NB-LRR proteins were found in the *Rps1-k* locus for resistance to *Phytophthora sojae* (Gao et al. 2005). Further research showed that the *Rps1*-k region is composed largely of repetitive sequences, but two full-length *Rps1*-k genes were identified and named *Rps1*-k-1 and *Rps1*-k-2 (Gao and Bhattacharyya 2008). These two genes are located about 20 kb apart. Although most cultivars marketed in the United States carry a resistance allele at the *Rps1* locus, many of them also carry QTL that provide partial, non-race-specific resistance.

Recent discoveries about phenomena like the role that gene structure variation, and particularly copy number variation (CNV), have demonstrated that classical genetics models and concepts are sometimes inadequate for explaining phenotypes (Anderson et al. 2014; Cook et al. 2012). Investigation of CNV at the *Rhg1* locus associated with SCN resistance has been particularly fascinating, and the findings show a very different situation from that of the *Rps1*-k locus described above. Tandem duplication of 31.2-kb interval containing five genes is associated with different levels of resistance (Cook et al. 2012, 2014). An allele consisting of seven to ten copies provides very high levels of resistance, whereas alleles with three copies provide moderate resistance, and a single copy is associated with susceptibility (Cook et al. 2014). Differential methylation of some DNA regions was also found to be correlated with relative resistance to SCN. Although high numbers of sequence polymorphisms were found within each *Rhg1* repeat in SCN-resistant lines, few of these have an effect on the protein-coding sequences. Subsequent investigations identified four distinct sequences of the 31.2-kb unit, and up to three of these can occur in certain *Rhg1* alleles (Lee et al. 2015). Lee et al. (2015) also found that the resistance phenotype was affected by the repeat sequence as well as the total number of copies. When SCN infestation occurs, the *Rhg1* locus appears to increase upregulation of several stress-related genes, including some that encode homologs of enzymes that increase levels of reactive oxygen species and proteins associated with the unfolded protein response (Kandoth et al. 2011). Genes involved in apoptic cell death, a hypersensitive response, and salicylic acid-mediated defence signalling are among the defence-related genes that are upregulated within the syncytia that result from SCN parasitism.

5 Pathogen genomics

Advances in DNA sequencing technologies and studies have resulted in proposals for substantial changes in taxonomic classifications of a number of important pathogens, and have permitted identification to the species level of pathogens present in diseased plant tissue. While the revisions to taxonomy based on DNA sequence instead of host specificity and pathogen morphology should unquestionably be improvements, changes in species and even genus names can result in some confusion, and can make it difficult to determine which organism was actually involved in some older studies. Some researchers are quick to embrace and use the revised scientific name of a pathogen, while others might not even be aware of the proposed revisions. A related dilemma is that some resistance genes were assigned names based on the initials of the genus and species of the pathogen at the time the gene was identified. One such example is the *Rdc/Rdm* family of genes that condition resistance to the SSC pathogen, which was recently found to be identical to *Diaporthe aspalathi*. The series of four resistance genes were initially named *Rdc* because the pathogen was assumed at the time to be *D. phaseolus* var. *caulivora*. Some researchers began to use the abbreviation *Rdm* for the genes after the fungus causing SSC was found to be different from the one causing northern stem canker, which was renamed *D. meridionalis*, so there is already some potential for confusion from that change. Another example of the influence that phylogenetic analyses have had on pathogen taxonomy is the previously mentioned proposal from Uzuhashi et al. (2010) to split *Pythium* spp. into four new genera.

6 Disease resistance from *G. soja* and perennial *Glycine* species

Glycine soja accessions are relatively easy to cross with *G. max*, and have been exploited to some degree as a source of SCN resistance genes. *G. soja* accession PI 468916 was found to have two resistance QTL in regions of the genome not previously reported to have nematode resistance genes (Kabelka et al. 2005; Wang et al. 2001). Kim et al. (2011) combined these two resistance QTL from PI 468916 with SCN resistance loci from the *G. max* accessions PI 88788 and PI 437654, and observed increased resistance to three nematode isolates.

Although much more difficult to transfer to *G. max* genetic backgrounds, genes from perennial *Glycine* species potentially offer diversity unavailable in cultivated soybean. The subgenus *Glycine* of the genus *Glycine* comprises 26 perennial diploid and tetraploid species native to Australia that represent a tertiary gene pool relative to domesticated soybean (Singh and Nelson 2014). Species have been identified with resistance to a variety of pathogens and pests, including *Phakopsora pachyrhizi* (Hartman et al. 1992; Schoen et al. 1992), *Septoria glycines* (Lim and Hymowitz 1987) and *Heterodera glycines* (Bauer et al. 2007, Riggs et al. 1998). Hartman et al. (2000) identified accessions of *G. tabacina* that were partially resistant to SSR and accessions of *G. tomentella* with partial resistance to SDS. Although recovery of fertile plants from crosses between *G. max* and perennial species is a difficult and painstaking process, Sing and Nelson (2014) successfully developed promising backcross lines and populations from crosses with *G. tomentella*. This success demonstrates that with sufficient dedication and patience, it should be possible to recover fertile and agronomically competitive lines from crosses between *G. max* and perennial *Glycine* accessions.

7 Transgene-mediated disease resistance

Soybean cultivars genetically engineered with disease resistance genes from other species could in theory be useful for reducing yield losses. The commercialization of such cultivars is impeded, however, by the comparative difficulty and expense of developing transgenic soybean plants and the high cost of getting them through the regulatory hurdles required for commercialization. To allow recovery of costs associated with commercialization approval, a transgene would presumably need to provide immunity or near-immunity against all or most races of a major pathogen and remain effective for a number of growing seasons. A chapter by Klink et al. (2013) provides an interesting discussion on the possibilities and approaches that could be used to develop transgenic resistance to SCN.

Sixteen soybean lines engineered with an *oxdc* decarboxylase transgene showed delayed white mould disease development and a 61–96% reduction in disease severity relative to a non-transgenic control (Cunha et al. 2010). The transgenic lines presumably had an enhanced ability to degrade oxalic acid secreted by the fungus, thereby reducing pathogenesis. Transgene-mediated resistance could be effective and durable in a situation like this in which the transgene would be expected to be effective against all isolates or populations of the pathogen, since it targets a compound involved in pathogenesis of an organism not known to have different races. Soybean plants expressing a receptor-like kinase from the soybean *Rfs2* resistance locus exhibited a high level of resistance to SDS and increased partial resistance to SCN (Srour et al. 2011).

8 Summary and future trends

Virtually all modern commercial soybean cultivars marketed in the United States and Canada carry resistance genes that protect them against at least some of the major pathogens likely to be encountered in the targeted planting area. This illustrates the success that soybean breeders in both the public and private sectors have had in developing high-yielding, agronomically elite soybean cultivars with effective and relatively durable resistance to economically important diseases like *Phytophthora* root and stem rot and the SCN. The durability of single widely used resistance genes has been variable, however, and the overall efficacy of some has eroded with time. The durability of the *rxp* gene for resistance to BP shows that certain genes are not easily defeated by the pathogens that they interact with. At the other extreme, an example would be the soybean rust resistance genes *Rpp1* and *Rpp3*, which were defeated within a few growing seasons in Brazil, even though selection pressure should not have been sufficient to accelerate this event. Resistance mediated by genes at the *rhg1* and *Rps1-k* loci has been widely used for decades and has held up relatively well, but some nematode and *Phytophthora sojae* populations, respectively, have succeeded in overcoming both types of resistance to some extent.

The number and diversity of disease resistance genes potentially available to breeders will continue to increase as new sources of resistance are discovered and as alternative approaches to mapping and identifying resistance genes or gene complexes are employed. An example is the successful use of GWAS with SNPs discovered through re-sequencing of germplasm accessions to identify *Phytophthora* resistance genes (Li et al. 2016; Schneider et al. 2016). Vuong et al. (2015) conducted a GWAS with 553 PIs to search for genomic regions associated with resistance to SCN HG Type 0. They confirmed six QTL previously mapped in biparental populations, including two whose locations corresponded with those of the *rps1* and *Rps4* loci, and also detected eight novel QTL.

Xu et al. (2013) employed a novel combination of methods and technologies to identify genes underlying QTL for southern root-knot nematode resistance in accession PI 438489B. A whole-genome re-sequencing approach was used to sequence 240 recombinant inbred lines (RILs) to an average depth of 0.19×. SNPs were identified and validated, and the parental source of each was inferred before they were used to genotype the RIL population. The genome was divided into 'bins' that contained one to ten genes each, and the bins were used as molecular markers to construct a linkage map with which three nematode resistance QTL were identified. A bin spanning a major resistance QTL on chromosome 10 was found to encompass three functional genes. A gene encoding a pectin methylesterase inhibitor was suspected to be involved in resistance to southern root-knot nematode. This type of clever combination of methods and currently available technologies should allow more effective exploitation of novel resistance genes from unique sources and should provide a better understanding of how resistance to diseases can be enhanced and made more durable.

While few soybean breeders are likely to question the need to incorporate and deploy a greater diversity of disease resistance genes in new soybean cultivars destined for major production regions, there is often a large gap between identified sources of resistance and the numbers of genes that are actually being deployed in contemporary cultivars. This problem is due in part to a concern about linkage drag resulting from the inadvertent co-introgression of undesirable genes at loci linked to the target resistance gene locus. Since many highly resistant soybean germplasm accessions from other countries have agronomically undesirable traits or are susceptible to other pathogens, the frequency of resistant progeny lines with a good agronomic trait package is likely to be low. Unless a soybean grower has had persistent and obvious problems with a specific disease or pest, he or she is likely to choose which cultivars to plant primarily based on yield performance, with pest resistance a secondary concern. In the competition for greater market share in seed sales, many private sector soybean breeders are likely to hesitate to make crosses with agronomically inferior disease-resistant lines. Public sector breeders could make some important contributions in the form of 'pre-breeding' to transfer unique alternative resistance genes from exotic germplasm lines into genetic backgrounds that are adapted to the targeted production area. This effort can obviously be facilitated by the use of MAS using molecular markers tightly linked to the gene or QTL of interest. MAS has become a critical part of private sector breeding programmes and is used to some extent in most public soybean breeding programmes.

Beyond biological limitations, patents on genes, germplasm and selection methods are an impediment to incorporating novel resistance genes into new cultivars and stacking resistance genes or QTL to enhance the durability of resistance. This is particularly an issue if MAS is used to develop enhanced resistant germplasm lines, though many phenotyping assays have also been patented. MAS for genes or QTL conditioning resistance traits that are difficult, tedious and/or expensive to assess phenotypically is particularly attractive, but both the procedure and the genes or QTL are likely to have been patented. One of the first patents for using MAS in soybean breeding was for the use of markers to select for SCN resistance (Webb et al. 1995). Countless patents have been issued for the use of MAS to develop lines with resistance to SDS and many other diseases (Lightfoot et al. 2001). Patent applications are often approved even if data to support the claims are vague or non-existent, and when key references from peer-reviewed journals about previous research on the subject are not mentioned in the patent application. The situation is probably unavoidable, but it hinders potential progress that could be made in developing soybean cultivars with broader and more durable resistance.

9 Where to look for further information

The 5th edition of the *Compendium of Soybean Diseases and Pests* from the American Phytopathological Society (Hartman et al. 2015c) is an excellent source of information about a wide range of pathogens and pests and the diseases that they cause on soybean. The 3rd edition of the American Society of Agronomy book titled *Soybeans: Improvement, Production, and Uses* (edited by H.R. Boerma and J.E. Specht, and published in 2004) has two chapters on soybean pathogens and one on nematodes. Those are good sources for basic background information and for additional references. Other chapters about genetics and genomics might be of interest to some readers.

10 References

Arahana, V. S., Graef, G. L., Specht, J. E., Stedman, J. R. and Eskridge, K. M. (2001), Identification of QTLs for resistance to *Sclerotinia sclerotiorum* in soybean. *Crop Sci.* 41(1), 180–8.

Arelli, P. R., Sleper, D. A., Yue, P. and Wilcox, J. A. (2000), Soybean reaction to races 1 and 2 of *Heterodera glycines*. *Crop Sci.* 40(3), 824–6.

Ashfield, T., Bocian, A., Held, D., Henk, A. D., Marek, L. F., Danesh, D., Peñuela, S., Mesem, K., Lightfoot, D. A., Young, N. D., Shoemaker, R. C. and Innes, R. W. (2003), Genetic and physical localization of the soybean *Rpg1-b* resistance gene reveals a complex locus containing several tightly linked families of NBS-LRR genes. *Mol. Plant Microbe Interact.* 16(9), 817–26.

Ashfield, T., Danzer, J. R., Held, D., Clayton, K., Keim, P., Saghai Maroof, Webb, D. M. and Innes, R. W. (1998), *Rpg1*, a soybean gene effective against all races of bacterial blight, maps to a cluster of previously identified disease resistance genes. *Theor. Appl. Genet.* 96, 1013–21.

Athow, K. L. and Probst, A. H. (1952), The inheritance of resistance to frogeye leaf spot of soybeans. *Phytopathology* 42, 660–2.

Athow, K. L., Probst, A. H., Kartzman, C. P. and Laviolette (1962), A newly identified physiological race of *Cercospora sojina* on soybean. *Phytopathology* 52, 712–14.

Bachman, M. S., Tamulonis, J. P., Nickell, C. D. and Bent, A. F. (2001), Molecular markers linked to brown stem rot resistance genes, *Rbs1* and *Rbs2*, in soybean. *Crop Sci.* 41(2), 526–35.

Backman, P. A., Weaver, D. B. and Morgan-Jones, G. (1985), Soybean stem canker: An emerging disease problem. *Plant Dis.* 69(8), 641–7.

Baker, W. A., Weaver, D. B., Qiu, J. and Pace, P. F. (1999), Genetic analysis of frogeye leaf spot resistance in PI54610 and Peking soybean. *Crop Sci.* 39(4), 1021–5.

Bates, G. D., Rothrock, C. S. and Rupe, J. C. (2008), Resistance of the soybean cultivar Archer to Pythium damping-off and root rot caused by several *Pythium* spp. *Plant Dis.* 92(5), 763–6.

Bauer, S., Hymowitz, T. and Noel, G. (2007), Soybean cyst nematode resistance derived from *Glycine tomentella* in amphiploid (*G. max* × *G. tomentella*) hybrid lines. *Nematropica* 37, 277–85.

Bernard, R. L. and Cremeens, C. R. (1988), Registration of 'Williams 82' soybean. *Crop Sci.* 28(6), 1027.

Boerma, H. R. and Hussey, R. S. (1992), Breeding plants for resistance to nematodes. *J. Nematol.* 24, 242–52.

Boerma, H. R., Monteros, M. J., Ha, B.-K., Wood, E. D., Phillips, D. V., Walker, D. R. and Missaoui, A. M. (2011), Registration of Asian soybean rust-resistant soybean germplasm G01-PR16. *J. Plant Reg.* 5(1), 1–5.

Boerma, H. R. and Phillips, D. V. (1983), Genetic implications of the susceptibility of Kent soybean to *Cercospora sojina. Phytopathology* 74, 1666–8.

Boerma, H. R. and Specht, J. E. (Eds) (2004), *Soybeans: Improvement, Production, and Uses*, 3rd Ed. American Society of Agronomy, Madison, Wisconsin, USA.

Boland, G. J. and Hall, R. (1987), Evaluating soybean cultivars for resistance to *Sclerotinia sclerotiorum* under field conditions. *Plant Dis.* 71, 934–6.

Bradley, C. A., Hartman, G. L., Mueller, D. S., Hoffman, D. D., Nickell, C. D. and Pedersen, W. L. (2005), Genetic analysis of partial resistance to *Rhizoctonia solani* in the soybean cultivar 'Savoy'. *Can. J. Plan Pathol.* 27, 137–42.

Bradley, C. A., Hartman, G. L., Nelson, R. L., Mueller, D. S. and Pedersen, W. L. (2001), Response of ancestral soybean lines and commercial cultivars to Rhizoctonia root and hypocotyl rot. *Plant Dis.* 85(10), 1091–5.

Brar, H., Swaminathan, S. and Bhattacharyya, M. K. (2011), The *Fusarium virguliforme* toxin FvTox1 causes foliar sudden death syndrome-like symptoms in soybean. *Mol. Plant-Microbe Interact.* 24, 1179–88.

Bromfield, K. R. (1984), *Soybean rust.* Monogr. 11. American Phytopathological Society, St. Paul, MN.

Bromfield, K. R. and Hartwig, E. G. (1980), Resistance to soybean rust and mode of inheritance. *Crop Sci.* 20(2), 254–5.

Bromfield, K. R., Melching, J. S. and Kingsolver, C. H. (1980), Virulence and aggressiveness of *Phakopsora pachyrhizi* isolates causing soybean rust. *Phytopathology* 70(1), 17–21.

Brown, E. A., Minor, H. C. and Calvert, O. H. (1987), A soybean genotype resistant to Phomopsis seed decay. *Crop Sci.* 27(5), 895–8.

Brucker, E., Carlson, S., Wright, E., Niblack, T. and Diers, B. (2005), *Rhg1* alleles from soybean PI 437654 and PI 88788 respond differently to isolates of *Heterodera glycines* in the greenhouse. *Theor. Appl. Genet.* 99, 811–18.

Buzzell, R. I. and Anderson, T. R. (1982), Plant loss response of soybean cultivars to *Phythophthora megasperma* f.sp. *glycinea* under field conditions. *Plant Dis.* 66(12), 1146–8.

Carter Jr., T. E., Nelson, R. L., Sneller, C. H. and Cui, Z. (2004), Genetic diversity in soybean, pp. 303–416. In *Soybean: Improvement, Production, and Uses,* H. R. Boerma and J. E. Specht (Eds). Am Soc. Agron, Madison, Wisconsin.

Chen, Y. and Wang, D. (2005), Two convenient methods to evaluate soybean for resistance to *Sclerotinia sclerotiorum.* *Plant Dis.* 89(12), 1268–72.

Chiesa, M. A., Pioli, R. N. and Morandi, E. N. (2009), Specific resistance to soybean stem canker conferred by the *Rdm4* locus. *Plant Pathol.* 58(6), 1032–8. doi:10.1111/j.1365-3059.2009.02145.X.

Chiesa, M. A., Pioli, R. N., Cambursano, M. V. and Morandi, E. N. (2013), Differential expression of distinct soybean resistance genes interacting with Argentinian isolates of *Diaporthe phaseolorum* var. *meridionalis*. *Eur. J. Plant Pathol.* 135, 351–62.

Chun, D., Kao, L. B. and Lockwood, J. L. (1987), Laboratory and field assessment of resistance in soybean to stem rot caused by *Sclerotinia sclerotiorum*. *Plant Dis.* 71(9), 811–15.

Concibido, V. C., Denny, R. L., Boutin, S. R., Hautea, R., Orf, J. H. and Young, N. D. (1994), DNA marker analysis of loci underlying resistance to soybean cyst nematode (*Heterodera glycines* Ichinoe). *Crop Sci.* 34(1), 240–6.

Concibido, V. C., Denny, R., Lange, D., Danesh, D., Orf, J. and Young, N. (1997), Genome mapping on soybean cyst nematode resistance genes in 'Peking', PI 90763, and PI 88788 using DNA markers. *Crop Sci.* 37(2), 258–64.

Concibido, V. C., Diers, B. W. and Arelli, P. R. (2004), A decade of QTL mapping for cyst nematode resistance in soybean. *Crop Sci.* 44(4), 1121–31.

Cook, D. E., Bayless, A. M., Wang, K., Song, Q., Jiang, J. and Bent, A. F. (2014), Distinct copy number, coding sequence and locus methylation patterns underlie *Rhg1*-mediated soybean resistance to soybean cyst nematode. *Plant Physiol.* 165, 630–47. doi:10.1104/pp.114.235952.

Cook, D. E., Lee, T. G., Guo, X., Melito, S., Wang, K., Bayless, A. M., Wang, J. Hughes, T. J., Willis, D. K. and Clemente, T. and et al. (2012), Copy number variation of multiple genes at *Rhg1* mediates nematode resistance in soybean. *Science* 338, 1206–9.

Cunha, W. G., Tinoco, M. L.P., Pancoti, H. L., Ribeiro, R. E. and Aragão, F. J. L. (2010), High resistance to *Sclerotinia sclerotiorum* in transgenic soybean plants transformed to express an oxalate decarboxylase gene. *Plant Pathol.* 59, 654–60. doi:10.1111/j.1365-3059.2010.02279.x.

Demirbas, A., Rector, B., Hohnes, D. G., Fioritto, R. J., Graef, G. L., Cregan, P. B., Shoemaker, R. C. and Specht, J. E. (2001), Simple sequence repeat markers linked to the soybean genes for Phytophthora resistance. *Crop Sci.* 41(4), 1220–7.

Diers, B. W., Kim, K.-S., Frederick, R. D., Hartman, G. L., Unfried, J., Shultz, S. and Cary, T. (2014), Registration of eight soybean germplasm lines resistant to soybean rust. *J. Plant Reg.* 8, 96–101.

Diers, B. W., Kopisch-Obuch, F. J., Hoffman, D. D., Hartman, G. L., Pedersen, W. L., Grau, C. R. and Wang, D. (2006), Registration of AxN-1-55 soybean germplasm with partial resistance to Sclerotinia stem rot. *Crop Sci.* 46(3), 1403–4.

Diers, B. W., Mansur, L., Imsande, J. and Shoemaker, R. C. (1992), Mapping phytophthora resistance loci in soybean with restriction fragment length polymorphism markers. *Crop Sci.* 32(2), 377–83.

Diers, B. W., Skorupska, H. T., Rao-Arelli, A. P. and Cianzio, S. R. (1997), Genetic relationships among soybean plant introductions with resistance to soybean cyst nematodes. *Crop Sci.* 37(6), 1966–72.

Dorrance, A. E., Berry, S. A., Anderson, T. R. and Meharg, C. (2008), Isolation, storage, pathotype characterization, and evaluation of resistance for *Phytophthora sojae* in soybean. *Plant Health Progr.* doi:10.1094/PHP-2008-0118-01-DG.

Dorrance, A. E. and Grünwald, N. (2009), *Phytophthora sojae*: Diversity among and within populations, pp. 197–212. In *Oomycetes Genetics and Genomics: Diversity, Interactions, and Research Tools,* K. Lamour and S. Kamoun (Eds). John Wiley & Sons, Inc., Hoboken, NJ.

Dorrance, A. E. and Schmitthenner, A. F. (2000), New sources of resistance to *Phytophthora sojae* in the soybean plant introductions. *Plant Dis.* 84, 1303–8.

Ellis, M. L., McHale, L. K., Paul, P. A., St. Martin, S. K. and Dorrance, A. E. (2013), Soybean germplasm resistant to Pythium irregular and molecular mapping of resistance quantitative trait loci derived from the soybean accession 424354. *Crop Sci.* 53, 1008–21.

Elshire, R. J., Glaubitz, J. C., Sun, Q., Poland, J. A., Kawamoto, K., Buckler, E. S. and Mitchell, S. E. (2011), A robust, simple genotyping-by-sequencing (GBS) approach for high diversity species. *PLoS One* 6, e19379.

Gao, X., Jackson, T. A., Hartman, G. L. and Niblack, T. L. (2006), Interactions between soybean cyst nematode and *Fusarium solani* f. sp. *glycines* based on greenhouse factorial experiments. *Phytopathology* 96, 1409–15.

Garcia, A., Calvo, É.S., Kiihl, R. A. S., Harada, A., Hiromoto, D. M. and Vieira, L. G. E. (2008), Molecular mapping of soybean rust (*Phakopsora pachyrhizi*) resistance genes: discovery of a novel locus and alleles. *Theor. Appl. Genet.* 117, 545–53.

Gelin, J. R., Arelli, P. R. and Rojas-Cifuentes, P. R. (2006), Using independent culling to screen plant introductions for combined resistance to soybean cyst nematode and sudden death syndrome. *Crop Sci.* 46, 2081–3.

Gizlice, Z., Carter Jr., T. E. and Burton, J. W. (1994), Genetic base for North American public soybean cultivars released between 1947 and 1988. *Crop Sci.* 34(5), 1143–51.

Glover, K. D., Wang, D., Arelli, P. R., Carlson, S. R., Cianzio, S. R. and Diers, B. W. (2004), Near isogenic lines confirm a soybean cyst nematode resistance gene from PI 88788. *Crop Sci.* 44(3), 936–41.

Gomes, R. R., Glienke, C., Videira, S. I. R., Lombard, L., Groenewald, J. Z. and Crous, P. W. (2013), Diaporthe: a genus of endophytic, saprobic and plant pathogenic fungi. *Persoonia* 31, 1–41

Goradia, L., Hartman, G. L. and Daniel, S. L. (2009), Evaluation of glyphosate-tolerant soybean cultivars for resistance to bacterial pustule. *Eur. J. Plant Pathol.* 124, 331–5.

Gordon, S. G., Martin, S. K. S. and Dorrance, A. E. (2006), *Rps*8 maps to a resistance gene rich region on soybean molecular linkage group F. *Crop Sci.* 46(1), 168–73.

Gore, M. A., Hayes, A. J., Jeong, S. C., Yue, Y. G., Buss, G. R. and Saghai Maroof, M. A. (2002), Mapping tightly linked genes controlling potyvirus infection at the *Rsv1* and *Rpv1* region in soybean. *Genome* 45, 592–9.

Grau, C. R., Dorrance, A. E., Bond, J. and Russing J. S. (2004), Fungal diseases, pp. 679–763. In *Soybeans: Improvement, Production and Uses,* Boerma, H. R. and Specht, J. E. (Eds), Agronomy Monograph No. 16. American Society of Agronomy, Crop Science Society of America, Soil Science Society of America.

Grau, C. R. and Hartman, G. L. (2015), Sclerotinia stem rot, pp. 59–62. In *Compendium of Soybean Diseases and Pests,* G. L. Hartman, J. C. Rupe, E. J. Sikora, L. L. Domier, J. A. Davis and K. L. Steffey (Eds). APS Press, St. Paul, MN.

Grau, C. R., Radke, V. L. and Gillespie, F. L. (1982), Resistance of soybean cultivars to *Sclerotinia sclerotiorum. Plant Dis.* 66, 506–8.

Guo, B., Sleper, D. A., Arelli, P. R., Shannon, J. G. and Nguyen, H. T. (2005), Identification of QTLs associated with resistance to soybean cyst nematode races 2, 3 and 5 in soybean PI 90763. *Theor. Appl. Genet.* 111, 965–71.

Guo, B., Sleper, D. A., Nguyen, H. T., Arelli, P. R. and Shannon, J. G. (2006), Quantitative trait loci underlying resistance to three soybean cyst nematode populations in soybean PI 404198A. *Crop Sci.* 46(1), 224–33.

Guo, X., Wang, D., Gordon, S. G., Helliwell, E., Smith, T., Berry, S. A., St. Martin, S. K. and Dorrance, A. E. (2008), Genetic mapping of QTLs underlying partial resistance to *Sclerotinia sclerotiorum* in soybean PI 391589A and PI 391589B. *Crop Sci.* 48(3), 1129–39.

Ha, B.-K., Bennett, J. B., Hussey, R. S., Finnerty, S. L. and Boerma, H. R. (2004), Pedigree analysis of a major QTL conditioning soybean resistance to southern root-knot nematode. *Crop Sci.* 44(3), 758–63.

Hansen, E. M., Wilcox, W. F. and Reeser, P. W. (2009), *Phytophthora rosacearum* and *P. sansomeana*, new species segregated from the *Phytophthora megasperma* 'complex'. *Mycologia* 101(1), 129–35.

Harris, D. K., Boerma, H. R., Hussey, R. S. and Finnerty, S. L. (2003), Additional sources of soybean germplasm resistant to two species of root-knot nematode. *Crop Sci.* 43(5), 1848–51.

Harris, D. K., Kendrick, M. D., King, Z. R., Pedley, K. F., Walker, D. R., Cregan, P. B., Buck, J. W., Phillips, D. V., Li, Z. and Boerma, H. R. (2015), Identification of unique genetic sources of soybean rust resistance from the USDA Soybean Germplasm Collection. *Crop Sci.* 55(5), 2161–76.

Hartman, G. L. (2015), Worldwide importance of soybean pathogens and pests, pp. 4–5. In *Compendium of Soybean Diseases and Pests.* The American Phytopathological Society, St. Paul, MN, USA.

Hartman, G. L., Bowen, C. R., Haudenshield, J. S., Fox, C. M., Cary, T. R. and Diers, B. W. (2015a), Evaluation of disease and pest damage on soybean cultivars released from 1923 through 2008 under field conditions in central Illinois. *Agron. J.* 107(6), 2373–80.

Hartman, G. L., Gardner, M. E., Hymowitz, T. and Naidoo, G. C. (2000), Evaluation of perennial *Glycine* species for resistance to soybean fungal pathogens that cause Sclerotinia stem rot and sudden death syndrome. *Crop Sci.* 40(2), 545–9.

Hartman, G. L. and Hill, C. B. (2010), Diseases of soybean and their management, pp. 276–99. In *The Soybean* (G. Singh (Ed.)). Wallingford, CABI.

Hartman, G. L., Huang, Y. H., Nelson, R. L. and Noel, G. R. (1997), Germplasm evaluation of *Glycine max* for resistance to *Fusarium solani*, the causal organism of sudden death syndrome. *Plan Dis.* 76, 396–9.

Hartman, G. L., Leandro, L. F. and Rupe, J. C. (2015b), Sudden death syndrome, pp. 88–90. In *Compendium of Soybean Diseases and Pests*. The American Phytopathological Society, St. Paul, MN, USA.

Hartman, G. L., Rupe, J. C., Sikora, E. J., Domier, L. L., Davis, J. A. and Steffey, K. L. (2015c), *Compendium of Soybean Diseases and Pests*. The American Phytopathological Society, St. Paul, MN, USA.

Hartman, G. L., Wang, T. C. and Hymowitz, T. (1992), Sources of resistance to soybean rust in perennial *Glycine* species. *Plant Dis.* 76, 396–9.

Hartman. G. L., West, E. and Herman, T. (2011), Crops that feed the world 2. Soybean – worldwide production, use, and constraints caused by pathogens and pests. *Food Sec.* 3(1), 5–17.

Hartwig, E. E. (1986), Identification of a fourth major gene conferring resistance to soybean rust. *Crop Sci.* 26(6), 1135–6.

Hartwig, E. E. and Bromfield, K. R. (1983), Relationships among three genes conferring specific resistance to rust in soybeans. *Crop Sci.* 23(2), 237–9.

Hoffman, D. D., Diers, B. W., Hartman, G. L., Nickell, C. D., Nelson, R. L., Pedersen, W. L., Cober, E. R., Graef, G. L., Steadman, J. R., Grau, C. R., Nelson, B. D., del Rio, L. E., Anderson, T., Poysa, V., Rajcan, I., Helms, T. and Stienstra, W. C. (2002), Selected soybean plant introductions with partial resistance to *Sclerotinia sclerotiorum*. *Plant Dis.* 86, 971–80.

Huang, J., Guo, N., Li, Y., Sun, J., Hu, G., Zhang, H., Li, Y., Zhang, X., Zhao, J., Xing, H. and Qiu, L. (2016), Phenotypic evaluation and genetic dissection of resistance to *Phytophthora sojae* in the Chinese soybean mini core collection. *BMC Genetics* 17(1), 85.

Hughes, T. J., Chen, W. and Grau, C. R. (2002), Pathogenic characterization of genotypes A and B of *Phialophora gregata* f. sp. *sojae*. *Plant Dis.* 86, 729–35.

Hyten, D. L., Hartman, G. L., Nelson, R. L., Frederick, R. D., Concibido, V. C., Narvel, J. M., and Cregan, P. B. (2007), Map location of the *Rpp1* locus that confers resistance to soybean rust in soybean. *Crop Sci.* 47, 837–8.

Hyten, D. L., Smith, J. R., Frederick, R. D., Tucker, M. L., Song, Q. and Cregan, P. B. (2009), Bulk segregant analysis using the GoldenGate assay to locate the *Rpp3* locus that confers resistance to *Phakopsora pachyrhizi* (soybean rust) in soybean. *Crop Sci.* 49, 265–71.

Hyten, D. L., Song, Q., Zhu, Y., Choi, I.-Y., Nelson, R. L., Costa, J. M., Specht, J. E., Shoemaker, R. C. and Cregan, P. B. (2006), Impacts of genetic bottlenecks on soybean genome diversity. *PNAS* 103(45), 16666–71.

Impullitti, A. E. and Malvick, D. K. (2014), Anatomical response and infection of soybean during latent and pathogenic infection by Type A and B of *Phialophora gregata*. *PLoS ONE* 9(5), e98311.

Iquira, E., Humira, S. and Belzile, F. (2015) Association mapping of QTLs for sclerotinia stem rot resistance in a collection of soybean plant introductions using a genotyping by sequencing (GBS) approach. *BMC Plant Biology* 15, 5. doi:10.1186/s12870-014-0408-y.

Jackson, E. W., Fenn, P. and Chen, P. (2005), Inheritance of resistance to Phomopsis seed decay in soybean PI 80837 and MO/PSD-0259 (PI 562694). *Crop Sci.* 45, 2400–4.

Jackson, E. W., Fenn, C. D., Fenn, P. and Chen, P. (2009), Genetic mapping of resistance to Phomopsis seed decay in the soybean breeding line MO/PSD-0259 (PI 562694) and plant introduction 80837. *J. Hered.* 100, 777–83.

Kabelka, E. A., Carlson, S. R. and Diers, B. W. (2005), Localization of two loci that confer resistance to soybean cyst nematode from *Glycine soja* PI 468916. *Crop Sci.* 45(6), 2473–81.

Kabelka, E. A., Carlson, S. R. and Diers, B. W. (2006), *Glycine soja* PI 468916 SCN resistance loci's associated effects on soybean seed yield and other agronomic traits. *Crop Sci.* 46(2), 622–9.

Kadam, S., Vuong, T. D., Qiu, D., Meinhardt, C. G., Song, L., Deshmukh, R., Patil, G., Wan, J., Valliyodan, B., Scaboo, A., Shannon, J. G. and Nguyen, H. T. (2016), Genomic-assisted phylogenetic analysis and marker development for next generation soybean cyst nematode resistance breeding. *Plant Sci.* 242, 3342–350.

Kandoth, P. K., Ithal, N., Recknor, J., Maier, T., Nettleton, D., Baum, T. and Mitchum, M. G. (2011), The soybean *Rhg1* locus for resistance to the soybean cyst nematode *Heterodera glycines* regulates the expression of a large number of stress- and defense-related genes in degenerating feeding cells. *Plant Physiol.* 155(4), 1960–75.

Kelly, H. Y., Dufault, N. S., Walker, D. R., Isard, S. A., Schneider, R. W., Giesler, L. J., Wright, D. L., Marois, J. J. and Hartman, G. L (2015), From select agent to an established pathogen: The response to *Phakopsora pachyrhizi* (soybean rust) in North America. *Phytopathology* 105(7), 905–16.

Kendrick, M. D., Harris, D. K., Ha, B.-K., Hyten, D. L., Cregan, P. B., Frederick, R. D., Boerma, H. R. and Pedley, K. F. (2011), Identification of a second Asian soybean rust resistance gene in Hyuuga soybean. *Phytopathology* 101(5), 535–43.

Kim, C., Guo, H., Kong, W., Chandnani, R., Shuang, L.-S. and Paterson, A. H. (2016), Application of genotyping by sequencing technology to a variety of crop breeding programs. *Plant Science* 242, 14–22. doi:10.1016/j.plansci.2015.04.016.

Kim, H. K., Kim, K. H., Van, K., Kim, M. Y. and Lee, S.-H. (2010a), Fine mapping of a resistance gene to bacterial leaf pustule in soybean. *Theor. Appl. Genet.* 120, 1443–50.

Kim, H. S. and Diers, B. W. (2000), Inheritance of partial resistance to Sclerotinia stem rot in soybean. *Crop Sci.* 40(1), 55–61.

Kim, H. S., Hartman, G. L., Mananandhar, J. B., Graef, G. L., Steadman, J. R. and Diers, B. W. (2000), Reaction of soybean cultivars to Sclerotinia stem rot in field, greenhouse, and laboratory evaluations. *Crop Sci.* 40(3), 665–9.

Kim, H. S., Sneller, C. H. and Diers, B. W. (1999), Evaluation of soybean cultivars for resistance to Sclerotinia stem rot in field environments. *Crop Sci.* 39(1), 64–8.

Kim, M., Hyten, D. L., Bent, A. F. and Diers, B. W. (2010b), Fine mapping of the SCN resistance locus *rhg1-b* from PI 88788. *Plant Gen.* 3(2), 81–9.

Kim, M., Hyten, D. L., Niblack, T. L. and Diers, B. W. (2011), Stacking resistance alleles from wild and domestic soybean sources improves soybean cyst nematode resistance. *Crop Sci.* 51(3), 934–43.

King, Z. R., Harris, D. K., Wood, E. D., Buck, J. W., Boerma, H. R. and Li, Z. (2016), Registration of four near-isogenic lines of G00-3213 for resistance to Asian soybean rust. *J. Plant Reg.* 10, 189–94.

Kirkpatrick, M. T., Rupe, J. C., Rothrock, C. S. and Gbur, E. E. (2006), The effect of *Pythium ultimum* and soil flooding on two soybean cultivars. *Plant Dis.* 90(5), 597–602.

Klink, V. P., Matsye, P. E., Lawrence, K. S. and Lawrence, G. W. (2013). Engineered soybean cyst nematode resistance, Ch. 6. In *Soybean - Pest Resistance* (Prof. Hany El-Shemy (Ed.)). InTech. doi:10.5772/54514. Available from: http://www.intechopen.com/books/soybean-pest-resistance/engineered-soybean-cyst-nematode-resistance.

Koenning, S. R. (2015), Root-knot nematodes, pp. 98–100. In *Compendium of Soybean Diseases and Pests,* G. L. Hartman, J. C. Rupe, E. J. Sikora, L. L. Domier, J. A. Davis and K. L. Steffey (Eds). APS Press, St. Paul, MN.

Koenning, S. R. and Wrather, J. A. (2010), Suppression of soybean yield potential in the continental United States by plant diseases from 2006 to 2009. Online. *Plant Health Prog.* doi:10.1094/PHP-2010-1122-01-RS.

Kolander, T. M., Bienapfl, J. C., Kurle, J. E. and Malvick, D. K. (2012), Symptomatic and asymptomatic host range of *Fusarium virguliforme*, the causal agent of soybean sudden death syndrome. *Plant Dis.* 96(8), 1148–53. doi:10.1094/PDIS-08-11-0685-RE.

Kopisch-Obuch, F. J., McBroom, R. L. and Diers, B. W. (2005), Association between SCN resistance loci and reduced yield in soybean. *Crop Sci.* 45(3), 956–65.

Lee, T. G., Kumar, I., Diers, B. W. and Hudson, M. E. (2015), Evolution and selection of *Rhg1*, a copy-number variant nematode-resistance locus. *Mol. Ecol.* 24, 1774–91.

Lemos, N. G., Lucca e Braccini, A. de, Abelnoor, R. V., Oliveira, M. C. N. de, Suenaga, K. and Yamanaka, N. (2011), Characterization of genes *Rpp2, Rpp4,* and *Rpp5* for resistance to soybean rust. *Euphytica* 182(1), 53–64.

Li, L., Guo, N., Niu, J., Wang, Z., Cui, X., Sun, J., Zhao, T. and Xing, H. (2016), Loci and candidate gene identification for resistance to *Phytophthora sojae* via association analysis in soybean [*Glycine max* (L.) Merr.]. *Mol. Genet. Genom.* 291(3):1095–103.

Li, S., Chen, P. and Hartman, G. L. (2015). pp. 47–8. *Compendium of Soybean Diseases and Pests,* G. L. Hartman, J. C. Rupe, E. J. Sikora, L. L. Domier, J. A. Davis and K. L. Steffey (Eds). APS Press, St. Paul, MN.

Li, S., Smith, J. R., Ray, J. D. and Frederick, R. D. (2012), Identification of a new soybean rust resistance gene in PI 567102B. *Theor. Appl. Genet.* 125, 133–42.

Li, X. P., Han, Y. P., Teng, W. L., Zhang S. Z., Yu, K. F., Poysa, V., Anderson, T., Ding, J. J. and Li, W. B. (2010), Pyramided QTL underlying tolerance to Phytophthora root rot in mega-environments from soybean cultivars 'Conrad' and 'Hefeng 25'. *Theor. Appl. Genet.* 121(4), 651–8.

Lightfoot, D. A. (2015), Two decades of molecular marker-assisted breeding for resistance to soybean sudden death syndrome. *Crop Sci.* 55(4), 1460–84.

Lightfoot, D. A., Meksem, K. and Gibson, P. T. (2001), Soybean sudden death syndrome resistant soybeans, soybean cyst nematode resistant soybeans and methods of breeding and identifying resistant plants: DNA markers. U.S. Patent 6,300,541. 9 October.

Lim, S. M. and Hymowitz, T. (1987), Reactions of perennial wild species of genus *Glycine* to *Septoria glycines*. *Plant Dis.* 71, 891–3.

Liu, S., Kandoth, P. K., Warren, S. D., Yeckel, G., Heinz, R., Alden, J., Yang, C., Jamai, A., El-Mellouiki, T., Juvale, P. S., Hill, J., Baum, T. J., Cianzio, S., Whitham, S. A., Korkin, D., Mitchum, M. G. and Meksem, K. (2012), A soybean cyst nematode resistance gene points to a new mechanism of plant resistance to pathogens. *Nature* 492, 456–60.

Liu, M., Li, S., Swaminathan, S., Sahu, B. B., Leandro, L. F., Cardinal, A. J., Bhattracharyya, M. K., Song, Q., Walker, D. R. and Cianzio, S. R. (2016), Identification of a soybean rust resistance gene in PI 567104B. *Theor. Appl. Genet.* 129, 863–77.

Lu, X., Robertson, A. E., Byamukama, E. Z. and Nutter Jr., F. W. (2010), Evaluating the importance of stem canker of soybean in Iowa. *Plant Dis.* 94(2), 167–73. doi:10.1094/PDIS-94-2-0167.

Luckew, A. S., Leandro, L. F., Bhattacharyya, M. K., Nordman, D. J., Lightfoot, D. A. and Cianzio, S. R. (2013), Usefulness of 10 genomic regions in soybean associated with sudden death syndrome resistance. *Theor. Appl. Genet.* 126, 2391–403.

Ludwig, A. and Tenhaken, R. (2000), Defence gene expression in soybean is linked to the status of the cell death program. *Plant Mol. Biol.* 44, 209–18.

Luedders, V. D. and Dropkin, V. H. (1983), Effect of secondary selection on cyst nematode reproduction on soybeans. *Crop Sci.* 23(2), 263–4.

Luzzi, B. M., Boerma, H. R. and Hussey, R. S. (1997), Resistance to three species of root-knot nematode in soybean. *Crop Sci.* 27(2), 258–62.

Mahalingam, R. and Skorupska, H. T. (1995), DNA markers for resistance to *Heterodera glycines* race-3 in soybean cultivar 'Peking'. *Breed. Sci.* 45, 435–43.

Malvick, D. K., Grau, C. R. and Gray, L. E. (2015), Brown stem rot, pp. 65–7. In *Compendium of Soybean Diseases and Pests,* G. L. Hartman, J. C. Rupe, E. J. Sikora, L. L. Domier, J. A. Davis and K. L. Steffey (Eds). APS Press, St. Paul, MN.

Manandhar, J. B., Hartman, G. L. and Sinclair, J. B. (1988), Soybean germ plasm evaluation for resistance to *Cholletotrichum trunculatum*. *Plant Dis.* 72, 56–9.

Mayhew, W. L. and Caviness, C. E. (1994), Seed quality and yield of early-planted, short-season soybean genotypes. *Agron J.* 86(1), 16–19.

McCann, J., Luedders, V. D. and Dropkin, V. H. (1982), Selection and reproduction of soybean cyst nematodes on resistant soybeans. *Crop Sci.* 22(1), 78–80.

McHale, L. K., Haun, W. J., Xu, W. W., Bhaskar, P. B., Anderson, J. E., Hyten, D. L., Gerhardt, D. J., Jeddeloh, J. A. and Stupar, R. M. (2012), Structural variants in the soybean genome localize to clusters of biotic stress-response genes. *Plant Physiol.* 159, 1295–308.

McHale, L. K., Tan, X., Koehl, P. and Michelmore, R. W. (2006), Plant NBS-LRR proteins: adaptable guards. *Genome Biol.* 7(4), 212.

McLean, R. J. and Byth, D. E. (1976), Resistance of soybean to rust in Australia. *Aust. Plant Pathol. Soc. Newsl.* 5(3), 34–6.

Mengistu, A., Arelli, P. A., Bellaloui, N., Bond, J. P., Shannon, J. G., Wrather, A. J., Rupe, J. C., Chen, P., Little, C. R., Canaday, C. H., Newman, M. A. and Pantalone, V. R. (2012), Evaluation of soybean genotypes for resistance to three seed-borne diseases. Online. *Plant Health Prog.* doi:10.1094/PHP-2012-0321-02-RS.

Mengistu, A., Arelli, P., Bond, J., Nelson, R., Rupe, J., Shannon, G. and Wrather, A. (2013), Identification of soybean accessions resistant to *Macrophomina phaseolina* by field screening and laboratory validation. Online. *Plant Health Prog.* doi:10.1094/PHP-2013-0318-01.RS.

Mengistu, A., Arelli, P., Bond, J. P., Shannon, J. G., Wrather, J. A., Rupe, J. C., Chen, P., Little, C., Canaday, C., Newman, M. and Pantalone, V. (2011a), Evaluation of soybean genotypes for resistance to charcoal rot. *Plant Health Prog.* doi:10.1094/PHP-2010-0926-01-RS.

Mengistu, A., Bond, J., Mian, R., Nelson, R., Shannon, G. and Wrather, A. (2011b), Identification of soybean accessions resistant to *Cercospora sojina* by field screening, molecular markers, and phenotyping. *Crop Sci.* 51(3), 1101–9.

Meyer, J. D. F., Silva, D. C. G., Yang, C., Pedley, K. F., Zhang, C., van de Mortel, M., Hill, J. H., Shoemaker, R. C., Abdelnoor, R. V., Whitman, S. A. and Graham, M. A. (2009), Identification and analyses of candidate genes for Rpp4-mediated resistance to Asian soybean rust in soybean. *Plant Physiol.* 150(5), 295–307.

Mian, M. A. R., Missaoui, A. M., Walker, D. R., Phillips, D. V. and Boerma. H. R. (2008), Frogeye leaf spot of soybean: A review and proposed race designations for isolates of *Cercospora sojina* Hara. *Crop Sci.* 48(1), 14–24.

Mian, M. A. R., Wang, T., Phillips, D. V., Alvernaz, J. and Boerma, H. R. (1999), Molecular mapping of the *Rcs3* gene for resistance to frogeye leaf spot of soybean. *Crop Sci.* 39(6), 1687–91.

Miles, M. R., Frederick, R. D. and Hartman, G. L. (2006), Evaluation of soybean germplasm for resistance to *Phakopsora pachyrhizi*. *Plant Health Prog.* doi:10.1094/PHP-2006-0104-01-RS.

Miles, M. R., Morel, W., Ray, J. D., Smith, J. R., Frederick, R. D. and Hartman, G. L. (2008), Adult plant evaluation of soybean accessions for resistance to *Phakopsora pachyrhizi* in the field and greenhouse in Paraguay. *Plant Dis.* 92(1), 96–105.

Minor, H. C., Brown, E. A. and Zimmerman, M. S. (1995), Developing soybean varieties with genetic resistance to *Phomopsis* spp. *J. Am. Oil Chem. Soc.* 72(12), 1431–4.

Mueller, D. S., Hartman, G. L., Nelson, R. L. and Pedersen, W. L. (2002), Evaluation of *Glycine max* germ plasm for resistance to *Fusarium solani* f. sp. *glycines*. *Plant Dis.* 86(7), 741–6.

Mueller, D. S., Nelson, R. L., Hartman, G. L. and Pedersen, W. L. (2003) Response of commercially developed soybean cultivars and the ancestral soybean lines to *Fusarium solani* f. sp. *glycines*. *Plant Dis.* 87(7), 827–31.

Muyolo, N. G, Lipps, P. E. and Schnitthenner, A. F. (1993), Anastomosis grouping and variation in virulence among isolates of *Rhizoctonia solani* associated with dry bean and soybean in Ohio and Zaire. *Phytopathology* 83, 438–44.

Narvel, J. M., Jakkula, L. R., Phillips, D. V., Wang, T., Lee, S.-H. and Boerma, H. R. (2001), Molecular mapping of *Rxp* conditioning reaction to bacterial pustule in soybean. *J. Hered.* 92(3), 267–70.

Nelson, B. D., Helms, T. C. and Kural, I. (1991), Effect of temperature and pathogen isolate on laboratory screening of soybean resistance to *Sclerotinia sclerotiorum*. *Can. J. Plant Sci.* 71, 347–52. https://doi.org/10.4141/cjps91-049

Niblack, T. L., Arelli, P. R., Noel, G. R., Opperman, C. H., Orf, J. H., Schmidt, D. P., Shannon, J. G. and Tylka, G. L. (2002), A revised classification system for genetically diverse populations of *Heterodera glycines*. *J. Nematol.* 34, 279–88.

Niblack, T. L., Colgrove, A. L., Colgrove, K. and Bond, J. P. (2008), Shift in virulence of soybean cyst nematode is associated with use of resistance from PI 88788. *Plant Health Prog.* doi:10.1094/PHP-2008–0118-01-RS.

Niblack, T. L. and Riggs, R. D. (2015), Soybean cyst nematode, pp. 100–4. In *Compendium of Soybean Diseases and Pests,* G. L. Hartman, J. C. Rupe, E. J. Sikora, L. L. Domier, J. A. Davis and K. L. Steffey (Eds). APS Press, St. Paul, MN.

Niblack, T. L., Tylka, G. L. and Riggs, R. D. (2004), Nematode pathogens of soybean, pp. 821–51. In *Soybeans: Improvement, Production and Uses,* Boerma, H. R. and Specht, J. E. (Eds), Agronomy Monograph No. 16. American Society of Agronomy, Crop Science Society of America, Soil Science Society of America.

Nickell, C. D., Lim, S. M., Eathington, S. and Warsaw, R. (1994), Registration of soybean germplasm line LL89-605, resistant to brown stem rot and bacterial blight. *Crop Sci.* 34(4), 1134.

Oerke, E. C. (2006), Crop losses to pests. *J. Agric. Sci.* 144, 31–43.

Okubura, P. A., Dickman, M. B. and Blechl, A. E. (2014), Molecular and genetic aspects of controlling soilborne pathogens *Rhizoctonia* and *Pythium*. *Plant Sci.* 228, 61–70.

Pace, P. F., Weaver, D. B. and Ploper, D. L. (1993), Additional genes for resistance to frogeye leaf spot race 5 in soybean. *Crop Sci.* 33(6), 1144–5.

Paris, R. L., Mengistu, A., Tyler, J. M. and Smith, J. R. (2006), Registration of soybean germplasm line DT97-4290 with moderate resistance to charcoal rot. *Crop Sci.* 46(5), 2324–5.

Patzoldt, M. E., Grau, C. R., Stephens, P. A., Kurtzeil, N. C., Carlson, S. R. and Diers, B. D. (2005), Location of a quantitative trait locus providing brown stem rot resistance in the soybean cultivar Bell. *Crop Sci.* 45(4), 1241–8.

Paul, C., Frederick, R. D., Hill, C. B., Hartman, G. L. and Walker, D. R. (2015), Comparison of pathogenic variation among *Phakopsora pachyrhizi* isolates collected from the United States and international locations, and identification of soybean genotypes resistant to the U.S. isolates. *Plant Dis.* 99(8), 1059–69.

Paul, C., Hartman, G. L, Marois, J. J., Wright, D. L. and Walker, D. R. (2013), First report of *Phakopsora pachyrhizi* adapting to soybean genotypes with *Rpp1* or *Rpp6* resistance genes in field plots in the United States. *Plant Dis.* 97(10), 1379.

Pawlowski, M. L., Hill, C. B. and Hartman, G. L. (2015), Resistance to charcoal rot identified in ancestral soybean germplasm. *Crop Sci.* 55(3), 1230–5.

Perez, P. T., Diers, B. D., Lundeen, P., Tabor, G. M. and Cianzio, S. R. (2010), Genetic analysis of new sources of soybean resistance to brown stem rot. *Crop Sci.* 50(6), 2431–9.

Pham, A.-T., Harris, D. K., Buck, J., Hoskins, A., Serrano, J., Abdel-Haleem, H., Cregan, P., Song, Q., Boerma, H. R. and Li, Z. (2015), Fine mapping and characterization of candidate genes that control resistance to *Cercospora sojina* K. Hara in two soybean germplasm accessions. *PLoS ONE* 10(5), e0126753. doi:10.1371/journal.pone.0126753.

Phillips, D. V. and Boerma, H. R. (1982), Two genes for resistance to race 5 of *Cercospora sojina* in soybeans. *Phytopathology* 72, 764–6.

Pierozzi, P. H. B., Ribeiro, A. S., Moreira, Laperuta, L. D. C., Rachid, B. F., Lima, W. F., Arias, C. A. A., Oliveira, M. F. and Toledo, J. F. F. (2008), New soybean (*Glycine max* Fabales, Fabaceae) sources of qualitative genetic resistance to Asian soybean rust caused by *Phakopsora pachyrhizi* (Uredinales, Phakopsoraceae). *Genet. Mol. Biol.* 31(2), 505–11.

Polzin, K. M., Lohnes, D. G., Nickell, C. D. and Shoemaker, R. C. (1994), Integration of *Rps2, Rmd*, and *Rj2* into linkage group J of the soybean molecular map. *J. Hered.* 85, 300–3.

Riggs, R. D., Wang, S., Singh, R. J. and Hymowitz, T. (1998), Possible transfer of resistance to *Heterodera glycines* from *Glycine tomentella* to soybean. Supp. *J. Nematol.* 30(4S), 547–52.

Rincker, K., Lipka, A. E. and Diers, B. W. (2016), Genome-wide association study of brown stem rot resistance in soybean across multiple populations. *Plant Genome* 9(2), doi:10.3835/plantgenome2015.08.0064.

Ross, J. P. (1986), Registration of eight soybean germplasm lines resistant to seed infection by *Phomposis* spp. *Crop Sci.* 26(1), 210–11.

Rosso, M. L., Rupe, J. C., Chen, P. and Mozzoni, L. A. (2008), Inheritance and genetic mapping of resistance to Pythium damping-off caused by *Pythium aphanidermatum* in 'Archer' soybean. *Crop Sci.* 48, 2215–22.

Rothrock, C. S., Avanzato, M. V. and Rupe, J. C. (2015), Pythium seed rot, damping-off and root rot, pp. 76–9. In *Compendium of Soybean Diseases and Pests,* G. L. Hartman, J. C. Rupe, E. J. Sikora, L. L. Domier, J. A. Davis and K. L. Steffey (Eds). APS Press, St. Paul, MN.

Rupe, J. C., Gbur, E. E. and Marx, D. M. (1991), Cultivar responses to sudden death syndrome of soybean. *Plant Dis.* 75(1), 47–50.

Rupe, J. C., Sutton, E. A., Becton, C. M. and Gbur Jr., E. E. (1999), Effect of soybean growth stage at the time of inoculation with *Diaporthe phaseolorum* var. *meridionales* on stem canker development and yield. *Plant Dis.* 83(6), 582–6.

Sandhu, D., Shallock, K. G., Rivera-Valez, N., Lundeen, P., Cianzio, S. and Bhattacharyya, M. (2005), Soybean Phytophthora resistance gene *Rps*8 maps closely to the *Rps*3 region. *J. Hered.* 96, 536–41.

Schmitthenner, A. F. (1985), Problems and progress in control of Phytophthora root rot of soybean. *Plant Dis.* 69, 362–8.

Schmitthenner, A. F. and Dorrance, A. E. (2015), Phytophthora root and stem rot, pp. 73–6. In *Compendium of Soybean Diseases and Pests,* G. L. Hartman, J. C. Rupe, E. J. Sikora, L. L. Domier, J. A. Davis and K. L. Steffey (Eds). APS Press, St. Paul, MN.

Schmutz, J., Cannon, S. B., Schlueter, J., Ma, J., Mitros, T., Nelson, W., Hyten, D. L., Song, Q., Thelen, J. J., Cheng, J., Xu, D., Hellsten, U., May, G. D., Yu, Y., Sakurai, T., Umezawa, T., Bhattahcaryya, M. K., Sandhu, D., Valliyodan, B., Lindquist, E., Peto, M., Grant, D., Shu, S., Goodstein, D., Barry, K., Futrell-Griggs, M., Abernathy, B., Du, J., Tian, Z., Zhu, L., Gill, N., Joshi, T., Libault, M., Sethuraman, A., Zhang, X. C., Shinozaki, K., Nguyen, H. T., Wing, R. A., Cregan, P., Specht, J., Grimwood, J., Roshkar, D., Stacey, G., Shoemaker, R. C. and Jackson, S. A. (2010), Genome sequence of the palaeopolyploid soybean. *Nature* 463, 178–83.

Schneider, R. W, Hollier, C. A., Whitam, H. K., Palm, M. E., McKenny, J. M., Hernandez, J. R., Levy, L. and DeVries-Paterson, R. (2005), First report of soybean rust caused by *Phakopsora pachyrhizi* in the continental United States. *Plant Dis.* 89, 774. https://doi.org/10.1094/pd-89-0774a

Schneider, R., Rolling, W., Song, Q., Cregan, P., Dorrance, A. E. and McHale, L. K. (2016), Genome-wide association mapping of partial resistance to *Phytophthora sojae* in soybean plant introductions from the Republic of Korea. *BMC Genomics* 17(607).

Schoen, D. J., Burdon, J. J. and Brown, A. H. D. (1992), Resistance of *Glycine tomentella* to soybean leaf rust *Phakopsora pachyrhizi* in relation to ploidy level and geographic distribution. *Theor. Appl. Genet.* 83, 827–32.

Siegel, S. P., Zhao, Y. F. and Bradley, C. A. (2008), Race characterization of *Pseudomonas savastanoi* pv. *glycinea* in Illinois. *Phytopathology* 98, S146.

Sikora, E. J., Allen, T. W., Wise, K. A., Bergstrom, G., Bradley, C. A., Bond, J., Brown-Rytlewski, D., Chilvers, M., Damicone, J., DeWolf, E., Dorrance, A., Dufault, N., Esker, P., Faske, T. R., Giesler, L., Goldberg, N., Golod, J., Gomez, I. R. G., Grau, C., Grybauskas, A., Franc, G., Hammerschmidt, R., Hartman, G. L., Henn, A., Hershman, D., Hollier, C., Isakeit, T., Isard, S., Jacobson, B., Jardine, D., Kemerait, B., Koenning, S., Langham, M., Malvick, D., Markell, S., Marois, J. J., Monfort, S., Mueller, D., Mueller, J., Mulroony, R., Newman, M., Osborne, L., Padgett, G. B., Ruden, B. E., Rupe, J., Schneider, R., Swartz, H., Shaner, G., Singh, S., Stromberg, E., Sweets, L., Tenuta, A., Vaiciunas, S., Yang, X. B., Young-Kelly, H. and Zidek, J. (2014), A coordinated effort to manage soybean rust in North America: a success story in soybean disease monitoring. *Plant Dis.* 98, 864–75. https://doi.org/10.1094/pdis-02-14-0121-fe

Silva, D. C. G, Yamanaka, N., Brogin, R. L., Arias, C. A. A., Nepomuceno, A. L., Mauro, A. D., Pereira, S. S., Nogueira, L. M., Passionotto, A. L. L. and Abdelnoor, R. V. (2008), Molecular mapping of two loci that confer resistance to Asian rust in soybean. *Theor. Appl. Genet.* 117, 57–63. https://doi.org/10.1007/s00122-008-0752-0

Sinclair, J. B. (1993), Phomopsis seed decay of soybeans – a prototype for studying seed disease. *Plant Dis.* 77(4), 329–34.

Singh, R. J. and Nelson, R. L (2014), Methodology for creating alloplasmic soybean lines by using *Glycine tomentella* as a maternal parent. *Plant Breed.* 133(5), 624–31. https://doi.org/10.1111/pbr.12196

Smith, S., Fenn, P., Chen, P. and Jackson, E. (2008), Inheritance of resistance to Phomopsis seed decay in PI 360841 soybean. *J. Hered.* 99(6), 588–92. https://doi.org/10.1093/jhered/esn037

Song, Q., Hyten, D. L., Jia, G., Quigley, C. V., Fickus, E. W., Nelson, R. L. and Cregan, P. B. (2013), Development and evaluation of SoySNP50K, a high-density genotyping array for soybean. *PloS One* 8(1), e54985. doi:10.1371/journal.pone.0054985.

Srour, A. Y., Afzal. A. J., Saini, N., Blahut-Beatty, L., Hemmati, N., Simmonds, D. H., Li, W., Liu, M., Town, C. D., Sharma, H., Arelli, P. and Lightfoot, D. A. (2012), The receptor like kinase at *Rhg1-a/Rfs2* caused pleiotropic resistance to sudden death syndrome and soybean cyst nematode as a transgene by altering signaling responses. *BMC Genomics* 13, 368–78.

Stewart, S., Robertson, A. E., Wickharmasinghe, D., Draper, M. A., Michel, A. and Dorrance, A. E. (2016), Population structure among and within Iowa, Missouri, Ohio, and South Dakota populations of *Phytophthora sojae*. *Plant Dis.* 100(2), 367–79. doi:10.1094/PDIS-04-15-0437-RE.

Stupar, R. M. and Specht, J. E. (2013), Insights from the soybean (*Glycine max* and *Glycine soja*) genome: Past, present, and future. *Adv. Agron.* 118, 177–204. https://doi.org/10.1016/b978-0-12-405942-9.00004-9

Sugimoto, T., Kato, M., Yoshida, S., Matsumoto, I. Kobayashi, T. Kaga, A., Hajika, M. Yamamoto, R., Watanabe, K. Aino, M. Matoh, T., Walker, D. R., Biggs, A. R. and Ishimoto, M. (2012), Pathogenic diversity of *Phytophthora sojae* and breeding strategies to develop Phytophthora-resistant soybeans. *Breed. Sci.* 61, 511–22. https://doi.org/10.1270/jsbbs.61.511

Sugimoto, T., Yoshida, S., Kaga, A., Hajika, M., Watanabe, K., Aino, M., Tatsuda, K., Yamamoto, R., Matoh, T., Walker, D. R., Biggs, A. R. and Isimoto, M. (2011), Genetic analysis and identification of DNA markers linked to a novel *Phytophthora sojae* resistance gene in the Japanese soybean cultivar Waseshiroge. *Euphytica* 182, 133–45. https://doi.org/10.1007/s10681-011-0525-8

Sun, J., Guo, N., Lei J., Li, L., Hu, G. and Xing, H. (2014), Association mapping for partial resistance to *Phytophthora sojae* in soybean (*Glycine max* (L.) Merr.). *J. Genet.* 93(2), 355–63. https://doi.org/10.1007/s12041-014-0383-y

Sun, S., Kim, M. Y., Van, K., Lee, Y.-W., Li, B. and Lee, S.-H. (2013), QTLs for resistance to Phomopsis seed decay are associated with days to maturity in soybean (*Glycine max*). *Theor. Appl. Genet.* 126, 2029–38. https://doi.org/10.1007/s00122-013-2115-8

Tamulonis, J. P., Luzzi, B. M., Hussey, R. S., Parrott, W. A. and Boerma, H. R. (1997a), DNA marker analysis of loci conferring resistance to peanut root-knot nematode in soybean. *Theor. Appl. Genet.* 95:664–70. https://doi.org/10.1007/s001220050610

Tamulonis, J. P., Luzzi, B. M., Hussey, R. S., Parrott, W. A. and Boerma, H. R. (1997b), DNA markers associated with resistance to Javanese root-knot nematode in soybean. *Crop Sci.* 37(3), 783–8.

Tolin, S. A. and Lacey, G. H. (2004), Viral, bacterial, and phytoplasmal diseases of soybean, pp. 765–819. In *Soybeans: Improvement, Production, and Uses*, 3rd ed., Boerma, H. R. and Specht, J. E. (Eds), Agronomy Monograph no. 16. American Society of Agronomy, Madison, Wisconsin, USA.

Tooley, P. W. and Grau, C. R. (1984), Field characterization of rate-reducing resistance to *Phytophthora megasperma* f. sp. *glycinea* in soybean. *Phytopathology* 74, 1201–8.

Triwitayakorn, K., Njiti, V. N., Iqbal, M. J., Yaegashi, S., Town, C. and Lightfoot, D. A. (2005), Genomic analysis of a region encompassing *QRfs1* and *QRfs2*: genes that underlie soybean resistance to sudden death syndrome. *Genome* 48, 125–38. doi:10.1139/G04-103.

Tucker, D. M., Saghai Maroof, M. A., Mideros, S., Skoneczka, J. A., Nabati, D. A., Buss, G. R., Hoeschele, I., Tyler, B. M., St. Martin, S. K. and Dorrance, A. E. (2010), Mapping quantitative trait loci for partial resistance to *Phytophthora sojae* in a soybean interspecific cross. *Crop Sci.* 50(2), 628–35.

Twizeyimana, M., Hill, C. B., Pawlowski, M., Paul, C. and Hartman, G. L. (2012), A cut-stem inoculation technique to evaluate soybean for resistance to *Macrophomina phaseolina*. *Plant Dis.* 96(8), 1210–15. https://doi.org/10.1094/pdis-02-12-0126-re

Tylka, G. L. and Marett, C. C. (2014), Distribution of the soybean cyst nematode, *Heterodera glycines*, in the United States and Canada: 1954 to 2014. *Plant Health Prog.* Online. Plant Management Network. doi:10.1094/php-BR-0006.

Tylka, G. L. and Mullaney (2015), Soybean cyst nematode-resistant soybean varieties for Iowa. Iowa State University, Extension and Outreach.

Uzuhashi, S., Tojo, M. and Kakishima, M. (2010), Phylogeny of the genus *Pythium* and description of new genera. *Mycoscience* 51, 337–65. https://doi.org/10.1007/s10267-010-0046-7

van de Mortel, M., Recknor, J. C., Graham, M. A., Nettleton, D., Dittman, J. D., Nelson, R. T., Godoy, C. V., Abdelnoor, R. V., Almeida, A. M. R. and Baum, T. J. (2007), Distinct biphasic mRNA changes in response to Asian soybean rust infection. *Mol. Plant Microbe Interact.* 20, 887–99. https://doi.org/10.1094/mpmi-20-8-0887

Vuong, T. D, Hoffman, D. D., Diers, B. W., Miller, J. K., Steadman, J. R. and Hartman, G. L. (2004), Utilization of the cut stem inoculation method to evaluate soybean, dry bean, and sunflower for resistance to *Sclerotinia sclerotiorum*. *Crop Sci.* 44(3), 777–83.

Vuong, T. D., Sleper, D. A., Shannon, J. G. and Nguyen, H. T. (2010), Novel quantitative trait loci for broad-spectrum resistance to soybean cyst nematode (*Heterodera glycines* Ichinohe) in soybean PI 567654. *Theor. Appl. Genet.* 121, 1253–66. https://doi.org/10.1007/s00122-010-1385-7

Vuong, T. D., Sonah, H., Meinhardt, C. G., Deshmukh, R., Kadam, S., Nelson, R. L., Shannon, J. G. and Nguyen, H. T. Genetic architecture of cyst nematode resistance revealed by genome-wide association study in soybean. *BMC Genomics* 16, 593. doi:10.1186/s12864-015-1811-y.

Walker, D. R., Boerma, H. R., Phillips, D. V., Schneider, R. W., Buckley, J. B., Ship, E. R., Mueller, J. D., Weaver, D. B, Sikora, E. J., Moore, S. H., Hartman, G. L., Miles, M. R., Harris, D. K., Wright, D. L., Marois, J. J. and Nelson, R. L. (2011), Evaluation of USDA soybean germplasm accessions for resistance to soybean rust in the southern United States. *Crop Sci.* 51(2), 678–93.

Walker, D. R., Harris, D. K., King, Z. R., Li, Z., Boerma, H. R., Buckley, J. B., Weaver, D. B., Sikora, E. J., Shipe, E. R., Mueller, J. D., Buck, J. W., Schneider, R. W., Marois, J. J., Wright, D. L. and Nelson, R. L. (2014a), Evaluation of soybean germplasm for resistance to *Phakopsora pachyrhizi* populations in the southeastern United States, 2009–2012. *Crop Sci.* 54(5), 1673–89.

Walker, D. R., Harris, D. K., King, Z. R., Li, Z., Phillips, D. V., Buck, J. W., Nelson, R. L. and Boerma, H. R. (2014b), Soybean germplasm accession seedling reactions to soybean rust isolates from Georgia. *Crop Sci.* 54(4), 1433–47.

Wang, D., Arelli, P. R., Shoemaker, R. C. and Diers, B. W. (2001), Loci underlying resistance to Race 3 of soybean cyst nematode in *Glycine soja* plant introduction 468916. *Theor. Appl. Genet.* 103, 561–6. https://doi.org/10.1007/pl00002910

Webb, D. M. inventor; Pioneer Hi-Bred International, Inc., assignee (2003), Quantitative trail loci associated with cyst nematode resistance and uses thereof. U.S. Patent 6,538,175. Date issued 25 March 2003.

Webb, D. M., Baltazar, B. M., Rao-Arelli, A. P., Schupp, J., Clayton, K., Keim, P. and Beavis, W. D. (1995), Genetic mapping of soybean cyst nematode race-3 resistance loci in soybean PI 437.654. *Theor. Appl. Genet.* 91, 574–81. https://doi.org/10.1007/bf00223282

Wegulo, S. N., Yang, X. B. and Martinson, C. A. (1998), Soybean cultivar responses to *Sclerotinia sclerotiorum* in field and controlled environment studies. *Plant Dis.* 82(11), 1264–70. https://doi.org/10.1094/pdis.1998.82.11.1264

Weng, C., Yu, K. Anderson, T. and Poysa, V. (2001), Mapping gene conferring resistance to Phytophthora root rot of soybean, *Rps*1a and *Rps*7. *J. Hered.* 92, 442–6. https://doi.org/10.1093/jhered/92.5.442

Weng, C., Yu, K., Anderson, T. R. and Poysa, V. (2007), A quantitative trait locus influencing tolerance to Phytophthora root rot in the soybean cultivar 'Conrad'. *Euphytica* 158, 81–6. https://doi.org/10.1007/s10681-007-9428-0

Whitaker, J. (2015), 2015 Georgia soybean production guide. (http://www.caes.uga.edu/commodities/fieldcrops/soybeans/documents/2015GeorgiaSoybeanProductionGuide-website.pdf).

Willmot, D. B. and Nickell, C. D. (1989), Genetic analysis of brown stem rot resistance in soybean. *Crop Sci.* 29(3), 672–4.

Wise, K. A. and Newman, M. E. (2015), Frogeye leaf spot, pp. 43–5. In *Compendium of Soybean Diseases and Pests,* G. L. Hartman, J. C. Rupe, E. J. Sikora, L. L. Domier, J. A. Davis and K. L. Steffey (Eds). APS Press, St. Paul, MN.

Wrather, J. A., Anderson, T. R., Arsyad, D. M., Gai, J., Ploper, L. D., Porta-Puglia, A., Ram, H. H. and Yorinori, J. T. (1997), Soybean disease loss estimates for the tope 10 soybean producing countries in 1994. *Plant Dis.* 81(1), 107–10. https://doi.org/10.1094/pdis.1997.81.1.107

Wrather, J. A. and Koenning, S. R. (2006), Estimates of disease effects on soybean yields in the United States 2003 to 2005. *J. Nematol.* 38(2), 173–80.

Wrather, J. A. and Koenning, S. R. (2009), Effects of diseases on soybean yields in the United States 1996 to 1997. Online. *Plant Health Prog.* doi:10.1094/PHP-2009-0401-01-RS.

Wrather, A., Shannon, G., Balardin, R., Carregal, L., Escobar, R., Gupta, G. K., Ma, Z., Morel, W., Ploper, D., and Tenuta, A. (2010), Effect of diseases on soybean yield in the top eight producing countries in 2006. Online. *Plant Health Prog.* doi:1094/PHP-2010–0125-01-RS.

Wu, X., Blake, S., Sleper, D. A., Shannon, J. G., Cregan, P. and Nguyen, H. T. (2009), QTL, additive and epistatic effects for SCN resistance in PI 437654. *Theor. Appl. Genet.* 118, 1093–105. https://doi.org/10.1007/s00122-009-0965-x

Xu, X., Zeng, L., Tao, Y., Vuong, T., Wan, J., Boerma, R., Noe, J., Li, Z., Finnerty, S., Pathan, S. M., Shannon, J. G. and Nguyen, H. T. (2013), Pinpointing genes underlying the quantitative trait loci for root-knot nematode resistance in paleopolyploid soybean by whole genome resequencing. *PNAS* 110(33), 13469–74. https://doi.org/10.1073/pnas.1222368110

Yamanaka, N., Yamaoka, Y., Kato, M., Lemos, N., Passianotto, A. L. de L., dos Santos, J. V. M, Benitez, E. R., Abdelnoor, R. V., Soares, R. M. and Suenaga, K. (2010), Development of classification criteria for resistance to soybean rust and differences in virulence among Japanese and Brazilian rust populations. *Trop. Plant Pathol.* 35(3), 153–62. https://doi.org/10.1590/s1982-56762010000300003

Yang, H.-C. and Hartman, G. L. (2015a), Anthracnose, pp. 31–4. In *Compendium of Soybean Diseases and Pests,* G. L. Hartman, J. C. Rupe, E. J. Sikora, L. L. Domier, J. A. Davis and K. L. Steffey (Eds). APS Press, St. Paul, MN.

Yang, H.-C. and Hartman, G. L. (2015b), Methods and evaluation of soybean genotypes for resistance to *Colletotrichum truncatum. Plant Dis.* 99(1), 143–8. https://doi.org/10.1094/pdis-03-14-0228-re

Yang, X. B. and Hartman, G. L. (2015), Rhizoctonia damping-off and root rot, pp. 80–2. In *Compendium of Soybean Diseases and Pests,* G. L. Hartman, J. C. Rupe, E. J. Sikora, L. L. Domier, J. A. Davis and K. L. Steffey (Eds). APS Press, St. Paul, MN.

Yates, J. L., Hussey, R. S., Finnerty, S. L. and Boerma, H. R. (2010), Three soybean plant introductions possess unique resistance to peanut root-knot nematode. *Crop Sci.* 50(1), 118–22.

Yorinori, J. T., Paiva, W. M., Frederick, R. D., Costamilan, L. M., Bertagnoli, P. F., Hartman, G. L., Godoy, C. V. and Nunes Jr., J. (2005), Epidemics of soybean rust (*Phakopsora pachyrhizi*) in Brazil and Paraguay from 2001 to 2003. *Plant Dis.* 89, 675–7. https://doi.org/10.1094/pd-89-0675

Yuan, J., Wen, Z., Gu, C. and Wang, D. (2014), Introduction of high throughput and cost effective SNP genotyping platforms in soybean. *Plant Genet. Genom. Biotechnol.* 2(1), 90–4. https://doi.org/10.5147/pggb.2014.0144

Zeng, F., Arnao, E., Zhang, G., Olaya, G., Wullschleger, J., Sierotzky, H, Ming, R., Bluhm, B. H., Bond, J. P., Fakhoury, A. M. and Bradley, C. A. (2015), Characterization of quinone outside inhibitor fungicide resistance in *Cercospora sojina* and development of diagnostic tools for its identification. *Plant Dis.* 99(4), 544–50. https://doi.org/10.1094/pdis-05-14-0460-re

Zhao, Y. F. (2015), Bacterial blight, pp. 18–19. In *Compendium of Soybean Diseases and Pests,* G. L. Hartman, J. C. Rupe, E. J. Sikora, L. L. Domier, J. A. Davis and K. L. Steffey (Eds). APS Press, St. Paul, MN.

Chapter 6

Advances in pest-resistant varieties of soybean

Shichen Zhang and Dechun Wang, Michigan State University, USA

1 Introduction
2 Mechanisms of HPR
3 Insect resistance in the soybean germplasm
4 Breeding challenges
5 *Bacillus thuringiensis* (Bt) soybean
6 Case study: developing aphid-resistant varieties
7 Conclusion
8 Future trends
9 Where to look for further information
10 References

1 Introduction

Soybean has been one of the most important crops in the United States as it has multiple uses including animal feed, cooking oil, biofuel, human protein source and so on. In 2016, the United States ranked first in world soybean production (4.31 billion bushels) with 2.03 billion bushels exported (SoyStats, 2016). However, soybean production has been threatened by an increasing number of insect pests including native species as well as invasive species.

In the United States, soybean has been traditionally attacked by 36 insect species, but only eight of these have been considered economically important pests (Lambert and Tyler, 1999). These important insect pests are in the orders of Lepidoptera, Coleoptera and Hemiptera. Lepidopteran and coleopteran pests cause yield losses by chewing on foliage or reproductive structures, whereas hemipteran pests jeopardize soybean production by sucking plant sap through their stylet-feeding mechanism.

Corn earworm (*Helicoverpa zea*), soybean looper (*Pseudoplusia includens*) and velvetbean caterpillar (*Anticarsia gemmatalis*) are the three major foliage-feeding lepidopterans that cause significant yield losses in the Southeast and Delta regions (Boerma and Walker, 2005). Larvae of these lepidopterans cause defoliation through direct foliage feeding. Larvae of corn earworm, however, also feed on soybean flowers, pods and developing

http://dx.doi.org/10.19103/AS.2017.0034.21

seeds. Thus, corn earworm has been considered as the most important lepidopteran pest in the southern states (Parrott et al., 2008).

In the order of Coleoptera, Mexican bean beetle (*Epilachna varivestis*) and bean leaf beetle (*Cerotoma trifurcata*) have been considered important soybean pests (Turnipseed and Kogan, 1987). Both larvae and adult Mexican bean beetles chew on foliage, sometimes causing significant defoliation that decreases soybean yield. Likewise, adult bean leaf beetles damage soybean production from defoliation. However, the larvae of bean leaf beetles feed on soybean roots and nodules, occasionally resulting in yield reduction. In addition, bean leaf beetles are also known to transmit *Bean pod mottle virus* and facilitate fungal invasion through the wounds (Turnipseed and Kogan, 1987).

Currently, the major hemipteran pests of soybean in the United States include the stink bug complex from the Pentatomidae family, soybean aphids from the Aphididae family and kudzu bugs from the Plataspidae family. Three stink bug species, including southern green stink bug (*Nezara viridula*), green stink bug (*Acrosternum hilare*) and brown stink bug (*Euschistus servus*), used to constitute the stink bug complex that damaged soybeans in the southern United States (McPherson and McPherson, 2000). The stink bug complex composition recently changed to include one newly established species, redbanded stink bug (*Piezodorus guildinii*), and an exotic species, brown marmorated stink bug (*Halyomorpha halys*). The redbanded stink bug did not cause significant economic losses until 2009, when it became the most destructive stink bug species in the southern United States (Musser et al., 2010; Temple et al., 2011). The brown marmorated stink bug, native to Asia, was first collected in Pennsylvania in 1998 and had been reported from 38 states by 2012 (Leskey et al., 2012). The brown marmorated stink bug population exploded in 2010 and caused severe yield losses (>50%) to soybean production in several Mid-Atlantic states (Leskey et al., 2012). Soybean aphids (*Aphis glycine*), native to China and first reported in Wisconsin in 2000, had spread to 30 states and 3 Canadian provinces by 2009 (Ragsdale et al., 2011). Kudzu bug (*Megacopta cribraria*), native to Asia, was first detected in Georgia in 2009 (Suiter et al., 2010) and has become an important economic pest in the southeastern United States (Ruberson et al., 2013).

Hemipteran pests damage soybean yield mainly through their direct feeding on plant sap with their piercing-sucking mouthparts. Stink bug nymphs and adults primarily attack the pods and seeds of soybean, often causing 'green soybean syndrome' (delayed senescence) through their feeding (Harbach et al., 2016). In contrast, kudzu bugs rarely feed on pods; adults and nymphs primarily feed on the main stem and petioles, leaving brown or purplish lesions (Dhammi et al., 2016). Soybean aphids feed on the whole plant with a preference for new growing parts. Additionally, virus transmitted by aphids often cause secondary yield loss (Hill et al., 2001; Clark and Perry, 2002). Both kudzu bugs and aphids secrete honeydew that leads to the growth of a fungus called sooty mould, causing additional yield loss by blocking the light reception of the leaves (Malumphy, 1997; Lemos Filho and Paiva, 2006).

To control these insect pests, insecticide application on soybean increased from 0.03 million pounds in 2000 to 4.7 million pounds in 2008 (Fernandez-Cornejo et al., 2014). Although insecticide application is effective in protecting soybean yield, this control method increases agricultural input, causes environmental contamination and jeopardizes beneficial insects such as natural enemies and pollinators (Ohnesorg et al., 2009; Lundin et al., 2015). Sometimes, the extensive use of insecticides leads to the appearance of pesticide-resistant insect populations, resurgence of primary pests and secondary pests.

An alternative way of controlling insect pests is to employ the natural host-plant resistance (HPR) that can provide soybean with economically and environmentally friendly season-long protection. Thus, HPR is an important component of integrated pest management (IPM) which aims at limiting the usage of pesticides. By identifying resistant germplasm and the underlying resistant genes, plant breeders have been able to incorporate genetic resistances into elite cultivars. The most successful example of utilizing HPR in soybean was the discovery and incorporation of the trait of dense pubescence that confers the resistance against potato leafhopper (*Empoasca fabae*) (Johnson and Hollowell, 1935). The use of pubescent soybean cultivars has suppressed potato leafhopper population to a non-pest status over the past 80 years in the United States (Boethel, 1999).

There are three categories of HPR, including antibiosis, antixenosis and tolerance (Painter, 1958). Antibiosis inhibits the insect biological and/or reproductive process through toxic plant secondary metabolites. Antixenosis morphologically or biochemically deters pests based on a 'non-preference' behaviour. Tolerance refers to the ability of the plant to maintain its yield under a moderate amount of damage from the pest. However, these categories are sometimes not mutually exclusive.

HPRs against different insect pests have been identified. However, one major drawback of employing HPR is the potential lack of its durability. Besides continually seeking for new resistance sources, several methods of sustaining HPR have been practised, including resistance gene pyramiding, variety mixture and resistance gene deployment based on biotype distribution. This chapter will review the recent research advances on improving soybean with HPR in North America, and discuss challenges of developing highly insect-resistant cultivars with competitive yield performance.

2 Mechanisms of HPR

2.1 Pubescence-related resistance

The damage caused by potato leafhoppers is described as 'hopperburn'; damaged leaves curl with yellow necrotic margins and stunting under heavy infection. Johnson and Hollowell (1935) reported that 'hairy' soybeans with normal pubescence showed little to no 'hopperburn' whereas glabrous soybeans showed severe 'hopperburn'. Years later, researchers studied this pubescence-related resistance more closely and uncovered that density, length and erect orientation of trichomes are positively correlated with potato leafhopper resistance (Singh et al., 1971; Broersma et al., 1972; Turnipseed, 1977). A plant introduction (PI) from Japan, PI 229358, has high levels of resistance to multiple insects including potato leafhopper because of its long and erect trichomes (Turnipseed, 1977). A few quantitative trait loci (QTLs) associated with pubescence density have been identified in the soybean genome and listed on SoyBase (Grant et al., 2010). The incorporation of pubescence trait into commercial varieties has relegated the potato leafhopper to a non-pest status in the United States (Boethel, 1999).

Pubescence-related resistance has also shown to be influenced by pubescence characteristics in studies of whiteflies (*Bemisia argentifolii*) and lepidopteran pests. Soybean genotypes with dense pubescence showed preference FOR oviposition from multiple insects such as whiteflies, bean flies (*Ophiomyia phaseoli*), pod borers (*Maruca vitrata*), corn earworms, velvetbean caterpillar and soybean looper (Beach and Todd 1988; Lambert and Kilen, 1989; McAuslane, 1996). However, the presence of pubescence inhibits the development of the larvae of lepidopteran defoliators (Lambert and Kilen,

1989; Beach and Todd, 1988). In addition, Hulburt et al. (2004) reported that soybean genotypes with sharp pubescence tip had significantly reduced defoliation from soybean looper, corn earworm and beet armyworm (*Spodoptera exigua*), and reduced weight gain of corn earworm. Lam and Pedigo (2001) reported that densely pubescent soybean isolines had significantly lower feeding on pods from bean leaf beetle and that feeding increased when the trichomes on pods were removed. However, dense pubescence did not provide protection against soybean aphids (Hill et al., 2004).

2.2 Biochemical resistance

Allelochemicals are secondary plant metabolites that are not essential for plant growth and development. Allelochemicals with negative allelopathic effects are known as important for plant defence against herbivory (Stamp, 2003). Phytoalexins are antibiotic metabolites that are produced or enriched under biotic stress (Hart et al., 1983). Some flavonoids are important phytoalexins with antiherbivory effects in *Glycine* spp. (Burden and Norris, 1992). For example, the inducible resistance against Mexican bean beetles in PI 227687 was due to increased phenylpropanoid metabolism (total phenolic content) when this PI was challenged by Mexican bean beetles (Chiang et al., 1987). A similar result was reported by Hart et al. (1983) that glyceolin (a type of flavonoid) was functioning as a deterrent to Mexican bean beetles and some other coleopterans. Coumestrol, another isoflavonoid, was suggested as contributing to antixenosis resistance against Mexican bean beetles in the cultivar 'Davis' (Burdern and Norris, 1992; Caviness and Walters, 1966). Genistein and daidzin can cause greater mortality; lower initial larval and pupal weight; reduced growth; and elongated larval cycle to southern green stink bugs, cabbage loopers and velvetbean caterpillars (Sharma and Norris 1991; Hoffmann-Campo et al., 2001; Piubelli et al., 2003, 2005).

Proteinase inhibitors are known as being involved in plant defence to herbivory injury; they are peptides or proteins that inhibit activities of digestive enzymes in insect gut, thus adversely affecting protein digestion and impeding the growth of insects (Ryan, 1990). Broadway and Duffey (1986) reported both soybean trypsin inhibitor and potato proteinase inhibitor significantly reduced the growth of larval beet armyworm and larval corn earworm. The soybean trypsin-chymotrypsin inhibitors induced significant mortality and growth inhibition of the pea aphid (*Acyrthosiphon pisum*) and potato aphid (*Macrosiphum euphorbiae*) (Rahbé et al., 2003; Azzouz et al., 2005). Potato proteinase inhibitors I and II increased mortality among late instar aphids and reduced production of nymphs in feeding trials of cereal aphid species (*Diuraphis noxia, Schizaphis graminum, Rhopalosiphum padi*) (Tran et al., 1997).The insecticidal effects of protein inhibitors were demonstrated in several transgenic plants, enhancing HPR against lepidopteran and coleopteran pests (Hilder et al., 1987; Johnson et al., 1989; De Leo et al., 2001; Falco and Silva-Filho, 2003; Lecardonnel et al., 1999; Alfonso-Rubí et al., 2003).

3 Insect resistance in the soybean germplasm

3.1 HPR against coleopteran pests

A two-year field screening of soybean genotypes from maturity groups VII and VIII against Mexican bean beetles in 1968 and 1969 uncovered three PIs from Japan, PI 171451 ('Kosamame'), PI 229358 ('Soden-daizu') and PI 227687 ('Miyako White'), with strong antibiosis resistance (Van Duyn et al., 1971). All three PIs were later reported as highly resistant to bean leaf beetles and PI 227687 additionally exhibited a high level of resistance to striped blister beetles (*Epicauta vittata*) (Clark et al., 1972). In that study, two additional PIs (PI 165989 and PI 229321) were rated as highly resistant to bean leaf beetles. Layton et al. (1987) reported similar results, but surprisingly observed that the cultivar 'Davis', which was frequently used as the susceptible check in screening trials of soybean genotypes against lepidopterans, exhibited greater resistance than these three Japanese PIs. In an evaluation test of 15 commercial cultivars for resistance against bean leaf beetles, 'Davis' was proven as having the strongest resistance (Jensen, 1968).

Smith et al. (1975) registered a cultivar, 'Shore', with both antibiosis and antixenosis resistance against Mexican bean beetles. However, 'Shore' was not derived from any of the three resistant Japanese PIs. Kraemer et al. (1988) screened all soybean PIs available by 1980 in MGs VI, VII and VIII for resistance against Mexican bean beetles. In MG VIII, PI 417061 and PI 417136 were determined to have stronger resistance than PI 227687, which was the resistant check in that maturity group. Additionally, new resistance sources FC 31665, PI 379621, PI 416925 and PI 416937 were identified for the first time in MG VI. The same research group evaluated PIs in MGs III, IV and V for resistance to Mexican bean beetles and identified three PIs (360847, 248511 and 407820) with moderate resistance and three PIs (423901-1, 417235 and 200450) with high levels of resistance in MG IV and MG V, respectively (Kraemer et al., 1990).

Although multiple soybean accessions have been reported as resistant to bean leaf beetles and Mexican bean beetles, QTLs related to the resistance have not been reported. Sisson et al. (1976) reported that inheritances of resistance to Mexican bean beetles in PI 171451, PI 229358 and PI 227687 were quantitative. Recently, a genome-wide association (GWA) study using the public SoySNP50K genotyping data (Song et al., 2013) and phenotypic data in the Germplasm Resources Information Network database (ars-grin.gov) discovered a significant single-nucleotide polymorphism (SNP), ss715609849, on chromosome 11 that is highly associated with resistance against Mexican bean beetles (Chang and Hartman, 2017). PI 171451 was later reported as resistant to the Japanese beetle, which has not been reported as a threat to soybean production in the United States (Hammond and Cooper, 1989). The Japanese beetle resistance from cultivar 'Forrest' (which is also resistant to soybean cyst nematode) was reported as controlled by a mixture of major and minor QTLs (Yesudas et al., 2010).

3.2 HPR against lepidopteran pests

Foliar-feeding lepidopterans (corn earworm, soybean looper and velvetbean caterpillar) are economically important pests in the southern United States (Boethel, 1999). After PI 171451, PI 229358 and PI 227687 were discovered as resistant to beetles (Mexican bean beetle, bean leaf beetle and striped blister beetle) (Van Duyn et al., 1971; Clark et al., 1972), they were frequently included in germplasm screenings and exhibited resistance to multiple lepidopteran pests including soybean looper, corn earworm, velvetbean caterpillar, cabbage looper (*Trichoplusia ni*) and tobacco budworm (*Heliothis virescens*) (Kilen et al., 1977; Luedders and Dickerson, 1977; Lambert and Kilen, 1984a).

Insect resistances from the three Japanese PIs were estimated as quantitatively inherited, with a few major genes involved (Sisson et al., 1976; Rufener et al., 1989; Kenty et al., 1996). By using restriction fragment length polymorphisms, Rector et al. (1998, 1999, 2000) detected QTLs underlying corn earworm resistance in three segregating populations developed from crossing a susceptible cultivar ('Cobb') to PI 171451, PI 227687 and PI 229358, respectively. Each of these three Japanese PIs possesses both shared and unique QTLs.

A major QTL on linkage group (LG) M (QTL-M) was detected as conditioning resistance to corn earworm from both PI 171451 and PI 229358. QTL-M was estimated as accounting for 37% of the variation in antixenosis resistance in these two PIs and 22%/28% of the variation in antibiosis resistance in PI 229358/PI 171451 (Rector et al., 1998, 1999, 2000). Narvel et al. (2001) confirmed QTL-M with simple sequence repeat (SSR) markers in the 'Cobb' × PI 229358 mapping population used by Rector et al. (1998, 2000), and found that 13 of 15 insect-resistant genotypes (cultivars and breeding lines), derived from PI 229358 or PI 171451, had introgressed QTL-M with various extent of linkage drag. Years later, QTL-M was fine mapped to a 0.52-cM segment on LG M between markers Sat_258 and Satt702 (Zhu et al., 2006).

In addition to QTL-M, QTLs on other LGs were identified with resistance to corn earworm and/or other lepidopterans. A QTL on LG H (QTL-H) conditioning antixenosis resistance against corn earworm was shared among all three PIs (171451, 227687 and 229358), and accounted for the variation in resistance (R^2) ranging from 9 to 19%. QTL-H was also suggested as possibly resistant to soybean looper and velvetbean caterpillar (Narvel et al., 2001; Boerma and Walker 2005). Rector et al. (2000) identified a QTL conditioning antibiosis resistance to corn earworm on LG G (QTL-G) in the 'Cobb' × PI 229358 population and its R^2 was 19%. However, both QTLs H and G only provide insect resistance in the presence of QTL-M (Zhu et al., 2006). Minor insect-resistant QTLs on several other LGs (C2, D1b, F and B2) derived from these three PIs were also detected by Rector et al. (1998, 1999 and 2000).

Surprisingly, one QTL conferring antixenosis resistance on LG F (R^2 = 20%) and one QTL conferring antibiosis resistance on LG J (R^2 = 19%) were found in the susceptible parent, 'Cobb' (Rector et al., 1999, 2000). It is not unusual to see favourable alleles contributed by parents with unfavourable phenotype. Terry et al. (1999, 2000) reported that recombinant lines derived from a cross between two non-resistant parents ('Minsoy' from China and 'Noir 1' from Hungary) were comparable in insect resistance to the three resistant Japanese PIs (171451, 227687 and 229358). And QTLs underlying corn earworm resistance were detected in this RIL population with most of the resistant alleles derived from the 'Minsoy' parent (Terry et al., 1999, 2000). The insect-resistant QTLs identified in susceptible parents suggested that useful insect-resistant alleles may exist in phenotypically insect-susceptible elite germplasm and could be transferred to other elite cultivars with minimal linkage drag.

In a mapping population developed from a cross of PI 227687 × 'Cobb', Hulburt et al. (2004) identified a QTL conveying both antibiosis and antixenosis resistance to corn earworm on LG E near the classical gene *Pb*, which conditions pubescence tip morphology. In that study, the authors observed that the sharp pubescence tip significantly reduced defoliation (antixenosis resistance) from corn earworm, beet armyworm and soybean looper as well as the weight gain (antibiosis resistance) of corn earworm. Therefore, the authors suggested that the introgression of sharp pubescence tip trait into elite cultivars would help improve resistance against these lepidopteran pests.

Soybean resistance to corn earworm has been suggested as highly correlated with resistance to other soybean defoliators (All et al., 1989; Rowan et al., 1991). For example, QTL-M which confers resistance to corn earworm also provides resistance against soybean looper (Zhu et al., 2008). Therefore, QTLs resistant to corn earworm (Table 1)

Table 1 Reported QTLs conferring resistance to corn earworm

QTL name[‡]	Resistance source(s)	LG/Chr.	Flanking markers	R^2(%)	Resistance modality	Reference(s)
Corn earworm 1-1, 2-3, 4-1, 6-3, 9-4, 9-5	PI 229358 and PI 171451	M/7	Sat_258 to Satt702	22–37	Antibiosis and Antixenosis	Rector et al., 1998, 1999, 2000; Narvel et al., 2001; Zhu et al., 2006
Corn earworm 8-2	'Minsoy'	M/7	Satt567	7	Antibiosis	Terry et al., 2000
Corn earworm 1-2, 2-2, 3-2, 9-3	PIs 229358, 171451 and 227687	H/12	Sat_122 to Satt541	9–19	Antixenosis	Rector et al., 1998, 1999; Narvel et al., 2001; Zhu et al., 2006
Corn earworm 7-2	'Minsoy'	H/12	Satt302 to Satt353	5–9	Antibiosis	Terry et al., 2000
Corn earworm 1-3, 9-1	PI 229358	D1b/2	Satt141 to Satt290	10	Antixenosis	Rector et al., 1998, 1999; Narvel et al., 2001
Corn earworm 2-1	'Cobb'	F/13	B212V-1 to A757V-2	20	Antixenosis	Rector et al., 1999
Corn earworm 5-2*	PI 227687	F/13	Sat_090 to Sat_074	12	Antibiosis	Rector et al., 2000; Hulburt, 2001
Corn earworm 3-1	PI 227687	C2/6	A132T to A670T	11	Antixenosis	Rector et al., 1999
Corn earworm 8-3	'Minsoy'	C2/6	Satt365	7	Antibiosis	Terry et al., 2000
Corn earworm 5-1	PI 227687	B2/14	A343b to Satt126	12–20	Antibiosis	Rector et al., 2000; Hulburt, 2001
NA	PI 227687	B2/14	Sat_230 to Satt126	17	Antixenosis	Hulburt, 2001
Corn earworm 6-1, 9-2	PI 229358	G/18	Satt472 to Satt191	19	Antibiosis	Rector et al., 2000; Narvel et al., 2001; Zhu et al., 2006
Corn earworm 6-2	'Cobb'	J/16	A064V to K411T-1	19	Antibiosis	Rector et al., 2000
Corn earworm 7-4	'Noir 1'	J/16	A060_1	4.8	Antibiosis	Terry et al., 2000
Corn earworm 7-1	'Minsoy'	E/15	Sat_112	9–17	Antibiosis	Terry et al., 2000
Corn earworm 8-1	'Minsoy'	E/15	Satt575	20-26	Antibiosis & Antixenosis	Terry et al., 2000; Hulburt et al., 2004
Corn earworm 7-3	'Noir 1'	D2/17	L204_2	6	Antibiosis	Terry et al., 2000
Corn earworm 7-5	'Noir 1'	D1a/1	R013_2 to Satt507	7–8	Antibiosis	Terry et al., 2000

[‡] Names listed on SoyBase (Grant et al., 2010).

* Source of resistant allele was incorrectly identified by Rector et al. (2000).

NA: not included in SoyBase.

Table 2 Soybean accessions with reported resistance to multiple important insect pests

Accession	MG	Insect[¶,*]	References
PI 196156	III	CEW, MBB(4)	Joshi, 1977
PI 567543C	III	SBA, KB	Mensah et al., 2005; Bray, 2015
PI 567598B	III	SBA, KB	Mensah et al., 2005; Bray, 2015
PI 567597C	III	SBA, KB	Mensah et al., 2005
PI 229319	IV	CEW, MBB(3)	Joshi, 1977
PI 89010	IV	CEW, MBB(4)	Joshi, 1977
PI 96786	V	CEW, PLH(1), SBL(4)	Joshi, 1977
PI 340051	V	CEW, SBL(4)	Joshi, 1977
PI 371610	V	CEW, SBL(4)	Joshi, 1977
PI 548439	V	CEW, PLH(1), SBL(4)	Joshi, 1977
FC 31665	VI	MBB(2), PLH(1)	Kraemer et al., 1988
PI 229358	VII	CEW, PLH(1), MBB(1), BLB, SBL(2), VBC(2)	Van Dyn et al., 1971; Clark et al., 1972; Kilen et al., 1977; Lambert and Kilen, 1984a; Kraemer et al., 1988
PI 171451	VII	CEW, PLH(1), BLB, MBB(1), JB, TBW	Van Dyn et al., 1971; Clark et al., 1972; Kraemer et al., 1988; Lambert and Kilen, 1984b
PI 227687	VIII	CEW, PLH(1), MBB(2), BLB, SBB, SBL(2), VBC(1)	Van Dyn et al., 1971; Clark et al., 1972
PI 209837	VIII	MBB(2), SBL(4),VBC(3), PLH(1)	Gray et al., 1985
FC 31592	VIII	MBB(3), SBL(3),VBC(3), PLH(2)	Gray et al., 1985
PI 417061	VIII	MBB(1), SBL(1)	Kraemer et al., 1988
PI 417136	VIII	MBB(2), SBL(3), VBC(2)	Kraemer et al., 1988

¶ Abbreviations: CEW – corn earworm; MBB – Mexican bean beetle; SBA – soybean aphid; KB – kudzu bug; BLB – bean leaf beetle; PLH – potato leafhopper; SBL – soybean looper; VBC – velvetbean caterpillar; JB – Japanese beetle; TBW – tobacco budworm; SBB – striped blister beetle.

* The number in the parentheses indicates the resistance level where (1) 0–20% defoliation, (2) 21–40% defoliation, (3) 41–60% defoliation, (4) 61–80% defoliation and (5) 81–100% defoliation, according to ars-grin.gov.

likely provide soybeans with resistance to multiple defoliating insect pests. Additionally, Chang and Hartman (2017) identified a single SNP (ss715592245) that may harbour resistance to soybean looper and a SNP (ss715629577) that is associated with resistance against velvetbean caterpillar in a GWA study. Additional soybean accessions resistant to lepidopteran pests and/or coleopteran pests are included in Table 2.

3.3 HPR against hemipteran pests

Direct feeding on developing pods and seeds from stink bugs not only causes soybean yield losses, but also decreases seed quality as it changes the protein profiles, fatty acid profiles and seed germination rate (Miner, 1966; Thomas et al., 1974; Todd and Turnipseed,

1974; Todd et al., 1973; Jensen and Newsom, 1972). In addition, the feeding injury creates an infection court for some soybean pathogens that can cause additional yield losses (McPherson et al., 1994).

Several soybean genotypes have been evaluated for resistance to stink bugs (Miner, 1966; Jones and Sullivan, 1978, 1979; Gilman et al., 1982). Among these genotypes, PI 171444 was identified as highly resistant to the southern green stink bug (Gilman et al., 1982) with both antibiosis and antixenosis resistance (Kester et al., 1984). The three insect-resistant Japanese PIs (171451, 227687 and 229358) that were found resistant to coleopteran and lepidopteran pests were also reported as effective against the southern green stink bug. PI 227687 was suggested as a promising resistance source because its shorter pod development stage limits the damages from pod feeders (Jones and Sullivan 1979). Five additional genotypes, V00-0742, V00-0842, V99-1685, V00-0870 and PI 558040, were reported as new resistance sources to the southern green stink bug (Campos et al., 2010). Seventeen genotypes (including PI 171451, PI 227687, PI 229358 and IAC-100) were reported as resistant to redbanded stink bugs, with fourteen of them conditioning antibiosis resistance (Silva et al., 2013). Pioneer 95Y20, a cultivar developed by Pioneer, was also reported as resistant to redbanded stink bugs (Davis et al., 2012).

A stink-bug-resistant cultivar, 'IAC-100', with resistance derived from PI 229358 and PI 274454 was released in Brazil (Rosseto, 1989, Carrao-Panizzi and Kitamura, 1995). The resistance mechanisms in IAC-100 were uncovered as shorter pod filling period, normal leaf senescence under stress (no 'green soybean syndrome'), abscission of young damaged pods and replacement with new pods (Rossetto et al., 1995). In general, plant maturity is a major factor for the resistance to stink bugs; early maturing genotypes have been observed to contain less damage as they avoid the peak of the pest population (Gilman et al., 1982). The stink bug resistance was estimated as quantitatively inherited in genotypes IAC-100, IAC-74 2832 and IAC-78 2318 (Rossetto et al., 1995; Souza and Toledo, 1995). However, QTL studies are needed to reveal loci underlying stink bug resistance from these cultivars.

The kudzu bug, an invasive species that feeds on kudzu, recently has become a serious pest for soybean production in the southern United States (Seiter et al., 2013; Dhammi et al., 2016). The successful establishment of kudzu (*Pueraria montana*, an invasive plant species) throughout the southern United States has allowed kudzu bug populations to overwinter and subsequently move to soybean. Adults and nymphs gather in large groups and suck sap from a host plant, weakening and stunting it. Severe feeding on leaf sap can cause extensive defoliation. Bray (2015) reported that Benning insect-resistant NILs with QTLs M, G, H and E were not preferred by kudzu bug, and that PI 567336A was confirmed as carrying antibiosis resistance against kudzu bug. As shown in Table 2, several other PIs (567543C, 567598B and 567597C) reported as resistant to soybean aphids were also observed as resistant to kudzu bugs (Bray, 2015). Fritz et al. (2016) reported that three cultivars (N7103, Vance and Nitrasoy) showed high level of resistance to kudzu bugs, and that most Benning insect-resistant NILs (Zhu et al., 2006, 2007) displayed moderate resistance. Future studies are needed to elucidate QTLs/genes underlying these resistances.

As mentioned earlier, the incorporation of pubescence trait into commercial soybean varieties has suppressed the potato leafhopper to a non-pest status in the United States (Boethel, 1999). Another successful example of exploiting HPR in soybean is the discovery and introgression of *Rag* (resistance to *Aphis glycine*) genes into elite cultivars. The biggest challenge, however, is the short durability of single *Rag* genes against emerging biotypes (Table 4). The history of breeding for aphid-resistant soybeans will be discussed in Section 6.

4 Breeding challenges

The insect resistances in these three Japanese PIs (171451, 229358 and 227687) were reported as quantitatively inherited (Sisson et al., 1976; Rector et al., 1998, 1999, 2000; Kenty et al., 1996) and difficult to be fully transferred into elite cultivars. In addition, all three PIs are low-yielding and susceptible to some important diseases and nematodes (Lambert and Kilen, 1984a). The tight linkages between the low-yielding alleles and the insect-resistant alleles are possibly hard to break with traditional breeding. Additionally, agronomic traits (e.g. yield) generally are quantitatively inherited, meaning that large segregating populations are needed to recover recombinant lines with most alleles contributing to yield and insect resistance. Therefore, the nature of linkage drag and quantitative inheritance of insect resistances and agronomic traits has made it difficult to develop high-yielding soybeans with sufficient insect resistance. Three insect-resistant cultivars ('Larmar', 'Crockett' and 'Lyon') derived from the three Japanese PIs had been released in the early 1990s (Bowers, 1990; Hartwig et al., 1990, 1994). However, none of them were well accepted by growers because of inferior yield performance and insufficient insect resistances (Table 3) (Boethel, 1999; Lambert and Tyler, 1999).

Molecular markers play important roles in modern breeding programmes. Family-based QTL mapping and association mapping have been used to identify non-random associations between phenotype and the genotype at a marker locus. With the assistance of genetic markers, superior recombinants are identified as possessing the resistant alleles from the un-adapted parent and favourable alleles contributed by the elite parent; however, it is difficult to identify these superior recombinants through a traditional breeding process. This selection process is called marker-assisted

Table 3 Registered insect-resistant soybean varieties

Name	Insect(s)[¶]	Reference(s)
Lamar	SBL, VBC and CEW	Hartwig et al., 1990
Lyon	SBL	Hartwig et al., 1994
Crockett	MBB, VBC, SGSB	Bowers, 1990
Shore	MBB	Smith et al., 1975
IAC-100	SGSB, SBL, MBB	Rossetto et al., 1995; McPherson and Buss, 2007
Davis	MBB, BLB	Caviness and Walters, 1966; Burden et al., 1992
Forrest	JB	Yesudas et al., 2010
N7103	KB	Fritz et al., 2016
Vance	KB	Fritz et al., 2016
Nitrasoy	KB	Fritz et al., 2016

¶ Abbreviations: SBL – soybean looper; VBC – velvetbean caterpillar; CEW – corn earworm; MBB – Mexican bean beetle; SGSB – southern green stinkbug; BLB – bean leaf beetle; JB – Japanese beetle; KB – kudzu bug.

selection (MAS), which is practised routinely in modern breeding/'molecular breeding' programmes. MAS can efficiently assist the introgression of insect resistance into elite cultivars by reducing the use of insect bioassays and by reducing linkage drags. Furthermore, issues such as environmental effects and large screening trials can be avoided by MAS. MAS also offers the opportunity of pyramiding insect-resistant genes that have similar phenotypic effects which cannot be distinguished by traditional phenotypic selection.

5 *Bacillus thuringiensis* (Bt) soybean

The limited availability of elite cultivars with sufficient resistance to lepidopteran pests prompted attempts to genetically engineer soybean by introducing a Bt insecticidal protein gene. Specific Bt genes convey specific insecticidal action towards a certain insect order, for example, Bt *CryI* is specifically toxic to the order Lepidoptera. Stewart et al. (1996) first transformed 'Jack', a susceptible cultivar, with a synthetic Bt insecticidal crystal protein gene (*Bt crylAc*), and this fertile Bt soybean exhibited high levels of resistance to corn earworm, soybean looper, tobacco budworm and velvetbean caterpillar.

An ideal transgenic soybean should have a broad spectrum of resistance, yet it is hard to achieve that goal. Another challenge that Bt soybeans face is the development of resistance in the pest populations. Gene pyramiding permits different modes of action to be combined to gain durable resistance against multiple insects. Researchers have been developing and evaluating Bt soybean pyramided with different native resistance genes (Walker et al., 2002, 2004; Zhu et al., 2008; Ortega et al., 2016). Walker et al. (2004) reported that QTL-M enhances the effectiveness of *cry1AC* transgene against lepidopteran pests in Jack-Bt soybeans (Stewart et al., 1996). Zhu et al. (2008) evaluated 16 NILs of different combinations of the insect-resistant QTLs (QTL-M, QTL-H and QTL-G) from PI 229358 and the *cry1AC* for efficacy against corn earworm and soybean looper in a 'Benning' genetic background (Boerma et al., 1997). It was found that adding two other QTLs (QTL-H and QTL-G) in the pyramid of Bt/QTL-M (Walker et al., 2004) further enhanced the resistance against corn earworm (Zhu et al., 2008). Ortega et al. (2016) pyramided different QTLs identified from PI 229358 (QTL-M, QTL-H and QTL-G) and PI 227687 (QTL-E) with *cry1AC* in the 'Benning' genetic background. In addition to achieving an acceptable level of resistance to lepidopteran pests, the combination of QTL-E and M further increased protection against Bt-tolerant pests, presenting an opportunity to protect the durability of *cry1AC* transgene in Bt soybeans.

In practice, Monsanto's Intacta RR2 Pro Bt soybeans have been widely grown in South America since 2013. Three benefits are combined in this variety, including 1) Bt protection against velvetbean caterpillar, soybean looper, corn earworm, lesser cornstalk borer (*Elasmopalpus lignosellus*) and cotton bollworm (*Helicoverpa armigera*); 2) increased yield potential; and 3) tolerance to glyphosate herbicide (www.monsanto.com). No Bt soybean varieties are commercially available in the United States yet, but the upswing in insect activity in the southern states may change that in the future as the soybean varieties being planted in South America overlap with those grown in the Mid-South of the United States.

6 Case study: developing aphid-resistant varieties

The soybean aphid, *Aphis glycine* Matsumura, has become the most economically important insect pest in soybean production in North America after its first discovery in the Mid-west at 2000 (Hartman et al., 2001). Soybean aphids cause primary yield loss through direct feeding and/or secondary yield loss through transmitting viruses (e.g. *Soybean dwarf virus, Soybean mosaic virus, Potato virus, Alfalfa mosaic virus* and *Tobacco ring spot virus*) (Hill et al., 2001; Clark and Perry, 2002; Davis et al., 2005). In addition, aphids secrete honeydew that leads to the growth of sooty mould on foliage (Malumphy, 1997; Lemos Filho and Paiva, 2006). The estimated economic impact of aphids on soybean production was reported as $3.6–4.9 billion annually in North America (Kim et al., 2008). Researchers have conducted studies on different control strategies, such as insecticide control, cultural control, biological control and host genetic control. Among these control methods, host genetic control and biological control are of most interest because they are favourable and compatible in IPM; natural enemies can effectively keep aphid populations in check when the population size is under control by HPR in the early season.

Li et al. (2004) first reported three soybean genotypes, 'Dowling' (PI 548663), 'Jackson' (PI 548657) and PI 200538, which confer antibiosis resistance against the Illinois soybean aphid isolate (Hartman et al., 2001). Antibiosis resistance from 'Dowling' and 'Jackson' were determined to be controlled by a single dominant gene (*Rag1/Rag*) that was later mapped to the same location between markers Satt435 and Satt463 on LG M/Chr. 7 (Hill et al., 2006a, b; Li et al., 2007). A subsequent genetic allelism test was conducted among 1000 F_2 plants from 'Dowling' × 'Jackson' and no susceptible plant was observed, suggesting the genes were allelic (Hill et al., 2012). Later, with additional SNP markers developed, *Rag1* was refined to a 115-kb interval and two nucleotide-binding site leucine-rich repeat (NBS-LRR) genes were proposed as the candidates for *Rag1* (Kim et al., 2010a) according to the Williams 82 reference genome annotation on SoyBase (Grant et al., 2010). No yield drag was observed with *Rag1* (Kim and Diers, 2009), and cultivars with *Rag1* were commercially released to growers (McCarville, 2012). However, the resistance conferred by *Rag1/Rag* to the Illinois isolate was broken down by an Ohio isolate (Kim et al., 2008). Thus, the Illinois isolate was referred to as Biotype 1 and the Ohio isolate was referred to as Biotype 2 (Kim et al., 2008).

Mian et al. (2008a,b) discovered three PIs (243540, 567301B and 567324) exhibiting resistance against Biotypes 1 and 2, and mapped a single dominant gene, *Rag2*, controlling antibiosis resistance in PI 243540 to LG F/Chr.13 between markers Satt334 and Sct_033. In the biotype study by Kim et al. (2008), PI 200538 retained strong antibiosis resistance to Biotype 2; later, a mapping study by Hill et al. (2009) revealed the underlying gene is the same gene as *Rag2* (Mian et al., 2008b) since it resides at the same genomic location and confers identical resistance reactions to different biotypes. Fine mapping of *Rag2* in PI 200538 delimited it to a 54-kb region with one NBS-LRR gene as the candidate (Kim et al., 2010b). Fox et al. (2014) found that *Rag2* was significantly associated with resistance in 20 of the 21 F_2 populations derived from 21 newly identified aphid-resistant PIs, suggesting *Rag2* may be a major aphid resistance source in the USDA soybean germplasm collection. Years later, a major QTL conferring antixenosis resistance in PI 567301B was mapped near *Rag2* on LG F/Chr. 13 but referred as a different gene (*Rag5*) in that detached leaves of PI 567301B did not maintain resistance towards soybean aphids while detached leaves of PI 243540 (source of *Rag2*) did (Michel et al., 2010; Jun et al., 2012). Additionally, in PI 567301B, a minor QTL conferring antixenosis resistance was mapped near marker BARC-063283-18296 on LG A2/Chr. 8 (Jun et al., 2012). Jun et al. (2013) discovered three aphid-resistant QTLs in PI 567324; two major ones (QTL_13_1 and QTL_13_2) were located close to the previously reported *Rag2* locus (Mian et al., 2008a; Kim et al., 2010b) and *Rag4* locus (Zhang et al., 2009), respectively, and a minor one (QTL_6_1) was detected on chromosome 6, where no aphid-resistant QTL has been reported so far. The oligogenic resistance from PI 567324 is expected to provide broader and more durable resistance against soybean aphids compared to cultivars with monogenic resistance (Jun et al., 2013).

Hill et al. (2010) reported the discovery of Biotype 3, which readily colonized plants with *Rag2* in Indiana. Mensah et al. (2005) screened 2,147 soybean accessions from MG 0 to III and identified four MG III accessions (PI 567543C, PI 567597C, PI 567541B and PI 567598B) resistant to mixed soybean aphid biotypes in Michigan. PI 567541B and PI 567598B carry antibiosis resistance while PI 567543C and PI 567597C possess antixenosis resistance. These four PIs were included in the biotype studies by Kim et al. (2008) and Hill et al. (2010), and found resistant to multiple biotypes. Mensah et al. (2008) reported that aphid resistance in PI 567541B and PI 567598B were both controlled by two major recessive genes. In PI 567541B, one recessive gene was mapped to the *Rag1* region and was named *Rag1c* whereas the other recessive gene was mapped to a different region (Satt649-Satt348) from *Rag2* on LG F/Chr.13 and was named *Rag4* (Zhang et al., 2009). The broad antixenosis resistance provided by PI 567543C and PI 567597C was reported controlled by a single partially dominant gene, *Rag3* and *Rag3e* (mapped in close proximity on LG J/Chr.16), respectively (Zhang et al., 2010; Du, 2016). Bales et al. (2013) reported the antibiosis resistance in PI 567598B were contributed by two recessive genes (*Rag1b* and *Rag3*) that were mapped to the previously identified *Rag1* (Li et al., 2007; Kim et al., 2010a) and *Rag3* (Zhang et al., 2010) regions.

Interestingly, in addition to *Rag3*, *Rag3e* and *Rag3*, a few more aphid-resistant genes were mapped to this shared genomic region. Zhang et al. (2013) detected a single dominant gene, *Rag3b*, conferring antibiosis resistance against mixed biotypes of soybean aphid in PI 567537. Du (2016) mapped a partially dominant gene, *Rag3d*, in PI 567585A, which confers antibiosis resistance against multiple biotypes. Another partially dominant gene, *Rag3c*, controlling moderate antibiosis resistance was identified in *Glycine soja* 85-32 (Zhang et al., 2017a). According to the Williams 82 reference genome annotation on SoyBase (Grant et al., 2010), this shared region is enriched with NBS-LRR genes. Fine mapping studies have been conducted and delimiting these Rag genes (*Rag3*, *Rag3b*, *Rag3c*, *Rag3d*, *Rag3e* and *Rag3*) to similar or different genomic locations (Bales, 2013; Du, 2016; Zhang et al., 2017b); it is possible that some of these Rag genes are different NBS-LRR genes while others are different alleles of the same NBS-LRR gene(s).

Isoflavone may play a role in plant defence against soybean aphids. Meng et al. (2011) mapped two aphid-resistant QTLs to the same locations of QTLs previously reported as associated with high isoflavone content in a Chinese soybean cultivar, 'Zhongdou 27'. One of the QTLs (qRa_1) was mapped to Satt470 on LG A2/Chr. 8 and explained a large portion of phenotypic variance. The second QTL (qRa_2) was mapped to Satt144 on LG

Table 4 Soybean aphid biotypes and their virulence to soybean resistance genes

		Aphid-resistance gene and the resistance source									
Biotype	State first found	*Rag* *Jackson*	*Rag1* *Dowling*	*Rag2 PIs* *243540, 200538*	*Rag1* + *Rag2* *Pyramid*	*Rag3* PI 567543C	*Rag1b, Rag3* PI 567598B	*Rag1c, Rag4* PI 567541B	*Rag3e* PI 567597C	*Rag5* PI 567301B	Reference
1	Illinois	–	–	–	–	–	–	–	–	–	Hill et al. 2004
2	Ohio	+	+	–	–	–	–	–	–	–	Kim et al. 2008
3	Indiana	NA	–/+*	+	–	–	–	–/+¶	–	NA	Hill et al. 2010
4	Wisconsin	NA	+	+	+	–	+	+	–	NA	Alt and Ryan-Mahmutagic 2013

+ Implies the aphid biotype readily colonizes on the soybean plants with the *Rag* gene(s).
– Implies the aphid biotype cannot colonizes on the soybean plants with the *Rag* gene(s).
NA represents 'Not Available' in the literature.
* Soybeans with *Rag1* were resistant to Biotype 3 in no-choice tests but susceptible to Biotype 3 in choice tests.
¶ PI 567541B was moderate resistant to Biotype 3 in choice tests but susceptible to Biotype 3 in no-choice tests.

F/Chr.13. The authors suggested that greater isoflavone concentration could help protect soybean from aphid attacks as some members of isoflavones have been reported as having antibiosis effects on some soybean pests. For example, genistein and daidzin are toxic to southern green stink bugs, cabbage loopers and velvetbean caterpillars (Sharma and Norris, 1991; Hoffmann-Campo et al., 2001; Piubelli et al., 2003, 2005). A Brazilian soybean cultivar, IAC-100, has been reported as having high isoflavone (daidzin and genistein) concentrations (Carrao-Panizzil and Kitamura, 1995) and antibiotic resistance against stink bugs (Rosseto, 1989).

Recently, Alt and Ryan-Mahmutagic (2013) reported that a new soybean aphid biotype, Biotype 4, discovered in Wisconsin readily colonized the previously known resistant soybean genotypes, including 'Dowling', PI 243540, PI 200538, *Rag1/Rag2* stacked material, PI 567541B and PI 567598B. Among the tested genotypes, PI 567543C and PI 567597C remained resistant to Biotype 4 (Alt and Ryan-Mahmutagic, 2013). The four reported aphid biotypes (Table 4) and unknown virulent biotypes put breeders under pressure to continually find new resistance sources. Kim et al. (2014) detected a possible new allele or gene at the *Rag1* region in PI 587732, conferring antibiosis resistance against Biotype 2. Nurden et al. (2010) reported that the antixenosis resistance against Biotype 2 in PI 71506 was controlled by a single dominant gene that is distinct from *Rag1*. This unknown gene is therefore worthwhile to study in the future to understand its location in the soybean genome and to provide additional resistance. A single dominant gene controlling antibiosis resistance was mapped to an interval different from *Rag2*, *Rag4*, *Rag5* on LG F/Chr. 13 and therefore referred to as a new aphid-resistant gene, R_P746 (Xiao et al., 2014).

In addition to qRa_1 (Meng et al., 2011) and the minor QTL from PI 567301B (Jun et al., 2012), two more loci were identified on LG A2/Chr. 8 with resistance to soybean aphids. The antixenosis resistance QTL, [*Rag6*]_*P203*, was delimited to a 192-kb interval between SSR_08_75 and SSR_08_88 from a Chinese soybean line P203 (Xiao et al., 2013). Zhang et al. (2017a, b) refined the antibiosis-resistant QTL, *Rag6*, from *G. soja* 85-32 to a 49-kb interval between markers MSUSNP08-100 and MSUSNP08-101, which is 2 Mb away from [*Rag6*]_*P203*. According to the William 82 reference genome (Schmutz et al., 2010), all four aphid-resistant QTLs on chromosome 8 are located distantly from each other (Meng et al., 2011; Jun et al., 2012; Xiao et al., 2013; Zhang et al., 2017a,b), providing additional genetic variability of aphid resistance for soybean breeding programmes.

In addition to the continuous discovery of new resistance sources, pyramiding of different *Rag* genes (Table 5) with the assistance of flanking markers could provide cultivars with broader and more durable resistance. Wiarda et al. (2012) investigated aphid development on soybeans with *Rag1* alone, *Rag2* alone and both genes combined, and discovered soybeans with both genes more resistant to aphids. A similar investigation also confirmed that pyramiding *Rag1* and *Rag2* provides yield protection from soybean aphids in North America (McCarville et al., 2014). A pool of advanced breeding lines pyramided with different Rag genes was developed at Michigan State University (Table 6), showing strong and durable resistance against Michigan soybean aphid populations over the years (unpublished data).

Gene deployment based on biotype distribution is an effective method to combat different aphid populations geographically. Therefore, the knowledge regarding distribution of the different biotype distribution is important. Cooper et al. (2015) studied the geographic distribution of aphid biotypes at 11 locations during 2008 to 2010 and

Table 5 Known SBA resistance genes

Gene	Source(s)	LG/Chr.	Flanking markers	R^{2c}	Resistance modality	Reference(s)
Rag/Rag1	PI 548657 (Jackson) and PI 548663 (Dowling)	M/7	Satt435-Satt463[a] SNPKS9-3-SNPKS5[b]	NA[d]	Primarily antibiosis	Hill et al., 2006a,b; Li et al., 2007; Kim et al., 2010a
Rag1b	PI 567598B	M/7	Satt567-Sattt435	14.0–35.5%	Antibiosis	Bales et al., 2013; Mensah 2008
Rag1c	PI 567541B	M/7	Satt299-Satt435	44.7–87.7%	Antibiosis	Mensah 2008; Zhang et al., 2009
Rag2	PI 200538, PI 243540	F/13	Satt334-Sct_033[a] SNP46169.7-SNP21A[b]	NA	Antibiosis	Hill et al., 2009; Kim et al., 2010b; Mian et al., 2008a, b
Rag3	PI 567543C	J/16	Sat_339-Satt414	74.3–90.4%	Antixenosis	Mensah et al., 2005; Zhang et al., 2010
Rag3	PI 567598B	J/16	Satt285-Satt414	28.4–45.8%	Antibiosis	Bales et al., 2013; Mensah 2008
Rag3b	PI 567537	J/16	Sat_399-Satt654	78.9–87.4%	Antibiosis	Zhang et al., 2013
Rag4	PI 567541B	F/13	Satt649-Satt348	0.9–9.2%	Antibiosis	Mensah 2008; Zhang et al., 2009
Rag5	PI 567301B	F/13	BARCSOYSSR_13_1131-BARCSOYSSR_13_1148	75–91%	Antixenosis	Jun et al., 2012; Mian et al., 2008a
[Rag6]_P203	P203	A2/8	SSR_08_75-SSR_08_88	NA	Antixenosis	Xiao et al. 2013
R_P746	P746	F/13	BARCSOYSSR_13_1278-BARCSOYSSR_13_1363	NA	Antibiosis	Xiao et al., 2014
qRa_1	Zhongdou 27	A2/8	Satt470	25–35%	Antibiosis	Meng et al., 2011
qRa_2	Zhongdou 27	F/13	Satt144	7–11%	Antibiosis	Meng et al., 2011
QTL_6_1	PI 567324	C2/6	BARCSOYSSR_06_0998	4.4–11.6%	Antixenosis	Jun et al., 2013
QTL_13_1	PI 567324	F/13	BARCSOYSSR_13_1139	42.7–70.6	Antixenosis	Jun et al., 2013
QTL_13_2	PI 567324	F/13	Satt649	2.1–13.1%	Antixenosis	Jun et al., 2013

[a] Markers flanking the initial mapping region.
[b] Markers flanking the fine mapping region.
[c] R^2 represents the percentage of phenotypic variation explained by a QTL.
[d] NA represents 'Not Available' in the literature.

Table 6 Advanced breeding lines pyramided with different *Rag* genes

Advanced breeding lines	Stacked Rag gene(s)	Pedigree information
E14912	*Rag3, Rag1b*	[LD01-5907 × (Titan × PI 567598B)] × LD02-4485
E13369	*Rag4, Rag1c*	E07051 × {[E00003 × (SDX00R-039-42 × PI 567541B] × E00003)}
E14902	*Rag3c, Rag6*	(Skylla × LD01-7323) × [E00003 × (Jiyu 71 × G. soja85-32)]
E13901	*Rag3, Rag4, Rag1c*	{(Skylla x PI 567598B) x [Skylla x (SDX00R-039-42 x PI 567541B)]} x E07051
E14919	*Rag3, Rag4, Rag1b, Rag1c*	[E00003 × (SDX00R-039-42 × PI 567541B)] × [LD01-5907 × (Titan × PI 567598B)]
E13903	*Rag3, Rag4, Rag1c, Rag2*	{[Skylla × PI 567598B] × [Skylla × (SDX00R-039-42 × PI 567541B)]} × LD08-12430a

All the advanced breeding lines listed in the table show antibiosis resistance to soybean aphids.

developed a panel of host differentials (indicator lines) to characterize aphid biotypes. According to this study, aphid populations had been diverse and dynamic across the United States and Canada. Additionally, PI 567598B and PI 567541B were identified as the most resistant and durable genotypes against aphid populations. The authors inferred the high level of resistance was due to two recessive genes working together in these two PIs. Deploying different *Rag* gene(s) according to the soybean aphid biotype distribution could avoid genetic vulnerability of a certain resistant cultivar over a large geographic area.

7 Conclusion

It is widely recognized that HPR provides great benefits such as suppressing pest populations, reducing economic input and protecting the ecosystem. This chapter briefly reviewed the identified HPRs against different insect pests in soybean (Tables 1–6 and Fig. 1). Three accessions from Japan, PI 171451 ('Kosamame'), PI 229358 ('Soden-daizu') and PI 227687 ('Miyako White'), were discovered as resistant to multiple insect pests of soybean, and have been frequently used as donors in various breeding programmes to build insect-resistant cultivars. Additional resistance sources were also discovered by several research groups. Four insect-resistant varieties were released but did not gain popularity among growers because of the insufficient resistance level and inferior yield performance that were due to the quantitative inheritance and linkage drag. The advent of DNA markers has enabled breeders to pyramid the QTLs underlying insect resistance in an elite cultivar with minimal linkage drag. Bt soybeans have been developed to supplement the natural HPR and widely grown in South America. However, a common drawback of HPR and Bt resistance is the loss of durability, which is due to the emergence of new biotypes. For instance, some of the Rag genes have lost resistance to the newly discovered soybean aphid biotype(s). To sustain HPR, several practices should be considered, including gene pyramiding, gene deployment based on biotype distribution and variety mixture.

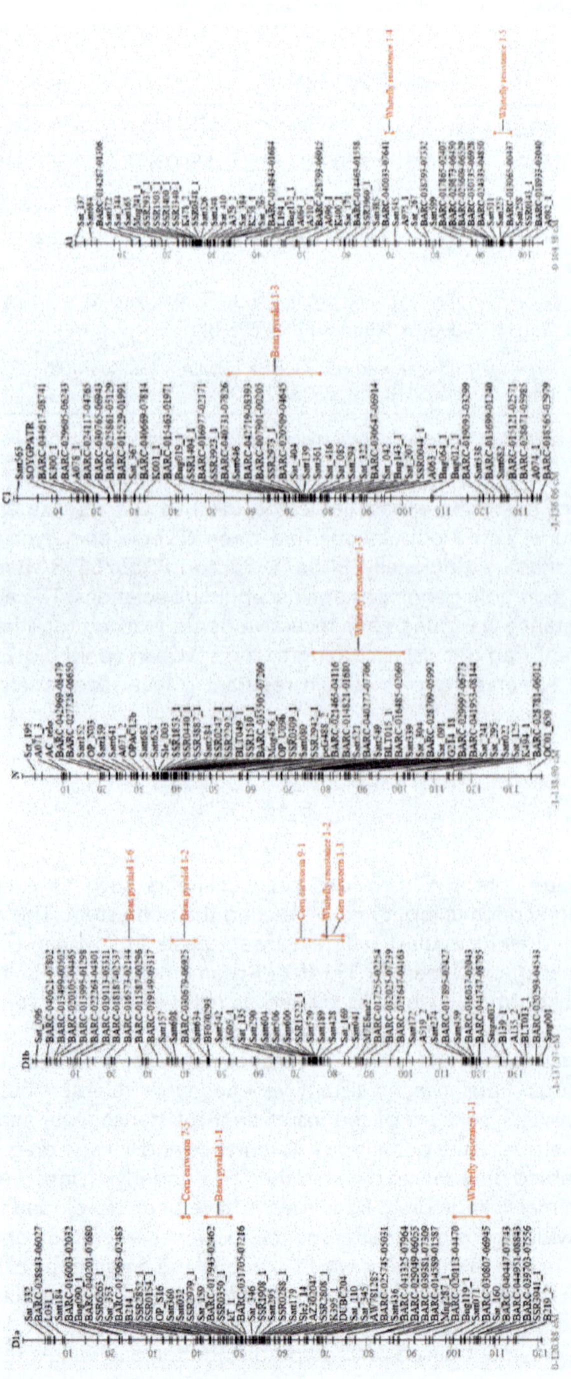

Figure 1 QTLs resistant to insect pests. Note: This map only shows the insect-resistant QTLs listed on SoyBase. Bean pyralid (BP; *Lamprosema indicata* Fabricius) is one of the major leaf-feeding insects that affect soybean crops in central and southern China.

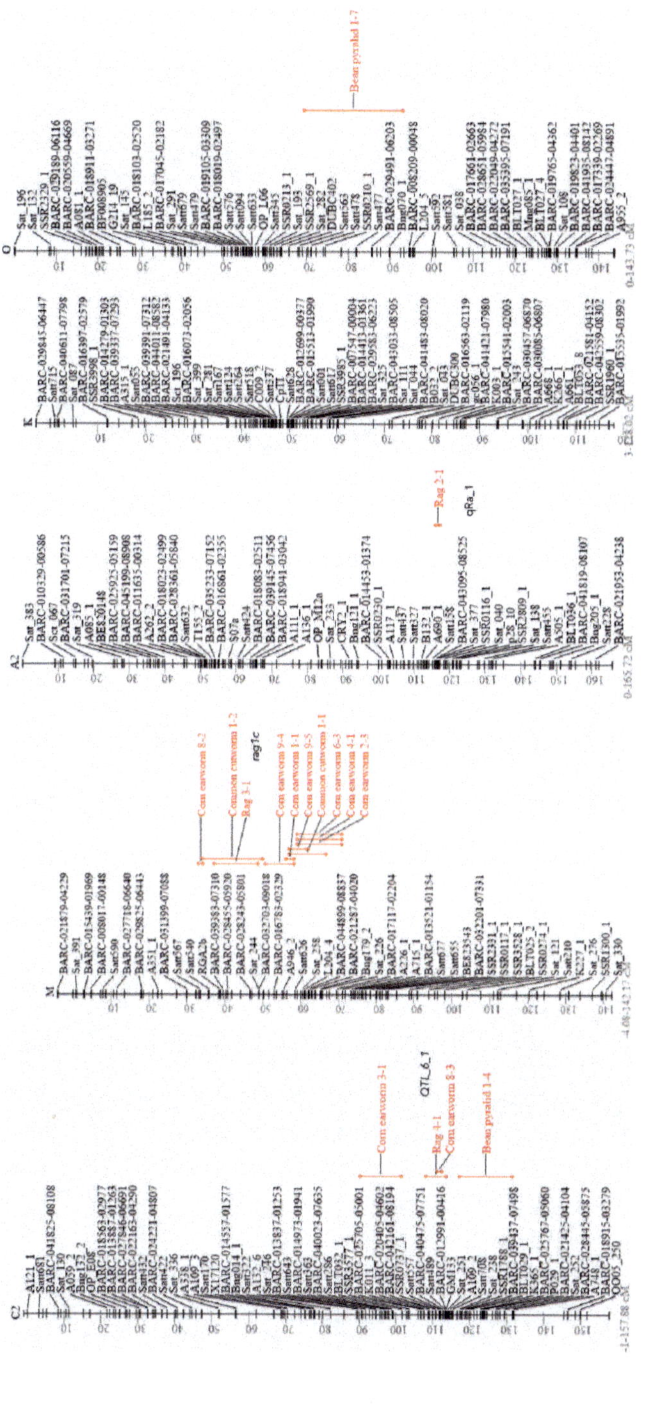

Figure 1 (Continued).

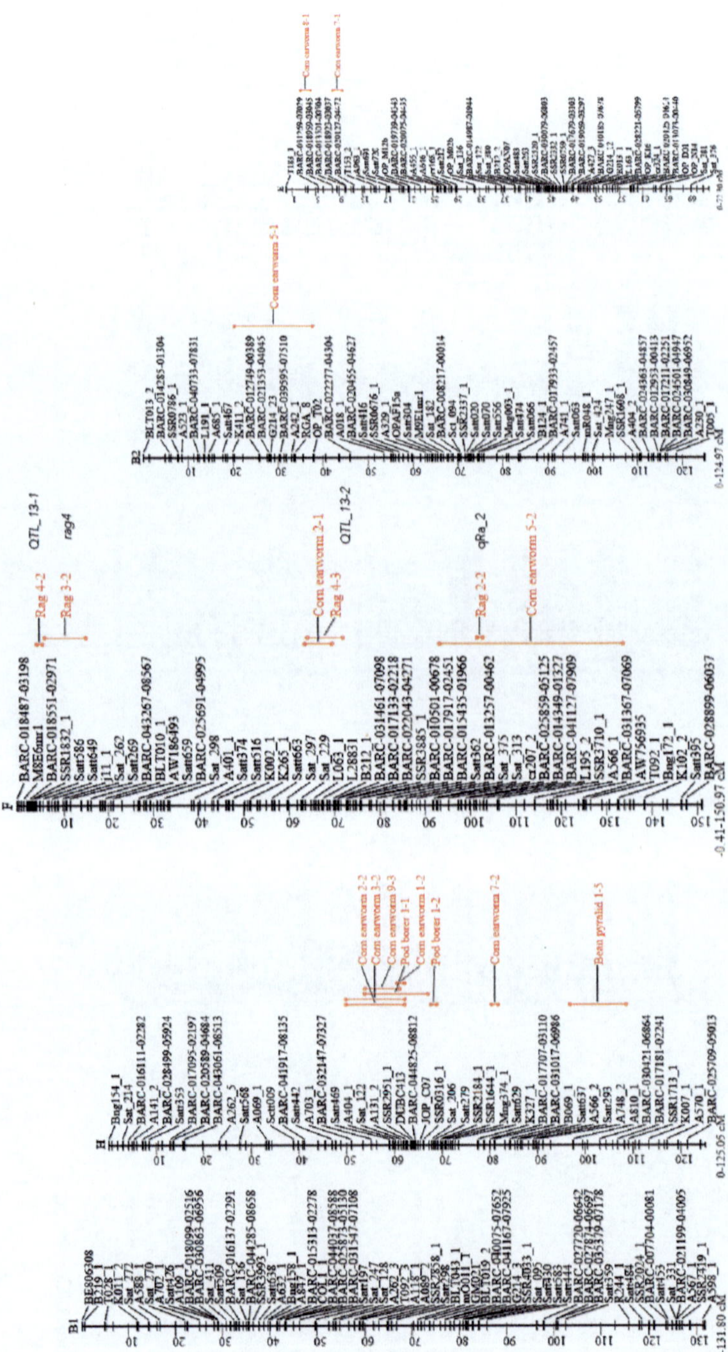

Figure 1 (Continued).

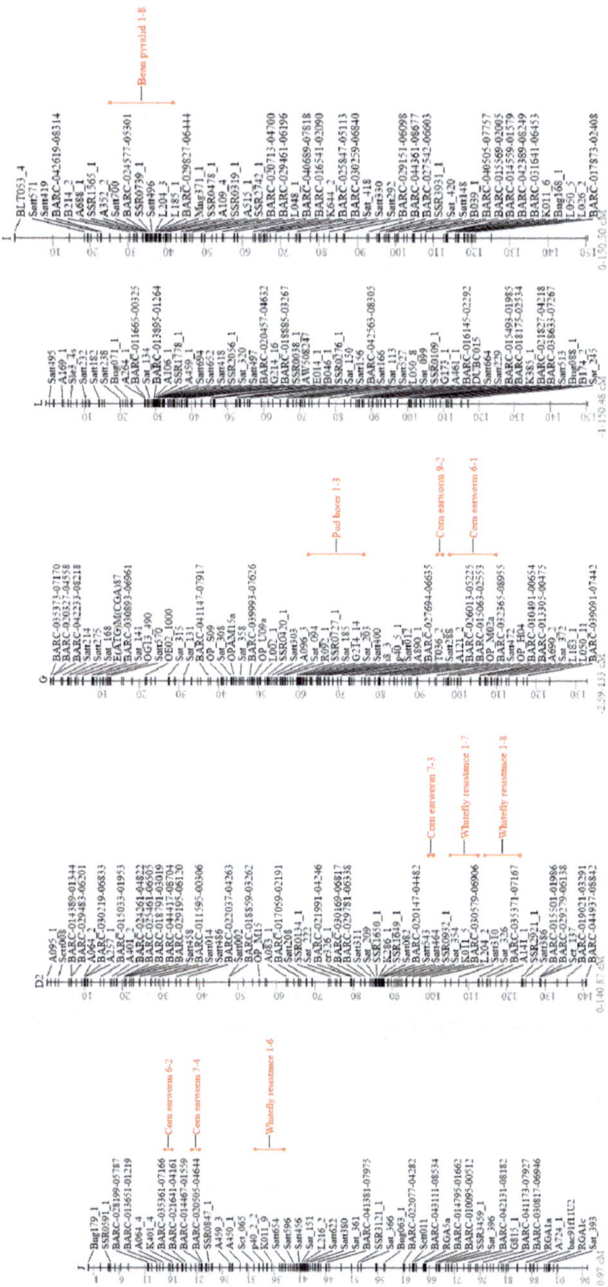

Figure 1 (Continued).

8 Future trends

Tolerant cultivars should be emphasized in breeding programmes. Unlike antixenosis and antibiosis, tolerance often involves plant features (e.g. rapid foliage recovery) which are not involved in plant–insect interactions. That means new virulent biotypes are less likely to emerge in a cropping system based on tolerant cultivars. Therefore, tolerant cultivars pyramided with additional resistance genes should be developed to control pests and to hamper the development of resistance in the pest populations.

Molecular biotype-specific markers would be helpful to diagnose aphid populations before planting and therefore deploy resistance genes geographically. This 'gene deployment based on biotype distribution' method would help sustain HPR and avoid genetic vulnerability by a wide growth of a certain resistance cultivar.

Besides Bt proteins, many different classes of plant proteins have been discovered as toxic to insects and proposed as targets for genetic engineering (see Table 3 in Hilder and Boulter 1999). Among these toxic proteins, protease inhibitors and α-amylase inhibitors have shown success in controlling lepidopteran pests (Wasmann et al., 1994; Edmonds et al., 1996; Thomas et al., 1995; Carbonero et al., 1999). Moreover, these proteinase inhibitor transgenes could be pyramided with Bt genes to limit the selection of Bt-resistant insects. Bt proteins cannot affect phloem-feeding insects such as aphids (Raps et al., 2001; Lawo et al., 2009). Therefore, the expressions of protease inhibitors or RNAi are alternative transgenic strategies. However, the adaptation of such approaches requires knowledge of aphid biology as well as aphid–plant interactions.

9 Where to look for further information

soybase.org

SoyBase is a USDA-ARS database providing professionally curated genetics, genomics and related data resources for soybean. It is a comprehensive repository that contains genetic maps integrated with QTLs of more than 90 unique traits as well as the soybean reference genome (Williams 82) sequence and annotation information.

ars-grin.gov

The Germplasm Resources Information Network (GRIN) Web server, which is supported by USDA-ARS, provides germplasm information of different species as well as access to the national plant germplasm collection. Pedigree information and phenotypic traits (such as insect resistance, disease resistance and more) were recorded for each plant accession.

10 References

Alfonso-Rubí, J., Ortego, F., Castañera, P., Carbonero, P. and Díaz, I. (2003). Transgenic expression of trypsin inhibitor CMe from barley in indica and japonica rice, confers resistance to the rice weevil Sitophilus oryzae. *Transgenic Research*, 12(1), pp. 23–31.

All, J. N., Boerma, H. R. and Todd, J. W. (1989). Screening soybean genotypes in the greenhouse for resistance to insects. *Crop Science*, 29(5), pp. 1156–9.

Alt, J. and Ryan-Mahmutagic, M. (2013). Soybean aphid biotype 4 identified. *Crop Science*, 53(4), pp. 1491–5.

Azzouz, H., Cherqui, A., Campan, E., Rahbe, Y., Duport, G., Jouanin, L., Kaiser, L. and Giordanengo, P. (2005). Effects of plant protease inhibitors, oryzacystatin I and soybean Bowman-Birk Inhibitor, on the aphid *Macrosiphum euphorbiae* (Homoptera, Aphididae) and its parasitoid *Aphelinus abdominalis* (Hymenoptera, Aphelinidae). *Journal of Insect Physiology*, 51, pp. 75–86.

Bales, C., Zhang, G., Liu, M., Mensah, C., Gu, C., Song, Q., Hyten, D., Cregan, P. and Wang, D. (2013). Mapping soybean aphid resistance genes in PI 567598B. *Theoretical and Applied Genetics*, 126, pp. 2081–91.

Bales, C. (2013). Fine Mapping of Aphid Resistance Genes in Soybean Plant Introduction (PI) 567598B. Dissertation, Michigan State University, East Lansing, MI.

Beach, M. R. and Todd, J. W. (1988). Oviposition preference of the soybean looper (Lepidoptera: Noctuidae) among four soybean genotypes differing in larval resistance. *Journal of Economic Entomology*, 81(1), pp. 344–8.

Boerma, H. R., Hussey, R. S., Phillips, D. W., Wood, E. D., Rowan, G. B. and Finnerty, S. L. (1997). Registration of 'Benning' soybean. *Crop Science*, 37, p. 1982.

Boerma, H. R. and Walker, D. R. (2005). Discovery and utilization of QTLs for insect resistance in soybean. In: Mauricio, R. (Ed.), *Genetics of Adaptation*. Springer, Netherlands, pp. 181–9.

Boethel, D. J. (1999). Assessment of soybean germplasm for multiple insect resistance. In: Quisenbury, S. S. and Clement, S. L. (Eds), *Global Plant Genetic Resources for Insect-Resistant Crops*. CRC Press, Boca Raton, FL, pp. 101–29.

Bowers, G. R. (1990) Registration of 'Crockett' soybean. *Crop Science*, 30, p. 427.

Bray, A. L. (2015). Characterizing Soybean Resistance to the Kudzu Bug. Master's Thesis, University of Georgia, Athens, GA.

Broadway, R. M. and Duffey, S. S. (1986). Plant proteinase inhibitors: Mechanism of action and effect on the growth and digestive physiology of larval Heliothis zea and Spodoptera exiqua. *Journal of Insect Physiology*, 32(10), pp. 827–33.

Broersma, D. B., Bernard, R. L. and Luckmann, W. H. (1972). Some effects of soybean pubescence on populations of the potato leafhopper. *Journal of Economic Entomology*, 65(1), pp. 78–82.

Burden, B. J. and Norris, D. M. (1992) Role of the isoflavenoid coumestrol in the constitutive antixenosic properties of 'Davis' soybeans against and oligophagous insect, the Mexican bean beetle. *Journal of Chemical Ecology*, 18, pp. 1069–81.

Campos, M., Knutson, A., Heitholt, J. and Campos, C. (2010). Resistance to seed feeding by southern green stink bug, *Nezara viridula* (Linnaeus), in soybean, *Glycine max* (L.) Merrill. *Southwestern Entomologist*, 35, pp. 233–9.

Carbonero, P., Díaz, I., Vicente-Carbajosa, J., Alfonso-Rubi, J., Gaddour, K. and Lara, P. (1999). Cereal α-amylase/trypsin inhibitors and transgenic insect resistance. In: G. T. Scarascia Mugnozza, E. Porceddu and M. A. Pagnotta (Eds), *Genetics and Breeding for Crop Quality and Resistance*. Springer, Netherlands, pp. 147–58.

Carrao-Panizzil, M. C. and Kitamura, K. (1995). Isoflavone content in Brazilian soybean cultivars. *Japanese Journal of Breeding*, 45, pp. 295–300.

Caviness, C. E. and Walters, H. J. (1966). Registration of 'Davis' soybean. *Crop Science*, 6(5), p. 502.

Chang, H. X. and Hartman, G. L. (2017). Characterization of insect resistance loci in the USDA soybean germplasm collection using genome-wide association studies. *Frontiers in Plant Science*, 8, p. 670.

Chiang, H. S., Norris, D. M., Ciepiela, A., Shapiro, P. and Oosterwyk, A. (1987). Inducible versus constitutive PI 227687 soybean resistance to Mexican bean beetle, *Epilachna varivestis*. *Journal of Chemical Ecology*, 13, pp. 741–9.

Clark, A. J. and Perry, K. L. (2002). Transmissibility of field isolates of soybean viruses by *Aphis glycines*. *Plant Disease*, 86, pp. 1219–22.

Clark, W. J., Harris, F. A., Maxwell, F. G. and Hartwig, E. E. (1972). Resistance of certain soybean cultivars to bean leaf beetle, striped blister beetle, and bollworm. *Journal of Economic Entomology*, 65, pp. 1669–72.

Cooper, S. G., Concibido, V., Estes, R., Hunt, D., Jiang, G. L., Krupke, C., McCornack, B., Mian, R., O'Neal, M., Poysa, V. and Prischmann-Voldseth, D. (2015). Geographic distribution of soybean aphid biotypes in the United States and Canada during 2008–2010. *Crop Science*, 55, pp. 2598–608.

Davis, J. A., Radcliffe, E. B. and Ragsdale, D. W. (2005) Soybean aphid, *Aphis glycines* Matsumura, a new vector of potato virus Y in potato. *American Journal of Potato Research*, 82, pp. 197–201.

Davis, J. A., Brown, S., Richter, A. R. and Kamminga, K. L. (2012). Assessing resistance to *Piezodorus guildinii* (Westwood) in six soybean cultivars. In *2012 Soybean Breeder's Workshop Oral Presentation,* https://soybase.org/meeting_presentations/soybean_breeders_workshop/.

De Leo, F., Bonadé-Bottino, M., Ceci, L. R., Gallerani, R. and Jouanin, L. (2001). Effects of a mustardtrypsin inhibitor expressed in different plants on three lepidopteran pests. *Insect Biochemistry and Molecular Biology*, 31, pp. 593–602.

Dhammi, A., van Krestchmar, J. B., Ponnusamy, L., Bacheler, J. S., Reisig, D. D., Herbert, A., Del Pozo-Valdivia, A. I. and Roe, R. M. (2016). Biology, pest status, microbiome and control of kudzu bug (Hemiptera: Heteroptera: Plataspidae): A new invasive pest in the US. *International Journal of Molecular Sciences*, 17(9), p. 1570.

Du, W. (2016) Map and Fine Map Aphid Resistance Genes in Soybean Plant Introduction (PI) 567597C, 567585A and 567537. Dissertation, Michigan State University, East Lansing, MI.

Edmonds, H. S., Gatehouse, L. N., Hilder, V. A. and Gatehouse, J. A. (1996). The inhibitory effects of the cysteine protease inhibitor, oryzacystatin, on digestive proteases and on larval survival and development of the southern corn rootworm (Diabrotica undecimpunctata howardi). *Entomologia Experimentalis et Applicata*, 78(1), pp. 83–94.

Falco, M. C. and Silva-Filho, M. C. (2003). Expression of soybean proteinase inhibitors in transgenic sugarcane plants: Effects on natural defense against *Diatraea saccharalis. Plant Physiology and Biochemistry*, 41, pp. 761–6.

Fernandez-Cornejo, J., Nehring, R., Osteen, C., Wechsler, S., Martin, A. and Vialou A. (2014). Pesticide Use in U.S. Agriculture: 21 Selected Crops, 1960–2008. USDA Economic Research Service – EIB124.

Fox, C. M., Kim, K. S., Cregan, P. B., Hill, C. B., Hartman, G. L. and Diers, B. W. (2014). Inheritance of soybean aphid resistance in 21 soybean plant introductions. *Theoretical and Applied Genetics*, 127, pp. 43–50.

Fritz, B. J., Reisig, D. D., Sorenson, C. E., Del Pozo-Valdivia, A. I. and Carter, T. E. (2016). Host Plant Resistance to *Megacopta cribraria* (Hemiptera: Plataspidae) in diverse soybean germplasm maturity Groups V through VIII. *Journal of Economic Entomology*, 109(3), pp. 1438–49.

Gilman, D. F., McPherson, R. M., Newsom, L. D., Herzog, D. C. and Williams, C. (1982). Resistance in soybeans to the southern green stink bug. *Crop Science*, 22, pp. 573–6.

Grant, D., Nelson, R. T., Cannon, S. B. and Shoemaker, R. C. 2010. SoyBase, the USDA-ARS soybean genetics and genomics database. *Nucleic Acids Research*, 38(suppl_1), pp. D843–D846. doi:10.1093/nar/gkp798.

Gray, D. J., Lambert, L. and Ouzts, J. D. (1985). Evaluation of soybean plant introductions for resistance to foliar feeding insects. *Journal of Mississippi Academy of Sciences*, 30, pp. 67–82.

Hammond, R. B. and Cooper, R. L. (1989). Development and antibiosis of released soybean germplasm lines resistant to Mexican bean beetle (Coleoptera: Coccinellidae). *Journal of Economic Entomology*, 82, pp. 259–63.

Harbach, C. J., Allen, T. W., Bowen, C. R., Davis, J. A., Hill, C. B., Leitman, M., Leonard, B., Mueller, D., Padgett, G., Phillips, X. and Schneider, R. (2016). Delayed senescence in soybean: Terminology, research update, and survey results from growers. Online. *Plant Health Progress*, 17, pp. 76–83.

Hart, S. V., Kogan, M. and Paxton, J. D. (1983). Effect of soybean phytoalexins on the herbivorous insects Mexican bean beetle and soybean looper. *Journal of Chemical Ecology*, 9(6), pp. 657–72.

Hartman, G. L., Domier, L. L., Wax, L. M., Helm, C. G., Onstad, D. W., Shaw, J. T., Solter, L. F., Voegtlin, D. J., D'Arcy, C. J., Gray, M. E., Steffey, K. L., Isard, S. A. and Orwick, P. L. (2001) Occurrence and distrubution of *Aphis glycines* on soybeans in Illinois in 2000 and its potential control. *Plant Health Progress*. Available at https://naldc.nal.usda.gov/download/11854/ (accessed July 2017) doi:10.1094/PHP-2001-0205-1001-HN.

Hartwig, E. E., Lambert, L. and Kilen, T. C. (1990) Registration of 'Lamar' soybean. *Crop Science*, 30, p. 231.

Hartwig, E. E., Kilen, T. C. and Young, L. D. (1994). Registration of 'Lyon' soybean. *Crop Science*, 34, pp. 1412.

Hilder, V. A., Gatehouse, A. M., Sheerman, S. E., Barker, R. F. and Boulter, D. (1987). A novel mechanism of insect resistance engineered into tobacco. *Nature*, 330, pp. 160–3.

Hilder, V. A. and Boulter, D. (1999). Genetic engineering of crop plants for insect resistance – a critical review. *Crop Protection*, 18(3), pp. 177–91.

Hill, C. B., Li, Y. and Hartman, G. L. (2004). Resistance to the soybean aphid in soybean germplasm. *Crop Science*, 44(1), pp. 98–106.

Hill, C. B., Li, Y. and Hartman, G. L. (2006a). A single dominant gene for resistance to the soybean aphid in the soybean cultivar Dowling. *Crop Science*, 46(4), pp. 1601–5.

Hill, C. B., Li, Y. and Hartman, G. L. (2006b). Soybean aphid resistance in soybean Jackson is controlled by a single dominant gene. *Crop Science*, 46(4), pp. 1606–8.

Hill, C. B., Kim, K. S., Crull, L., Diers, B. W. and Hartman, G. L. (2009). Inheritance of resistance to the soybean aphid in soybean PI 200538. *Crop Science*, 49(4), pp. 1193–200.

Hill, C. B., Crull, L., Herman, T. K., Voegtlin, D. J. and Hartman, G. L. (2010). A new soybean aphid (Hemiptera: Aphididae) biotype identified. *Journal of Economic Entomology*, 103, pp. 509–15.

Hill, C. B., Chirumamilla, A. and Hartman, G. L. (2012). Resistance and virulence in the soybean-*Aphis glycines* interaction. *Euphytica*, 186(3), pp. 635–46.

Hill, J. H., Alleman, R., Hogg, D. B. and Grau, C. R. (2001). First report of transmission of Soybean mosaic virus and Alfalfa mosaic virus by Aphis glycines in the New World. *Plant Disease*, 85(5), p. 561.

Hoffmann-Campo, C. B., Harborne, J. B. and McCaffery, A. R. (2001). Pre-ingestive and post-ingestive effects of soya bean extracts and rutin on Trichoplusia ni growth. *Entomologia Experimentalis et Applicata*, 98(2), pp. 181–94.

Hulburt, D. J. (2001). Identifying Additional Insect Resistance Quantitative Trait Loci in Soybean Using Simple Sequence Repeats. M.S. Thesis, University of Georgia, Athens, GA.

Hulburt, D. J., Boerma, H. R. and All, J. N. (2004). Effect of pubescence tip on soybean resistance to lepidopteran insects. *Journal of Economic Entomology*, 97, pp. 621–7.

Jensen, R. L. (1968). Potential of Varietal Resistance for Control of Insect Pests of Soybeans. M.S. Thesis, Louisiana State Univ., Baton Rouge.

Jensen, R. L. and Newsom, L. D. (1972). Effect of stink bug-damaged soybean seeds on germination, emergence, and yield. *Journal of Economic Entomology*, 65(1), pp. 261–4.

Johnson, H. W. and Hollowell, E. A. (1935). Pubescent and glabrous characters of soybeans as related to resistance to injury by the potato leafhopper. *Journal of Agricultural Research*, 51, pp. 371–81.

Johnson, R., Narvaez, J., An, G. and Ryan, C. (1989). Expression of proteinase inhibitors I and II in transgenic tobacco plants: Effects on natural defense against *Manduca sexta* larvae. *Proceedings of the National Academy of Sciences*, 86(24), pp. 9871–5.

Jones, W. A. and Sullivan, M. J. (1978). Susceptibility of certain soybean cultivars to damage by stink bugs. *Journal of Economic Entomology*, 71(3), pp. 534–6.

Jones, W. A. and Sullivan, M. J. (1979). Soybean resistance to the southern green stink bug, *Nezara viridula*. *Journal of Economic Entomology*, 72(4), pp. 628–32.

Joshi, J. M. (1977). Research notes: Field screening of soybean germplasm (Maturity Group 00 to IV) against *H. zea* damage. *Soybean Genet Newsletter*, 4, pp. 46–56.

Jun, T. H., Mian, M. A. R. and Michel, A. P. (2012). Genetic mapping revealed two loci for soybean aphid resistance in PI 567301B. *Theoretical and Applied Genetics*, 124, pp. 13–22.

Jun, T. H., Mian, M. A. R. and Michel, A. P. (2013). Genetic mapping of three quantitative trait loci for soybean aphid resistance in PI 567324. *Heredity*, 111(1), pp. 16–22.

Kenty, M. M., Hinson, K., Quesenberry, K. H. and Wofford, D. (1996). Inheritance of resistance to the soybean looper in soybean. *Crop Science*, 36, pp. 1532–7.

Kester, K. M., Smith, C. M. and Gilman, D. F. (1984). Mechanisms of resistance in soybean (*Glycine max* [L.] Merrill) genotype PI171444 to the southern green stink bug, *Nezara viridula* (L.)(Hemiptera: Pentatomidae). *Environmental Entomology*, 13, pp. 1208–15.

Kilen, T. C., Hatchett, J. H. and Hartwig, E. E. (1977). Evaluation of early generation soybeans for resistance to soybean looper. *Crop Science*, 17(3), pp. 397–8.

Kim, C. S., Schaible, G., Garrett, L., Lubowski, R. and Lee, D. (2008). Economic impacts of the US soybean aphid infestation: A multi-regional competitive dynamic analysis. *Agricultural and Resource Economics Review*, 37(2), pp. 227–42.

Kim, K. S., Hill, C. B., Hartman, G. L., Mian, M. A. and Diers, B. W. (2008). Discovery of soybean aphid biotypes. *Crop Science*, 48(3), pp. 923–8.

Kim, K. S. and Diers, B. W. (2009). The associated effects of the soybean aphid resistance locus on soybean yield and other agronomic traits. *Crop Science*, 49(5), pp. 1726–32.

Kim, K. S., Bellendir, S., Hudson, K. A., Hill, C. B., Hartman, G. L., Hyten, D. L., Hudson, M. E. and Diers, B. W. (2010a). Fine mapping the soybean aphid resistance gene *Rag1* in soybean. *Theoretical and Applied Genetics*, 120, pp. 1063–71.

Kim, K. S., Hill, C. B., Hartman, G. L., Hyten, D. L., Hudson, M. E. and Diers, B. W. (2010b). Fine mapping of the soybean aphid-resistance gene *Rag2* in soybean PI 200538. *Theoretical and Applied Genetics*, 121, pp. 599–610.

Kim, K. S., Chirumamilla, A., Hill, C. B., Hartman, G. L. and Diers, B. W. (2014). Identification and molecular mapping of two soybean aphid resistance genes in soybean PI 587732. *Theoretical and Applied Genetics*, 127(5), pp. 1251–9.

Kraemer, M. E., Rangappa, M., Benepal, P. S. and Mebrahtu, T. (1988). Field evaluation of soybeans for Mexican bean beetle resistance. I. Maturity groups VI, VII, and VIII. *Crop Science*, 28(3), pp. 497–9.

Kraemer, M. E., Rangappa, M., Mebrahtu, T. and Benepal, P. S. (1990). Field evaluation of soybean for Mexican bean beetle resistance. II. Maturity Groups III, IV, and V. *Crop Science*, 30(2), pp. 374–7.

Lam, W. K. F. and Pedigo, L. P. (2001). Effect of trichome density on soybean pod feeding by adult bean leaf beetles (Coleoptera: Chrysomelidae). *Journal of Economic Entomology*, 94(6), pp. 1459–63.

Lambert, L. and Kilen, T. C. (1984a). Multiple insect resistance in several soybean genotypes. *Crop Science*, 24, pp. 887–90.

Lambert, L. and Kilen, T. C. (1984b) Influences of three soybean genotypes and their F1 intercrosses on the development of five insect species. *Journal of Economic Entomology*, 77, pp. 622–5.

Lambert, L. and Kilen, T. C. (1989). Influence and performance of soybean lines isogenic for pubescence type on oviposition preference and egg distribution by corn earworm (Lepidoptera: Noctuidae). *Journal of Entomological Science*, 24(3), pp. 309–16.

Lambert, L. and Tyler, J. (1999). Appraisal of insect-resistant soybeans. In: Webster, J. A. and Webster, B. R. (Eds), *Economic, Environmental, and Social Benefits of Insect Resistance in Field Crops*. Thomas Say Publ., Entomol. Soc. of America, Lanham, MD, pp. 131–48.

Lawo, N. C., Wäckers, F. L. and Romeis, J. (2009). Indian Bt cotton varieties do not affect the performance of cotton aphids. *PLoS ONE*, 4(3), pp. e4804.

Layton, M. B., Boethel, D. J. and Smith, C. M. (1987). Resistance to adult bean leaf beetle and banded cucumber beetle (Coleoptera: Chrysomelidae) in soybean. *Journal of Economic Entomology*, 80(1), pp. 151–5.

Lecardonnel, A., Chauvin, L., Jouanin, L., Beaujean, A., Prévost, G. and Sangwan-Norreel, B. (1999). Effects of rice cystatin I expression in transgenic potato on Colorado potato beetle larvae. *Plant Science*, 140(1), pp. 71–9.

Lemos Filho, J. P. D. and Paiva, É. A. S. (2006). The effects of sooty mold on photosynthesis and mesophyll structure of mahogany (Swietenia macrophylla King., Meliaceae). *Bragantia*, 65(1), pp. 11–17.

Leskey, T. C., Hamilton, G. C., Nielsen, A. L., Polk, D. F., Rodriguez-Saona, C., Bergh, J. C., Herbert, D. A., Kuhar, T. P., Pfeiffer, D., Dively, G. P. and Hooks, C. R. (2012). Pest status of the brown marmorated stink bug, *Halyomorpha halys* in the USA. *Outlooks on Pest Management*, 23(5), pp. 218–26.

Li, Y., Hill, C. B. and Hartman, G. L. (2004). Effect of three resistant soybean genotypes on the fecundity, mortality, and maturation of soybean aphid (Homoptera: Aphididae). *Journal of Economic Entomology*, 97(3), pp. 1106–11.

Li, Y., Hill, C. B., Carlson, S. R., Diers, B. W. and Hartman, G. L. (2007). Soybean aphid resistance genes in the soybean cultivars Dowling and Jackson map to linkage group M. *Molecular Breeding*, 19(1), 25–34.

Luedders, V. D. and Dickerson W. A. (1977). Resistance of selected soybean genotypes and segregating populations to cabbage looper feeding. *Crop Science*, 17, pp. 395–6.

Lundin, O., Rundlöf, M., Smith, H. G., Fries, I. and Bommarco, R. (2015). Neonicotinoid insecticides and their impacts on bees: A systematic review of research approaches and identification of knowledge gaps. *PLoS ONE*, 10(8), p. e0136928.

Malumphy, C. P. (1997). Morphology and anatomy of honeydew eliminating organs. In: Ben-Dov, Y. and Hodgson, C. J. (Eds) *World Crop Pests*. Elsevier Science B.V., Amsterdam, the Netherlands, 7, pp. 269–74.

McAuslane, H. J. (1996). Influence of leaf pubescence on ovipositional preference of Bemisia argentifolii (Homoptera: Aleyrodidae) on soybean. *Environmental Entomology*, 25(4), pp. 834–41.

McCarville, M., Hodgson, E. and O'Neal, M. (2012). Soybean Aphid- Resistant Soybean Varieties for Iowa. Iowa State University Extension and Outreach PM 3023, Ames, IA, USA.

McCarville, M. T., O'Neal, M. E., Potter, B. D., Tilmon, K. J., Cullen, E. M., McCornack, B. P., Tooker, J. F. and Prischmann-Voldseth, D. A. (2014). One gene versus two: A Regional study on the efficacy of single gene versus pyramided resistance for soybean aphid management. *Journal of Economic Entomology*, 107, pp. 1680–7.

McPherson, J. E. and McPherson, R. (2000). *Stink Bugs of Economic Importance in America North of Mexico*. CRC Press, Boca Raton, pp. 71–100.

McPherson, R. M., Todd J. W. and Yeargan K. V. (1994). Stink bugs. In: L. G. Higley and D. J. Boethel (Eds.), *Handbook of Soybean Insect Pests*. Entomological Society of America, Lanham, MD, pp. 87–90.

McPherson, R. M. and Buss, G. R. (2007). Evaluating lepidopteran defoliation resistance in soybean breeding lines containing the stink bug (Hemiptera: Pentatomidae) resistance IAC-100 cultivar in their pedigrees. *Journal of Economic Entomology*, 100(3), pp. 962–8.

Meng, F., Han, Y., Teng, W., Li, Y. and Li, W. (2011). QTL underlying the resistance to soybean aphid (Aphis glycines Matsumura) through isoflavone-mediated antibiosis in soybean cultivar 'Zhongdou 27'. *Theoretical and Applied Genetics*, 123, pp. 1459–65.

Mensah, C., DiFonzo, C., Nelson, R. L. and Wang, D. (2005). Resistance to soybean aphid in early maturing soybean germplasm. *Crop Science*, 45(6), pp. 2228–33.

Mensah, C. (2008). Inheritance of Aphid Resistance in PI 567541B and PI 567598B, Identification of Aphid Resistance QTL in PI 567598B, and a New Aphid Biotype in Michigan, Dissertation, Michigan State University, East Lansing, MI.

Mian, M. A. R., Hammond, R. B. and St Martin, S. K. (2008a). New plant introductions with resistance to the soybean aphid. *Crop Science*, 48(3), pp. 1055–61.

Mian, M. R., Kang, S. T., Beil, S. E. and Hammond, R. B. (2008b). Genetic linkage mapping of the soybean aphid resistance gene in PI 243540. *Theoretical and Applied Genetics*, 117(6), pp. 955–62.

Michel, A. P., Mian, M. R., Davila-Olivas, N. H. and Cañas, L. A. (2010). Detached leaf and whole plant assays for soybean aphid resistance: Differential responses among resistance sources and biotypes. *Journal of Economic Entomology*, 103(3), pp. 949–57.

Miner, F. D. (1966). Biology and control of stink bugs on soybeans. Ark. Agric. Exp. Stn. Bull. No. 708.

Musser F. R., Lorenz G. M., Stewart S. D. and Catchot A. L. (2010). 2009 Soybean insect losses for Mississippi, Tennessee, and Arkansas. *Midsouth Entomology*, 2010(3), pp. 48–54.

Narvel, J. M., Walker, D. R., Rector, B. G., All, J. N., Parrott, W. A. and Boerma, H. R. (2001). A retrospective DNA marker assessment to the development of insect resistant soybean. *Crop Science*, 41, pp. 1931–9.

Nurden, A. V., Scott, R. A., Hesler, L., Tilmon, K., Glover, K. and Carter, C. (2010). Inheritance of soybean aphid resistance from PI 71506. *Journal of Crop Improvement*, 24, pp. 400–16.

Ohnesorg, W. J., Johnson, K. D. and O'Neal, M. E. (2009). Impact of reduced risk insecticides on soybean aphid and their natural enemies. *Journal of Economic Entomology*, 102, pp. 1816–26.

Ortega, M. A., All, J. N., Boerma, H. R. and Parrott, W. A. (2016). Pyramids of QTLs enhance host–plant resistance and Bt-mediated resistance to leaf-chewing insects in soybean. *Theoretical and Applied Genetics*, 129(4), pp. 703–16.

Painter, R. H. (1958). Resistance of Plants to Insects. *Annual Review of Entomology*, 3, pp. 267–90.

Parrott, W., Walker, D., Zhu, S., Boerma, H. R. and All, J. (2008). Genomics of insect-soybean interactions. In Stacey, G. (Ed.), *Genetics and Genomics of Soybean*. Springer, New York, pp. 269–91.

Piubelli, G. C., Hoffmann-Campo, C. B., De Arruda, I. C., Franchini, J. C. and Lara, F. M., (2003). Flavonoid increase in soybean as a response to *Nezara viridula* injury and its effect on insect-feeding preference. *Journal of Chemical Ecology*, 29(5), pp. 1223–33.

Piubelli, G. C., Hoffmann-Campo, C. B., Moscardi, F., Miyakubo, S. H. and De Oliveira, M. C. N. (2005). Are chemical compounds important for soybean resistance to Anticarsia gemmatalis? *Journal of Chemical Ecology*, 31(7), pp. 1509–25.

Ragsdale, D. W., Landis, D. A., Brodeur, J., Heimpel, G. E. and Desneux, N. (2011). Ecology and management of the soybean aphid in North America. *Annual Review of Entomology*, 56, pp. 375–99.

Rahbé, Y., Ferrasson, E., Rabesona, H. and Quillien, L. (2003). Toxicity to the pea aphid Acyrthosiphon pisum of anti-chymotrypsin isoforms and fragments of Bowman–Birk protease inhibitors from pea seeds. *Insect Biochemistry and Molecular Biology*, 33(3), pp. 299–306.

Raps, A., Kehr, J., Gugerli, P., Moar, W. J., Bigler, F. and Hilbeck, A. (2001). Immunological analysis of phloem sap of Bacillus thuringiensis corn and of the non-target herbivore *Rhopalosiphum padi* (Homoptera: Aphididae) for the presence of Cry1Ab. *Molecular Ecology*, 10, pp. 525–33.

Rector, B. G., All J. N., Parrott W. A. and Boerma H. R. (1998). Identification of molecular markers associated with quantitative trait loci for soybean resistance to corn earworm. *Theoretical and Applied Genetics*, 96, pp. 786–90.

Rector, B. G., All, J. N., Parrott, W. A. and Boerma, H. R. (1999). Quantitative trait loci for antixenosis resistance to corn earworm in soybean. *Crop Science*, 39, pp. 531–8.

Rector, B. G., All, J. N., Parrott, W. A. and Boerma, H. R. (2000). Quantitative trait loci for antibiosis resistance to corn earworm in soybean. *Crop Science*, 40, pp. 233–8.

Rosseto, C. J. (1989). Breeding for resistance to stink bugs. In: Pascale, A. J. (Ed.), *Proceedings of the World Soybean Research Conference IV*, 5–9 March 1989, Buenos Aires, Argentina. Assoc. Argentina de la Soja Press, Buenos Aires, Argentina, pp. 2046–60.

Rossetto, C. J., Gallo, P. B., Razera, L. F., Bortoletto, N., Igue, T., Medina, P. F., Tisselli O. F., Aquilera, V., Veiga, R. F. A. and Pinheiro, J. B. (1995) Mechanisms of resistance to stink bug complex in the soybean cultivar 'IAC-100'. *Anais-Sociedade Entomologica Do Brasil*, 24(3), pp. 517–22.

Rowan, G. B., Boerma, H. R., All, J. N. and Todd, J. (1991). Soybean cultivar resistance to defoliating insects. *Crop Science*, 31(3), pp. 678–82.

Ruberson, J. R., Takasu, K., Buntin, G. D., Eger Jr., J. E., Gardner, W. A., Greene, J. K., Jenkins, T. M., Jones, W. A., Olson, D. M., Roberts, P. M. and Suiter, D. R. (2013). From Asian curiosity to eruptive American pest: *Megacopta cribraria* (Hemiptera: Plataspidae) and prospects for its biological control. *Applied Entomology and Zoology*, 48(1), pp. 3–13.

Rufener II, G. K., St. Martin, S. K., Cooper, R. C. and Hammond, R. B. (1989). Genetics of antibiosis resistance to Mexican bean beetle in soybean. *Crop Science*, 29, pp. 618–22.

Ryan, C. A. (1990). Protease inhibitors in plants: Genes for improving defenses against insects and pathogens. *Annual Review of Phytopathology*, 28(1), pp. 425–49.

Schmutz, J., Cannon, S. B., Schlueter, J., Ma, J., Mitros, T., Nelson, W., Hyten, D. L., Song, Q., Thelen, J. J., Cheng, J., Xu, D., Hellsten, U., May, G. D., Yu, Y., Sakurai, T., Umezawa, T., Bhattacharyya, M. K., Sandhu, D., Valliyodan, B., Lindquist, E., Peto, M., Grant, D., Shu, S., Goodstein, D., Barry, K., Futrell-Griggs, M., Abernathy, B., Du, J., Tian, Z., Zhu, L., Gill, N., Joshi, T., Libault,

M., Sethuraman, A., Zhang, X., Shinozaki, K., Nguyen, H. T., Wing, R. A., Cregan, P., Specht, J., Grimwood, J., Rokhsar, D., Stacey, G., Shoemaker, R. C. and Jackson, S. A. (2010). Genome sequence of the palaeopolyploid soybean. *Nature*, 463, pp. 178–83.

Seiter, N. J., Greene, J. K. and Reay-Jones, F. P. (2013). Reduction of soybean yield components by *Megacopta cribraria* (Hemiptera: Plataspidae). *Journal of Economic Entomology*, 106(4), pp. 1676–83.

Sharma, H. C. and Norris, D. M. (1991). Chemical basis of resistance in soy bean to cabbage looper, *Trichoplusia ni*. *Journal of Science Food and Agriculture*, 55, pp. 353–64.

Silva, J. P., Baldin, E. L., Souza, E. S., Canassa, V. F. and Lourenção, A. L. (2013). Characterization of antibiosis to the redbanded stink bug *Piezodorus guildinii* (Hemiptera: Pentatomidae) in soybean entries. *Journal of Pest Science*, 86(4), pp. 649–57.

Singh, B. B., Hadley, H. H. and Bernard, R. L. (1971). Morphology of pubescence in soybeans and its relationship to plant vigor. *Crop Science*, 11(1), pp. 13–16.

Sisson, V. A., Miller, P. A., Campbell, W. V. and Van Duyn, J. W. (1976). Evidence of inheritance of resistance to the Mexican bean beetle in soybeans. *Crop Science*, 16(6), pp. 835–7.

Smith, T. J., Camper Jr., H. M. and Schillinger, J. A. (1975). Registration of Shore soybean. *Crop Science*, 15, p. 100.

Song, Q., Hyten, D. L., Jia, G., Quigley, C. V., Fickus, E. W., Nelson, R. L. and Cregan, P. B. (2013). Development and evaluation of SoySNP50K, a high-density genotyping array for soybean. *PLoS ONE*, 8(1), p. e54985.

Souza, R. F. and Toledo, J. F. F. (1995). Genetic analysis of soybean resistance to stinkbug. *Brazilian Journal of Genetics*, 18(4), pp. 593–8.

SoyStats. (2016) http://soystats.com/.

Stamp, N. (2003). Out of the quagmire of plant defense hypotheses. *The Quarterly Review of Biology*, 78(1), pp. 23–55.

Stewart, C. N., Adang, M. J., All, J. N., Boerma, H. R., Cardineau, G., Tucker D. and Parrott, W. A. (1996). Genetic transformation, recovery, and characterization of fertile soybean transgenic for a synthetic *Bacillus thuringiensis* cry1Ac gene. *Plant Physiology*, 112, pp. 121–9.

Suiter, D. R., Eger, J. E., Gardner, W. A., Kemerait, R. C., All, J. N., Roberts, P. M., Greene, J. K., Ames, L. M., Buntin, G. D., Jenkins, T. M. and Douce, G. K. (2010). Discovery and distribution of *Megacopta cribraria* (Hemiptera: Heteroptera: Plataspidae) in northeast Georgia. *Journal of Integrated Pest Management*, 1(1), pp. F1–F4.

Temple, J., Davis, J. A., Hardke, J., Price, P., Micinski, S., Cookson, C., Richter, A. and Leonard, B. R. (2011). Seasonal abundance and occurrence of the redbanded stink bug in Louisiana soybeans. *Louisiana Agriculture*, 54, pp. 20–2.

Terry, L. I., Chase, K., Orf, J., Jarvik, T., Mansur, L. and Lark, K. G. (1999). Insect resistance in recombinant inbred soybean lines derived from non-resistant parents. *Entomologia Experimentalis et Applicata*, 91(3), pp. 467–78.

Terry, L. I., Chase, K., Jarvik, T., Orf, J., Mansur, L. and Lark, K. G. (2000). Soybean quantitative trait loci for resistance to insects. *Crop Science*, 40(2), pp. 375–82.

Thomas, G. D., Ignoffo, C. M., Morgan, C. E. and Dickerson, W. A. (1974). Southern green stink bug: Influence on yield and quality of soybeans. *Journal of Economic Entomology*, 67(4), pp. 501–3.

Thomas, J. C., Adams, D. G., Keppenne, V. D., Wasmann, C. C., Brown, J. K., Kanost, M. R. and Bohnert, H. J. (1995). *Manduca sexta* encoded protease inhibitors expressed in *Nicotiana tabacum* provide protection against insects. *Plant Physiology and Biochemistry*, 33(5), pp. 611–14.

Todd, J. W., Jellum, M. D. and Leuck, D. B. (1973). Effects of southern green stink bug damage on fatty acid composition of soybean oil. *Environmental Entomology*, 2(4), pp. 685–90.

Todd, J. W. and Turnipseed, S. G. (1974). Effects of southern green stink bug damage on yield and quality of soybeans. *Journal of Economic Entomology*, 67(3), pp. 421–6.

Tran, P., Cheesbrough, T. M. and Keickhefer, R. W. (1997). Plant proteinase inhibitors are potential anticereal aphid compounds. *Journal of Economic Entomology*, 90(6), pp. 1672–7.

Turnipseed, S. G. (1977). Influence of trichome variations on populations of small phytophagous insects in soybean. *Environmental Entomology*, 6, pp. 815–17.

Turnipseed, S. G. and Kogan, M. (1987) Integrated control of insects. In: J. R. Wilcox (Ed.), *Soybeans: Improvement, Production, and Uses*. ASA, CSSA and SSSA, Madison, pp. 779–817.

Van Duyn, J. W., Turnipseed, S. G. and Maxwell, J. D. (1971). Resistance in soybeans to the Mexican bean beetle. I. Sources of Resistance. *Crop Science*, 11(4), pp. 572–3.

Walker, D., Boerma, H. R., All, J. and Parrott, W. (2002). Combining cry1Ac with QTL alleles from PI 229358 to improve soybean resistance to lepidopteran pests. *Molecular Breeding*, 9, pp. 43–51.

Walker, D. R., Narvel, J. M., Boerma, H. R., All, J. N. and Parrott W. A. (2004) A QTL that enhances and broadens Bt insect resistance in soybean. *Theoretical and Applied Genetics*, 109, pp. 1051–7.

Wasmann, C. C., Echt, C., Dunn, R. L., Bohnert, H. J. and McCoy, T. J. (1994). Introduction and expression of an insect proteinase inhibitor in alfalfa (*Medicago sativa* L.). *Plant Cell Reporter*, 14, pp. 31–6.

Wiarda, S. L., Fehr, W. R. and O'Neal, M. E. (2012). Soybean aphid (*Hemiptera: Aphididae*) development on soybean with Rag1 alone, Rag2 alone, and both genes combined. *Journal of Economic Entomology*, 105(1), pp. 252–8.

Xiao, L., Hu, Y., Wang, B. and Wu, T. (2013). Genetic mapping of a novel gene for soybean aphid resistance in soybean (*Glycine max* [L.] Merr.) line P203 from China. *Theoretical and Applied Genetics*, 126(9), pp. 2279–87.

Xiao, L., Zhong, Y. P., Wang, B. and Wu, T. L. (2014). Mapping an aphid resistance gene in soybean [*Glycine max* (L.) Merr.] P746. *Genetics and Molecular Research*, 13(4), pp. 9152–60.

Yesudas, C. R., Sharma, H. and Lightfoot, D. A. (2010). Identification of QTL in soybean underlying resistance to herbivory by Japanese beetles (*Popillia japonica*, Newman). *Theoretical and Applied Genetics*, 121(2), pp. 353–62.

Zhang, G., Gu, C. and Wang, D. (2009). Molecular mapping of soybean aphid resistance genes in PI 567541B. *Theoretical and Applied Genetics*, 118, pp. 473–82.

Zhang, G., Gu, C. and Wang, D. (2010). A novel locus for soybean aphid resistance. *Theoretical and Applied Genetics*, 120, pp. 1183–91.

Zhang, G., Gu, C. and Wang, D. (2013). Mapping and validation of a gene for soybean aphid resistance in PI 567537. *Molecular Breeding*, 32, pp. 131–8.

Zhang, S., Zhang, Z., Bales, C., Gu, C., DiFonzo, C., Li, M., Song, Q., Cregan, P., Yang, Z. and Wang, D. (2017a). Mapping novel aphid resistance QTL from wild soybean, Glycine soja 85-32. *Theoretical and Applied Genetics* 130, pp. 1941–52 doi: https://doi.org/10.1007/s00122-017-2935-z.

Zhang, S., Zhang, Z., Wen, Z., Gu, C., An, Y., Bales, C., DiFonzo, C., Song, Q. and Wang, D. (2017b). Fine mapping of the soybean aphid resistance genes Rag6 and Rag3c from Glycine soja 85-32. *Theoretical and Applied Genetics* doi: https://doi.org/10.1007/s00122-017-2979-0.

Zhu, S., Walker, D. R., Boerma, H. R., All, J. N. and Parrott W. A. (2006). Fine mapping of a major insect resistance QTL in soybean and its interaction with minor resistance QTLs. *Crop Science*, 46, pp. 1094–9.

Zhu, S., Walker, D. R., Warrington, C. V., Parrott, W. A., All, J. N., Wood, E. D. and Boerma, H. R. (2007). Registration of four soybean germplasm lines containing defoliating insect resistance QTLs from PI 229358 introgressed into 'Benning'. *Journal of Plant Registrations*, 1(2), pp. 162–3.

Zhu, S., Walker, D. R., Boerma, H. R., All, J. N. and Parrott, W. A. (2008) Effects of defoliating insect resistance QTLs and a *cry1Ac* transgene in soybean near-isogenic lines. *Theoretical and Applied Genetics*, 116, pp. 455–63.

Chapter 7

Integrated weed management in soybean cultivation

Bob Hartzler, Iowa State University, USA

1 Introduction
2 Establishing weed-management goals
3 Management tactics: crop rotation, competitiveness of soybean and use of cover crops
4 Further management tactics
5 Interactions between weeds and other pest complexes
6 Factors limiting the adoption of diversified weed management
7 Conclusion
8 Where to look for further information
9 References

1 Introduction

1.1 Overview

Successful weed management is an essential component of soybean production. Unlike other pests (e.g. insects, pathogens, etc.), almost all soybean (*Glycine max*) fields have weed infestation that must be managed for farmers to sustain economic viability. Weed-control tactics are among the most disruptive practices associated with soybean production; thus, advances in sustainable soybean production must include development of improved systems for managing weeds.

Widespread adoption of herbicide-based weed control in industrialized countries has isolated weed-management practices from other aspects of soybean production. Integrated weed management involves a systems approach that considers contributions of plant breeding, cultural, mechanical and chemical practices in suppressing weed populations and minimizing the impacts of weeds on yield (Deytieux et al., 2012). Integrated weed management changes the focus from eliminating weed interference with soybean growth and yield, to managing the weed seed bank and population fluctuations. Shaner (2014) stated that simple weed-management systems are not sustainable and diversification is the only system that will provide long-term weed management. In addition, cropping systems that utilize integrated weed management improve most environmental indicators (e.g.

http://dx.doi.org/10.19103/AS.2017.0034.22

energy input, greenhouse gas emission, ecotoxicity) when compared with a conventional system (Deytieux et al., 2012). While the benefits of diversified weed management are well documented, market forces have favoured continued use of simplified weed-management schemes based on herbicides.

For the remainder of this introduction, we will discuss the impacts of herbicides on soybean production to show the importance of diversification in weed management.

1.2 Impacts of herbicides on soybean production

Industrialized forms of agriculture are utilized in many soybean-producing regions of the world. These production systems are heavily reliant on herbicides for weed management, and the effectiveness of these tools has greatly reduced the use of alternative control tactics. Advantages provided by herbicides include dramatic reductions in labour, adoption of no-till and other forms of conservation tillage and increases in farm size (Gianessi and Reigner, 2007). The introduction of herbicide-resistant soybean varieties in the late 1990s, particularly glyphosate-resistant varieties, further increased the reliance on herbicides for weed management (Green, 2012). The adoption and impact of genetically modified herbicide-resistant crops have been the focus of several reviews (Bonny, 2016; Sandermann, 2006; Benbrook, 2012; Owen and Zelaya, 2005).

One of the profound impacts of the heavy reliance on herbicides has been the evolution of herbicide-resistant weeds (Gould, 1991; Neve, 2007). The rapid rise of weed biotypes with multiple resistances is viewed as a threat to the sustainability of current production systems and world crop production (Yu and Powles, 2014; Délye et al., 2013). In response to the threat of herbicide-resistant weeds, a call for diversification in weed management has been a common theme among weed scientists (Harker and O'Donovan, 2013; Norsworthy et al., 2012; Shaner and Beckie, 2014).

2 Establishing weed-management goals

2.1 Establishing goals for weed management

For most soybean producers, the primary objective in managing weeds is minimizing the effects of weeds on crop yields. Economic thresholds are intended to assist producers in making better weed-management decisions (Coble and Mortensen, 1992; Neeser et al., 2004), and have been a fundamental component of many integrated pest-management programmes. Economic thresholds utilize predictions of the yield impact of weed infestations to determine economic returns from implementing control tactics. Weed infestations below the threshold level are allowed to remain in the field since the yield gained from control would not recoup the expense of controlling the weed.

While economic thresholds focus on weed densities' impact on crop yields following full-season competition, critical periods of competition consider the time of removal of weeds on their impact on crop yields (van Heemst, 1985). Critical periods of competition are based on the assumption that weeds can be allowed to compete with crops for a portion of the growing season without affecting crop yields. The availability of post-emergence herbicides that provided greater flexibility in application timing markedly increased interest in determining when weeds that emerge with the crop first begin to impact yield (Van Acker et al., 1993; Knezevic et al., 2003).

Although considerable effort has been invested in developing economic thresholds for weeds, these tools have not been accepted commercially (Czapar et al., 1995). The high variability in soybean yield response to weed infestations, patchy weed distributions and other costs associated with weeds (e.g. harvest efficiency, seed production, aesthetics, etc.) all contribute to the low-adoption rate of economic thresholds.

A limitation of most economic thresholds for weed management is that they take a short-term view of the costs of weeds on crop production (Jones and Medd, 2000). Economic optimum thresholds take into account the impact of weeds allowed to remain in the field on future management. They are a modified threshold that considers the impact of weed seed production by weeds allowed to remain in the field. The economic optimum threshold for two broadleaf weeds in soybean was 7.5- and 3.6-fold lower than the single-year economic threshold (Bauer and Mortensen, 1992). When taking into account the cost of seed produced by weeds, 'acceptable' densities may be low enough that subjective evaluations provide sufficient precision in determining when to implement control tactics (Hartzler, 1997). The evolution of multiple herbicide resistances in weeds has led to the recommendation of zero thresholds, or no-seed thresholds, for certain problematic weeds (Norsworthy et al., 2012, 2014). This approach establishes the goal of preventing survival of any weeds within a field.

2.2 Managing the weed seed bank

Most weed-management strategies are intended to either reduce the establishment of weeds or control weeds early in the season in order to protect crop yields. Since tactics used to control weeds (e.g. herbicides, tillage) generally are density independent, the number of weeds that survive is directly related to the magnitude of the seed bank (Gallandt, 2006; Deileman et al., 1999; Hartzler and Roth, 1993). Establishing a goal of minimizing the size of the weed seed bank will enhance the effectiveness of all weed-control tactics by reducing the likelihood of weeds surviving control strategies.

The magnitude of a field's seed bank is determined by the relative rate of seed inputs compared with seed losses (Buhler et al., 1997). Weeds that escape control within a field contribute the majority of seeds to that field's seed bank; thus, achieving high levels of weed control and enhancing crop competitiveness with weeds minimize contributions to the seed bank. Davis (2006) used population models to assess the value of managing seed banks of weeds encountered in different ecosystems. Enhancing seed losses from the seed bank strongly influenced population dynamics of summer annual weeds typical of soybean fields, whereas seed bank dynamics had much less effect on the success of biennial and perennial weeds common to less disturbed habitats.

Although the majority of weed seeds in a field's seed bank are produced by weeds escaping control within the field, external seed sources introduce new weed species into the fields. Mechanical dispersal is the primary means of movement for most important weeds of soybean. Farm equipment, including tillage implements, planters, sprayers and especially harvesters are capable of moving seed from infested fields to clean fields (Thill and Mallory-Smith, 1997).

Increases in herbicide-resistant weeds have led to research and development of tools intended to intercept and destroy weed seeds before they enter the seed bank (Walsh and Powles, 2007). These tools generally are coupled with the machine used to harvest the crop. Much of this research has been conducted in Australia where multiple herbicide-resistant weeds have threatened dryland production systems (Walsh and Powles, 2007;

Walsh et al., 2012). This strategy is effective only on weed species that retain the majority of seed on mother plants until crop harvest (Forcella et al., 1996; Walsh et al., 2014).

An alternative method of reducing the magnitude of the seed bank is to enhance seed predation and other sources of biological losses. A variety of seed predators, both invertebrates and vertebrates, consume significant quantities of weed seed, both prior to and following dispersal (Marino et al., 1997). Seed predation was shown to be important in regulating weed populations in production systems less reliant on herbicides (Westerman et al., 2005). An increased understanding of how management practices and the environment regulate the activity of seed predators is required to increase suppression of weed populations by predation (Davis and Raghu, 2010).

A characteristic of most weedy species is persistent seed. Soil-borne microbes influence the survival of seed within the soil profile (Kremer, 1993; Dalling et al., 2011). Different types of defence systems against microbial antagonists were predicted to occur for weed species with different dormancy types (Dalling et al., 2011). The authors stated that an understanding of these seed defence mechanisms could lead to the development of innovative management strategies that reduce weed seed persistence in the soil.

The high level of effectiveness of herbicides has led to a change in many producers' perspective for establishing weed-management goals. Weeds are commonly viewed as acceptable as long as they do not interfere with productivity in terms of yield or harvest efficiency. In order to develop more efficient and sustainable soybean-production systems, the focus of weed management must shift from protecting crop yields to managing the weed seed bank (Gallandt, 2006; Davis, 2006).

3 Management tactics: crop rotation, competitiveness of soybean and use of cover crops

3.1 Crop rotation

The benefits of crop rotation in managing weeds are well documented. A review of crop rotation experiments that examined weed communities reported that weed densities were lower within diverse rotations compared with monocultures in 78% of the experiments (Liebman and Dyck, 1992). Crop rotation influenced weed seed bank density more than the tillage system in a long-term study (Cardina et al., 2002).

Several factors contribute to improved suppression and management of weeds in diverse rotations. Rotations that include crops with a single life cycle (e.g. summer annuals) allow weeds with a similar life cycle to increase over time compared with rotations including crops with different life cycles (Blackshaw et al., 1994; Cardina et al., 2002). Diverse rotations, typically including crops with different growth and competitive characteristics, allow the use of a wider range of control tactics than simple rotations, and are produced using different cultural practices (e.g. row spacing, planting density, etc.).

A long-term study has compared weed management and population dynamics in the corn (*Zea mays*)–soybean rotation that dominates the United States Corn Belt to 3- and 4-year rotations, which include a small grain and forage crop (Davis et al., 2012; Liebman et al., 2014). Over an 8-year period, herbicide use was reduced by more than 80% in the 3- and 4-year rotations compared with the 2-year rotation. Although weed densities were higher in the diverse rotations than the conventional 2-year rotation, grain yields and profitability were equal or greater in the diverse rotations (Davis et al., 2012). Weed species responded differently in sensitivity to the three rotations (Liebman et al., 2014). Velvetleaf (*Abutilon theophrasti*) emergence was reduced, and seed mortality was increased by more than 70% in 3- and 4-year rotations compared with a 2-year rotation (corn–soybean), whereas giant foxtail (*Setaria faberi*) life-history processes were unaffected by the rotations. Velvetleaf seed production was more than 25-fold greater in the 4-year rotation than in the 2-year rotation. However, greater seed predation occurred in the 4-year rotation, and these losses along with other mortality events prevented the populations from increasing over time.

Crop rotations that include dissimilar crops result in weeds being exposed to multiple stresses and mortality events. Successful weeds are those that are best adapted to the interactions between the site characteristics, the crops produced and the forms of disturbances utilized (Lègére and Samson, 1999). Westerman et al. (2005) proposed that multiple stress factors placed on weeds in diverse cropping systems act as a buffer against rapid, uncontrollable weed population increases commonly observed in simple systems relying on a few control tactics for weed management. Although the multiple benefits of diverse rotations are well documented, current farm policies, existing infrastructure and economic viability of alternative systems discourage the types of rotations that enhance weed management in many areas of soybean production.

3.2 Enhancing the competitiveness of soybean

The ability of the soybean canopy to suppress weeds is one of the strongest weed-management tools available in the struggle against weeds. The competitiveness of a crop is influenced by its planting density and arrangement, planting date and inherent characteristics of the crop cultivar. The time required to develop a closed canopy is directly related to row spacing (Weber et al., 1966). More rapid development of the soybean canopy in narrow rows increased early season crop tolerance to weeds (Knezevic et al., 2003) and also reduced the establishment of late season weeds (Yelverton and Coble, 1991). Many factors influence grower decisions regarding row spacing, including equipment costs and speed of planting (Lambert and Lowenberg-DeBoer, 2003), yield potential (DeBruin and Pedersen, 2008) and incidence of other pest complexes (Grau and Radke, 1984; Buchanan et al., 2015). The importance placed on row spacing for weed management is influenced by the availability of effective weed-control tactics for the specific production system.

Prior to the development of modern herbicides, soybean seeding rates were generally higher than needed for optimum yields (DeBruin and Pedersen, 2008). Higher seeding rates increase competitiveness of soybean with weeds, therefore reducing crop yield losses and weed biomass (Weber and Staniforth, 1957). As soybean seed costs increased, recommended seeding rates were reduced in many production areas (Basol et al., 2013; DeBruin and Pedersen, 2008). As with row spacing, the importance of seeding rates for weed management varies among production systems. Place et al. (2009) stated that seeding rates greater than 550 000 seed/ha in 76 cm row spacing provided improved weed control, yields and profits in organic soybean systems. In contrast, seeding rates of 185 000 and 309 000 seed/ha provided at least 95% of the maximum yield and were more profitable than higher seeding rates (DeBruin and Pedersen, 2008). The value of higher seeding rates as a component of integrated weed management has been inconsistent. Increased seeding rates had little impact on weed control in systems where herbicides provided effective control (Bell et al., 2015; DeWerff et al., 2014). In contrast, seeding rates between 240 000 and 420 000 seeds/ha did not consistently affect weed density, but end-of-season weed biomass was inversely related to the soybean seeding rate (Arce et al., 2009).

Differences in competitiveness of cultivars have been reported in many crop species, including soybean (Callaway, 1992; Bussan et al., 1997). Traits that lead to rapid canopy development and capture of light resources (e.g. quick emergence and rapid early season vegetative growth, high leaf area index and profuse tillering or branching) have consistently been identified with increased crop competitiveness against weeds (Callaway, 1992; Wortmann, 1993; Jordan, 1993). Soybean height 6–7 weeks after emergence was found to have a strong correlation with weed suppressive ability and highly heritable (Jannink et al., 1992). Jordan (1993) stated that an improved understanding of any yield costs related to enhanced competitiveness in crops and the extent that the adoption of competitive cultivars would allow other weed-control tactics to be reduced is needed prior to undertaking development of breeding programmes for this trait.

Although competition is the most common interaction among plants growing in close proximity, other processes can interfere with crop growth (Fuerst and Putman, 1983). Selection of allelopathic cultivars has been evaluated in numerous crops (Jabran et al., 2015), but there are no reports in the literature of soybean cultivars demonstrating allelopathic properties. Plants growing in the presence of other plants undergo a variety of responses associated with alteration of the red: far red light ratio (Franklin, 2008). The shade-avoidance response has been linked to yield losses associated with the presence of weeds early in the growing season (Green-Tracewicz et al., 2001; Page et al., 2010). Horvath et al. (2015) identified several downregulated genes in soybean plants grown in the presence of weeds, and suggested the genes could serve as a target for manipulating weed tolerance.

3.3 Cover crops

Concerns over the environmental impacts of agriculture on soil and water quality and the increase in herbicide-resistant weeds have increased interest in cover crops (Schipanski, 2014). Cover crops may impact weed communities via competition for limited resources, release of allelochemicals, altering the physical and microbial characteristics of the soil and influencing seed predation (Birthisel et al., 2014; Conklin et al., 2002; Lou et al., 2016; Teasdale, 1996). Weed suppression by cereal rye (*Secale cereale*) is closely associated with the reduction in light at the soil surface, indicating the importance of the physical barrier provided by the mulch. The amount of rye biomass required to reduce Palmer amaranth emergence and light at the soil surface was 5200 and 5370 kg ha^{-1}, respectively (Webster et al., 2016).

Species used as cover crops vary widely in their ability to suppress weeds and provide other ecosystem services (Teasdale, 1996). Wyse (1994) stated that progress in using cover crops for weed management would be dependent on developing species specifically for this purpose. Characteristics for an ideotype of a spring-seeded cover crop in the north central region of the United States included rapid seedling emergence, horizontal leaf angle, maximum height of 10 cm, life cycle of five weeks or less and non-dormant seed (DeHaan et al., 1994). Genetic engineering was used to demonstrate the concept of developing a cover crop that would self-terminate with the arrival of warm temperatures in spring, therefore eliminating the need for tillage or herbicides to kill the cover crop, and also eliminate negative effects on crop growth (Stanislaus et al., 2002).

Mirsky et al. (2015) reviewed the role and challenges of cover crops in organic, no-till soybean production. In this system, cover crops are required to contribute more to weed management than in other production systems due to a lack of tillage or herbicides, and thus, the cover crops must be managed specifically to optimize performance. A threshold of 8000 kg ha^{-1} cover crop biomass was suggested for consistent weed suppression. Timing of cover crop termination is critical to maximize weed suppression, and varies among cover crop species (Keene et al., 2016). Wortman et al. (2013) reported that the cover crop termination method influenced weed biomass more than the cover crop species or mixtures used in organic production systems, illustrating the importance in appropriate management of cover crop residue to maximize the contribution of cover crops to weed management.

4 Further management tactics

4.1 Mechanical control

Tillage and other forms of mechanical disturbance of the soil or weeds were the primary control tactics used for weed management prior to the development of modern herbicides (Buhler, 1995). The effect of tillage on weeds varies according to the weed life stage present at the time of the disturbance, and can include alteration of the soil environment, positioning of weed seed within the soil profile, killing of emerged plants via mechanical disruption or desiccation and fragmentation of perennial reproductive structures. Herbicide-resistant weeds have led to a reduction in conservation tillage due to the inability to control weeds with herbicides alone (Price et al., 2011).

Tillage used for seedbed preparation can reduce weed densities faced during the cropping season by burying seeds at depths unfavourable for emergence (Pareja et al., 1985; Buhler and Mester, 1991). Tillage can be especially effective in reducing weed density in years following weed-control failures due to the large quantity of seeds deposited on the soil surface (Hartzler and Roth, 1993). Weed seed persistence generally increases with burial depth (Bekker et al., 1998); thus, deep tillage may only serve as a temporary solution as seeds are returned to near the surface with subsequent tillage.

A variety of tools are available to control weeds that emerge following planting. Tools such as rotary hoes and harrows are typically used within the first few weeks after planting. Selectivity is based primarily on the differential emergence depth of crop and weeds (Fogelberg and Gustavsson, 1998); thus, these tools must be used before weeds establish a well-developed root system that protects them from shallow soil disturbance. Inter-row cultivators are used later in the season to dislodge weeds growing between rows and can bury weeds within the row by soil movement (Gunsolus, 1990). In many soybean-producing regions, the use of narrow-row spacing (<38 cm) limits the use of inter-row cultivation. Flaming is used similarly as mechanical tools, but offers the advantage of being able to be operated when soils are too wet for tools that control weeds via soil disturbance. The energy requirement for flaming varies with weed species and size (Uloa et al., 2010).

Increased regulations on herbicides, spread of herbicide-resistant weeds and the increased adoption of organic production have led to the development of more efficient mechanical weed-control tools (Van Der Weide et al., 2008). Kunz et al. (2015) reported that automatic steering systems allow cultivators to be operated at twice the speed of manually operated tractors while maintaining the level of weed control without increasing soybean injury. Reducing the time required for mechanical cultivation is critical in order to increase its adoption since the increased time needed to cover fields with mechanical weed-control implements compared to herbicide application is a primary factor that has led to the decline in inter-row cultivation.

The next generation of mechanical weed-control tools likely will involve autonomous robotic systems (Slaughter et al., 2008). A variety of methods have been used to differentiate weeds from crops, including morphology, light reflectance and visual texture. A vision-guided cultivator removed up to 41% more weeds in broccoli (*Broccoli oleracea*) and lettuce (*Lactuca sativa*) than a conventional cultivator, and reduced the time requirement for hand-weeding by up to 45% (Lati et al., 2016).

4.2 Soil management

The role of soil quality/health on the sustainability of crop production has been an area of increased emphasis in recent decades (Doran et al., 2000; Karlen et al., 1997). Gallandt et al. (1999) stated that an improved understanding of the relationship between soil quality and weed population dynamics could lead to the development of cropping systems with enhanced weed-suppression capacity. Reductions in crop yield losses associated with interference, weed seed persistence in the soil and safe sites for seedling establishment were cited as opportunities to improve weed management by exploiting soil–weed relationships.

Supplemental applications of essential nutrients have long been used to enhance crop yields. Manipulating the quantity (Jornsgard et al., 1996), form (Davis and Liebman, 2001), timing (Paolini et al., 1999) or placement (Rasmussen, 2002) of fertilizers can influence availability and utilization of the nutrients by crops and weeds. Improved understanding of the mechanisms that influence availability, absorption and utilization of nutrients by both crops and weeds could enhance soybean competitiveness with weeds (Di Tomaso, 1995).

Organic materials such as manure and green manure crops are used in many cropping systems to fulfil nutritional needs of crops. These materials differ from synthetic fertilizers in that the release of nutrients occurs at a slower rate than with synthetic fertilizers (Eghball et al., 2002), and the release of organic compounds or increase in soil organic matter may influence soil microorganisms and plant growth (Conklin et al., 2002; Kennedy and Kremer, 1996). Increases in total soil biological activity or specific microorganisms have been associated with weed-suppressive soils (Kennedy, 1999; Kennedy et al., 1991; Kremer and Li, 2003; Hoagland et al., 2008).

Smith et al. (2009) proposed that changes in soil resources associated with cropping system diversity affect the ability of crops to tolerate weed interference. They speculated that diverse cropping systems create heterogeneous resource pools. Weeds utilize different resource pools than crops, resulting in less inter-specific competition for the resources, and thus less impact on crop growth and yield.

4.3 Biological control

The use of biological control agents to manage weeds has had more success for weeds of natural systems than those in agro-ecosystems (McFayden, 1998; Müller-Schärer et al., 2000). The need for rapid control of weeds to minimize impact on crop yields and the high disturbance levels that occur during soybean production reduce the opportunities for classical forms (inoculation) of biological control.

The bioherbicide (inoculative) approach to biological control of weeds has had some success in soybean production and other agronomic crops. Collego was a commercialized formulation of the fungus *Colletotrichum gloeosporioides* introduced in 1982 to control northern jointvetch (*Aeschynomene virginica*) in soybean and rice (*Oryza sativa*) (Templeton, 1986). Hallett (2005) stated that future success in this area is dependent on genetic manipulation of pathogens to enhance their virulence or the development of enhanced delivery systems. Co-application of a sub-lethal dose of glyphosate with a mycoherbicide enhanced the virulence of the pathogen and allowed lower doses of inoculum (Sharon et al., 1992). The authors stated that the development of specific compounds that predispose a weed to infection could improve the efficacy of mycoherbicides and increase their adoption.

4.4 Herbicides

The initiation of the chemical era of weed management is generally attributed to the discovery of the growth regulator herbicides in the 1940s. In 1960, less than 10% of the US soybean acres were treated with herbicide; this rose to more than 90% in 1990 (Pike et al., 1991). The popularity of most active ingredients during the early period of herbicidal weed control was relatively short-lived as either better products were introduced or, in some cases, weeds adapted to the herbicides and reduced their effectiveness.

The introduction of glyphosate-resistant soybean resulted in dramatic changes in weed-management programmes in countries where the technology was approved. The resistant soybeans were introduced in 1996 and by 2006, they were planted on more than 50 million hectares; approximately 96% of the soybean grown in the United States, Argentina and Brazil were glyphosate-resistant varieties (Dill et al., 2008). Changes associated with adoption of the technology included dramatic reductions in other herbicides and increased adoption of conservation tillage practices.

The rapid spread of herbicide-resistant weeds threatens the current production system in many soybean-producing regions (Yu and Powles, 2014; Délye et al., 2013). The lack of discovery of new modes of action in more than 20 years is limiting herbicide options (Duke, 2012). Since these production systems are facilitated by herbicidal control of weeds, many alternative practices either are difficult to incorporate into the system or farmers are reluctant to incorporate alternative strategies. The primary strategies used to manage herbicide resistance include 1) use of full herbicide rates, 2) rotating herbicide sites of action and 3) using multiple, effective sites of action (Owen, 2016). It has been stressed that relying on herbicides to manage herbicide resistance is a short-term solution, and that sustainable weed-management systems will require increased reliance on other management strategies (Beckie and Harker, 2017; Owen, 2016).

4.5 Scouting/mapping weed infestations

The majority of weed-management decisions are driven by arbitrary assessments of weed infestations, rather than strict threshold levels; thus, there has been relatively little effort at developing systematic scouting methods. The traditional method of weed counts within randomly placed quadrats in a field was stated to be appropriate, if weeds were randomly distributed within a field, but may fail to accurately assess patchy distributions of weeds (Gold et al., 1996). Clay and Johnson (2002) recommended flexible sampling systems based on previous experience and knowledge, and that sampling systems should consider how the data will be used. They emphasized that scouting should provide information regarding changes in both weed density and area infested, rather than static assessments.

In a ranking of herbicide-resistant weed-management strategies, scouting and surveys of weeds both before and after herbicide applications was ranked third in a list of ten strategies (Beckie and Harker, 2017). The authors emphasized the need for maintenance of databases over time to document fluctuations in weed communities. This information is essential for adjusting management programmes to avoid outbreaks of economic infestations of weeds.

Increases in both farm size and the percentage of rented farmland have resulted in a loss of much of the historical knowledge of weed infestations frequently used by farmers to guide development of weed-management programmes. Advances in technology have simplified data collection and management. Unmanned aerial vehicles (UAVs) provide a relatively inexpensive and rapid method of mapping weed infestations within fields (Rasmussen et al., 2013). A UAV was able to map early season weeds in maize with 86% accuracy (Peña et al., 2013). Knowledge related to the nature of weed communities within individual fields is essential for increasing the efficiency of weed management. Increased efforts in determining how best to utilize new technologies to map weed infestations are needed.

5 Interactions between weeds and other pest complexes

In conventional cropping systems, decisions related to weed management are often made in isolation from other aspects of crop production. However, most agronomic fields are infested with multiple categories of pests (weeds, insects and pathogens). Norris (2005) stated that both the weeds and tactics used to control them can have direct or indirect effects on other pest classes, and that development of integrated pest-management programmes requires a multi-disciplinary approach.

The presence of weeds in fields has been shown to influence the prevalence of other pests (Wisler and Norris, 2005). Several weed species have been identified as host to the soybean cyst nematode (*Heterodera glycines*), including winter annual species that could serve as sites for nematode reproduction during parts of the year when soybeans are not present (Venkatesh et al., 2000). Hepperly et al. (1980) isolated three pathogens of soybean on velvetleaf and stated that the importance of weeds serving as a source of inoculum for pathogens should not be underestimated. Volunteer corn in soybean fields was found to harbour western corn rootworm (*Diabrotica virgifera virgifera*), and that volunteer corn possessing the Bt trait could accelerate the development of Bt-resistant western corn rootworm (Krupke et al., 2009). Smith et al. (2016) presented evidence that seed treatments may influence weed seed banks in corn and soybean.

Relatively few studies have evaluated the interactions among pest complexes on crop growth and yield. The effects of three pests on soybean yield [weeds, soybean cyst nematode and corn earworm (*Helicoverpa zea*)] were reported to be primarily additive (Alston et al., 1993). Interactions between herbicide injury, soybean cyst nematode and defoliating insects on soybean yields resulted in the need to adjust management strategies based on the infestation levels of the different pests (Browde et al., 1994).

6 Factors limiting the adoption of diversified weed management

Conventional production systems for soybean, often referred to as industrial agriculture, have relied on a combination of mechanization and agrochemicals to replace labour (Woodhouse, 2010). Smaller-scale production systems were reported to be more efficient in energy use; however, small-scale systems have lower labour productivity and thus, lower earnings than large-scale, industrial systems. Globalization has led to the widespread adoption of industrial systems for soybean production (Masuda and Goldsmith, 2009), but the environmental impacts of these systems (Deytieux et al., 2012) and the ability to stay ahead of evolving weed problems (Duke, 2012; Shaner, 2014) have led to the call for changes in agricultural production systems.

An advantage of herbicide-based systems is the ease with which herbicides can be substituted within the system. An herbicide-based weed-management system that is no longer effective can be easily modified by adding or changing herbicide-active ingredients without the need for new knowledge, skills, equipment or additional labour. Kniss (2017) reported the herbicide area treatments, an estimate of the number of herbicide treatments applied to each soybean field, declined approximately 40% following the introduction of glyphosate-resistant soybean in the United States. However, the rise of glyphosate-resistant weeds resulted in herbicide area treatments increasing to a level higher than before the introduction of glyphosate-resistant soybean. Thus, the response to herbicide resistance has largely been a concomitant increase in herbicide use.

Diversified weed management requires greater knowledge and management skills than herbicide-based systems. Gunsolus (1990) stated that non-herbicidal weed control was as much art as science, and required highly specific management skills. Increased variability of yields and economic returns with non-herbicide-based systems compared with herbicide-based systems were reported to favour the use of herbicides, even if greater returns were expected in the non-herbicide system (Olson and Eidman, 1992). Uncertainty regarding the effectiveness of integrated pest management and the dominance of conventional systems were cited as barriers to widespread adoption of integrated management systems (Cowan and Gunby, 1996).

Changes in soybean weed management that reduce reliance on herbicides will likely be driven by continued evolution of herbicide resistance, as occurred in dryland production regions of Australia (Walsh and Powles, 2007). Heap (2014) stated that herbicide resistance will lead to diversification of both weed-management practices and cropping systems. More sustainable systems will change the focus of weed management from protecting crop yields to managing the weed seed bank and creating an environment less favourable for the establishment and growth of weeds. The primary challenge moving forward is replacing a relatively 'simple' system for farmers with one that relies on an integration of multiple strategies, where it is difficult for the producer to quantify the contribution of individual components.

7 Conclusion

The widespread adoption of herbicide-based weed management systems in soybean has increased labour productivity and earnings for farmers. Effective herbicides have allowed weed management decisions to be isolated from other aspects of soybean production. However, the lack of diversity in these systems allows weeds to quickly adapt, thereby threatening their sustainability. Integrated management requires greater knowledge of the weeds and how both cultural and weed management practices influence weed survival and reproduction. This knowledge must be used to design a production system that creates an environment less favourable for the establishment and survival of weeds, rather than simply creating a system that allows the use of tactics that provide high mortality events such as herbicide application and tillage.

8 Where to look for further information

A comprehensive reference on cultural, biological and mechanical approaches to weed management is *Ecological Management of Agricultural Weeds* (2001) by Liebman, M., Mohler, C.L, and Staver, C. P., Cambridge University Press, 532 pp. The authors state that their goal in writing the book was to show how insights into ecological processes can explain the successes and failures of weed management and how this knowledge can be used to develop better management strategies. Owen et al. (2014) reviewed the suite of tactics available for weed management in major agronomic crops of the United States and Canada. They provided an assessment of the adoption of integrated approaches to weed management, and the barriers preventing greater use of these systems.The leading journals focused on weed management research are *Weed Research* (European Weed Science Society) and *Weed Science* (Weed Science Society of America). Other journals that regularly feature articles on weed management research include *Pest Management Science* (Society of Chemical Industry), *Crop Protection* (International Association for the Plant Protection Sciences) and *Weed Technology* (Weed Science Society of America).

9 References

Alston, D. G., Schmitt, D. P., Bradley Jr., J. R. and Coble, H. D. (1993) 'Multiple pest interactions in soybean: Effects on *Heterodera glycines* egg populations and crop yield', *J. Nematol.* 25(1), 42–49.

Arce, G. D., Pedersen, P. and Hartzler, R. G. (2009) 'Soybean seeding rate effects on weed management', *Weed Technol.* 23(1), 17–22.

Basol, T. L., Wright, D. and Lenssen, A. W. (2013) 'Understanding soybean plant population recommendations for Iowa', *Agric. Envir. Exten. Public.* Book 180.

Bauer, T. A. and Mortensen, D. A. (1992) 'A comparison of economic and economic optimum thresholds for two annual weeds in soybeans', *Weed Technol.* 6(1), 228–35.

Beckie, H. J. and Harker, H. J. (2017) 'Our top 10 herbicide-resistant weed management practices', *Pest Manag. Sci.* 73(6), 1045–52.

Bekker, R. M., Bakker, J. P., Grandin, U., Kalamees, R., Milberg, P., Poschlod, P., Thompson, K. and Willems, J. H. (1998) 'Seed size, shape and vertical distribution in the soil: Indicators of seed longevity', *Funct. Ecol.* 12(5), 834–42.

Bell, H. D., Norsworthy, J. K., Scott, R. C. and Popp, M. (2015) 'Effect of row spacing, seeding rate, and herbicide program in glufosinate-resistant soybean on palmer amaranth management', *Weed Technol.* 29(3), 390–404.

Benbrook, C. (2012) 'Impacts of genetically engineered crops on pesticide use in the U. S. – the first sixteen years', *Environ. Sci. Eur.* 24, doi:10.1186/2190-4715-24-24.
Birthisel, S. K., Gallandt, E. R. and Jabbour, R. (2014) 'Habitat effects on second-order predation of the seed predator *Harpalus rufipes* and implications for weed seedbank management', *Biol. Control* 70(1), 65–72.
Blackshaw, R. E., Larney, F. O., Lindwall, C. W. and Kozub, G. C. (1994) 'Crop rotation and tillage effects on weed populations on the semi-arid Canadian prairies', *Weed Technol.* 8(2), 231–7.
Bonny, S. (2016) 'Genetically modified herbicide-tolerant crops, weeds, and herbicides: Overview and impact', *Environ. Manage.* 57(1), 31–48.
Browde, J. A., Pedigo, L. P., Owen, M. D. K. and Tylka, G. L. (1994) 'Soybean yield and pest management as influenced by nematodes, herbicides, and defoliating insects', *Agron. J.* 86(4), 601–8.
Buchanan, A. L., Zobel, E., Hinds, J., Rosario-Lebron, A. and Hooks, C. R. R. (2015) 'Can row spacing influence arthropod communities in soybean? Implications for early and late planting', *Environ. Entomol.* 44(3), 557–61.
Buhler, D. D. (1995) 'Influence of tillage systems on weed population dynamics and management in corn and soybean in the Central USA', *Crop Sci.* 35(5), 1247–58.
Buhler, D. D., Hartzler, R. G. and Forcella, F. (1997) 'Implications of weed seedbank dynamics to weed management', *Weed Sci.* 45(3), 329–36.
Buhler, D. D. and Mester, T. C. (1991) 'Effect of tillage systems on the emergence depth of giant (*Setaria faberi*) and green foxtail (*Setaria viridis*)', *Weed Sci.* 39(2), 200–3.
Bussan, A. J., Burnside, O. C., Orf, J. H. and Puettmann, K. J. (1997) 'Field evaluation of soybean (*Glycine max*) genotypes for weed competitiveness', *Weed Sci.* 45(1), 31–7.
Callaway, M. (1992) 'A compendium of crop varietal tolerance to weeds', *Am. J. Alter. Agric.* 7(4), 168–80.
Cardina, J., Herms, C. P. and Doohan, D. J. (2002) 'Crop rotation and tillage system effects on weed seedbanks', *Weed Sci.* 50(4), 448–60.
Clay, S. and Johnson, G. (2002) 'Scouting for weeds'. Online. *Crop Manage.* doi:10.1094/CM-2002-1206-01-MA.
Coble, H. D. and Mortensen, D. A. (1992) 'The threshold concept and its application to weed science', *Weed Technol.* 6(1), 191–5.
Conklin, A. E., Erich, M. S., Liebman, M., Lambert, D., Gallandt, E. R. and Halteman, W. A. (2002) 'Effects of red clover (*Tifolium pretense*) green manure and compost soil amendments on wild mustard (*Brassica kaber*) growth and incidence of disease', *Plant Soil* 238(2), 245–56.
Cowan, R. and Gunby, P. (1996) 'Sprayed to death: path dependence, lock-in and pest control strategies', *Econ. J.* 106(436), 521–42.
Czapar, G. F., Curry, M. P. and Gray, M. E. (1995) 'Survey of integrated pest management practices in central Illinois', *J. Prod. Agric.* 8(4), 483–6.
Dalling, J. W., Davis, A. S., Schutte, B. J. and Arnold, A. E. (2011) 'Seed survival in soil: interacting effects of predation, dormancy and the soil microbial community', *J. Ecol.* 99(1), 89–95.
Davis, A. S. (2006) 'When does it make sense to target the weed seed bank?', *Weed Sci.* 54(3), 558–65.
Davis, A. S., Hill, J. D., Chase, C. A., Johanns, A. M. and Liebman, M. (2012) 'Increasing cropping system diversity balances productivity, profitability and environmental health', *PLOS One* 7(10), doi.org/10.1371/journal.pone.0047149.
Davis, A. S. and Liebman, M. (2001) 'Nitrogen source influences wild mustard growth and competitive effect on sweet corn', *Weed Sci.* 49(4), 558–66.
Davis, A. S. and Raghu, S. (2010) 'Weighing abiotic influences on weed seed predation', *Weed Res.* 50(5), 402–12.
De Bruin, J. and Pedersen, P. (2008) 'Effect of row spacing and seeding rate on soybean yield', *Agron. J.* 100(3), 704–10.
De Haan, R. L., Wyse, D. L., Ehlke, N. J., Maxwell, B. D. and Putman, D. H. (1994) 'Simulation of spring-seeded smother plants for weed control in corn (*Zea mays*)', *Weed Sci.* 42(1), 35–43.

Délye, C., Jasieniuk, M. and Le Corre, V. (2013) 'Deciphering the evolution of herbicide resistance in weeds', *Trends Gen.* 29(11), 649–58.

DeWerff, R. P., Conley, S. P., Colquhoun, J. B. and Davis, V. M. (2014) 'Can soybean seeding rate be used as an integrated component of herbicide resistance management?', *Weed Sci.* 62(4), 625–36.

Deytieux, V., Nemecek, T., Knuchel, R. F., Gaillard, G. and Munier-Jolain, N. M. (2012) 'Is integrated weed management efficient for reducing environmental impacts of cropping systems? A case study based on life cycle assessment', *Eur. J. Agron.* 36(1), 55–65.

Dieleman, J. A., Mortensen, D. A. and Martin, A. R. (1999) 'Influence of velvetleaf (*Abutilon theophrasti*) and common sunflower (*Helianthus annuus*) density variation on weed management outcomes', *Weed Sci.* 47(1), 81–9.

Dill, G. M., Cajacob, C. A. and Padgette, S. R. (2008) 'Glyphosate-resistant crops: Adoption, use and future considerations', *Pest Manage. Sci.* 64(4), 326–31.

Di Tomaso, J. M. (1995) 'Approaches for improving crop competitiveness through the manipulation of fertilization strategies', *Weed Sci.* 43(3), 491–7.

Doran, J. W. and Zeiss, M. R. (2000) 'Soil health and sustainability: Managing the biotic component of soil quality', *Appl. Soil Ecol.* 15(1), 3–11.

Duke, S. O. (2012) 'Why have no new herbicide modes of action appeared in recent years?' *Pest Manage. Sci.* 68(4), 505–12.

Eghball, B., Wienhold, B. J., Gilley, J. E. and Eigenberg, R. A. (2002) 'Mineralization of manure elements', *J. Soil Water Conserv.* 57(6), 470–3.

Fogelberg, F. and Gustavsson, A. M. (1998) 'Resistance against uprooting in carrots (*Daucus carota*) and annual weeds: A basis for selective mechanical weed control', *Weed Res.* 38(3), 183–90.

Forcella, F., Peterson, D. H. and Barbour, J. C. (1996) 'Timing and measurement of weed shed in corn (*Zea mays*)', *Weed Technol.* 10(3), 535–43.

Franklin, K. A. (2008) 'Shade avoidance', *New Phytol.* 179(4), 930–44.

Fuesrt, E. P. and Putman, A. R. (1983) 'Separating the competitive and allelopathic components of interference', *J. Chem. Ecol.* 9(8), 937–44.

Gallandt, E. R. (2006) 'How can we target the weed seedbank?', *Weed Sci.* 54(3), 588–96.

Gallandt, E. R., Liebman, M. and Huggins, D. R. (1999) 'Improving soil quality: implications for weed management', *J. Crop Prod.* 2(1), 95–121.

Gianessi, L. P. and Reigner, N. P. (2007) 'The value of herbicides in U.S. crop production', *Weed Technol.* 21(2), 559–66.

Gold, H. J., Bay, J. and Wilkerson, G. G. (1996) 'Scouting for weeds, based on the negative binomial distribution', *Weed Sci.* 44(3), 504–10.

Gould, F. (1991) 'The evolutionary potential of crop pests', *Am. Scientist* 79, 496–507.

Grau, C. R. and Radke, V. L. (1984) 'Effects of cultivars and cultural practices on *Sclerotinia* stem rot of soybean', *Plant Dis.* 68(1), 56–8.

Green, J. M. (2012) 'The benefits of herbicide-resistant crops', *Pest Manage. Sci.* 68(10), 1323–31.

Green-Tracewicz, E., Page, E. R. and Swanton, C. J. (2011) 'Shade avoidance in soybean reduces branching and increases plant-to-plant variability in biomass and yield per plant', *Weed Sci.* 59(1), 43–9.

Gunsolus, J. L. (1990) 'Mechanical and cultural weed control in corn and soybeans', *Am. J. Alter. Agric.* 5(1), 114–19.

Hallett, S. G. (2005) 'Where are the bioherbicides', *Weed Sci.* 53(3), 404–15.

Harker, K. and O'Donovan, J. (2013) 'Recent weed control, weed management, and integrated weed management', *Weed Technol.* 27(1), 1–11.

Hartzler, R. G. (1997) 'Velvetleaf (*Abutilon theophrasti*) interference in soybean (*Glycine max*): A survey of yield loss estimates and management recommendations', *Crop Prot.* 16(5), 483–5.

Hartzler, R. G. and Roth, G. W. (1993) 'Effect of prior year's weed control on herbicide effectiveness in corn (*Zea mays*)', *Weed Technol.* 7(3), 611–14.

Heap, I. (2014) 'Global perspective of herbicide-resistant weeds', *Pest Manage. Sci.* 70(9), 1306–15.

Hepperly, P. R., Kirkpatrick, B. L. and Sinclair, J. B. (1980) '*Abutilon theophrasti:* wild host for three fungal parasites of soybean', *Phytopath.* 70(4), 307–10.

Hoagland, L., Carpenter-Boggs, L., Reganold, J. P. and Mazzola, M. (2008) 'Role of native soil biology in Brassicaceous seed meal-induced weed suppression', *Soil Biol. Biochem.* 40(7), 1689–97.

Horvath, D. P., Hansen, S. A., Moriles-Miller, J. P., Pierik, R., Yan, C., Clay, D. E., Scheffler, B. and Clay, S. A. (2015) 'RNAseq reveals weed-induced PIF3-like as a candidate target to manipulate weed stress response in soybean', *New Phytol.* 207(1), 196–210.

Jabran, K., Mahajan, G., Sardana, V. and Chauhan, B. S. (2015) 'Allelopathy for weed control in agricultural systems', *Crop Prot.* 72(1), 57–65.

Jannink, J L., Orf, J. H., Jordan, N. R. and Shaw, R. G. (2000) 'Index selection for weed suppressive ability in soybean', *Crop Sci.* 40(4), 1087–94.

Jones, R. E. and Medd, R. W. (2000) 'Economic thresholds and the case for longer term approaches to population management of weeds', *Weed Technol.* 14(2), 337–50.

Jordan, N. (1993) 'Prospects for weed control through crop interference', *Ecol. Appl.* 3(1), 84–91.

Jornsgard, B., Rasmussen, K., Hill, J. and Christiansen, J. L. (1996) 'Influence of nitrogen on competition between cereals and their natural weed populations', *Weed Res.* 36(6), 461–70.

Karlen, D. L., Mausbach, M. J., Doran, J. W., Cline, R. G., Harris, R. F. and Schuman, G. E. (1997) 'Soil quality: A concept, definition, and framework for evaluation', *Soil Sci. Soc. Am. J.* 61(1), 4–10.

Keene, C. L., Curran, W. S., Wallace, J. M., Ryan, M. R., Mirsky, S. B., VanGessel, M. J. and Barbercheck, M. E. (2016) 'Cover crop termination timing is critical in organic rotational no-till systems' *Agron. J.* 109(1), 272–82.

Kennedy, A. C. (1999) 'Soil microorganisms for weed management', *J. Crop Prod.* 2(1), 123–38.

Kennedy, A. C., Elliott, L. F., Young, F. L. and Douglas, C. L. (1991) 'Rhizobacteria suppressive to the weedy downy brome', *Soil Sci. Soc. Am. J.* 55(3), 722–7.

Kennedy, A. C. and Kremer, R. J. (1996) 'Microorganisms in weed control strategies', *J. Prod. Agric.* 9(4), 480–5.

Knezevic, S. Z., Evans, S. P. and Mainz, M. (2003) 'Row spacing influences the critical timing for weed removal in soybean (*Glycine max*)', *Weed Technol.* 17(4), 666–73.

Kniss, A. R. (2017) 'Long-term trends in the intensity and relative toxicity of herbicide use', *Nat. Comm.* 8, 14865, doi:10.1038/ncomms14865.

Kremer, R. J. (1993) 'Management of weed seed banks with microorganisms', *Ecol. Appl.* 3(1), 42–52.

Kremer, R. J. and Li, J. (2003) 'Developing weed-suppressive soils through improved soil quality management', *Soil Till. Res.* 72(2), 193–202.

Krupke, C., Marquardt, P., Johnson, W., Weller, S. and Conley, S. P. (2009) 'Volunteer corn presents new challenges for insect resistance management', *Agron. J.* 101(4), 797–9.

Kunz, C., Weber, J. F. and Gerhards, R. (2015) 'Benefits of precision farming technologies for mechanical weed control in soybean and sugar beet – comparison of precision hoeing with conventional mechanical weed control', *Agron. J.* 5(1), 130–42.

Lambert, D. M. and Lowenberg-DeBoer, J. (2003) 'Economic analysis of row spacing for corn and soybean', *Agron. J.* 95(3), 564–73.

Lati, R. N., Siemens, M. C., Rachuy, J. S. and Fennimore, S. A. (2016) 'Intrarow weed removal in broccoli and transplanted lettuce with an intelligent cultivator', *Weed Technol.* 30(3), 655–63.

Marino, P. C., Gross, K. L. and Landis, D. A. (1997) 'Weed seed loss due to predation in Michigan maize fields. *Agric. Ecosyst. Environ.* 66(3), 189–96.

Lègére, A. and Samson, N. (1999) 'Relative influence of crop rotation, tillage, and weed management on weed associations in spring barley cropping systems', *Weed Sci.* 47(1), 112–22.

Liebman, M. and Dyck, E. (1993) 'Crop rotation and intercropping strategies for weed management', *Ecol. Appl.* 3(1), 92–122.

Liebman, M., Miller, Z. J., Williams, C. L., Westerman, P. R., Dixon, P. M., Heggenstaller, A., Davis, A. S., Menalled, F. D. and Sundberg, D. N. (2014) 'Fates of *Setaria faberi* and *Abutilon theophrasti* seeds in three crop rotation systems', *Weed Res.* 54(3), 293–306.

Lou, Y., Davis, A. S. and Yannarell, A. C. (2016) 'Interactions between allelochemicals and the microbial community affect weed suppression following cover crop residue incorporation into soil', *Plant Soil* 399(1), 357–71.

Marino, P. C., Gross, K. L. and Landis, D. A. (1997) 'Weed seed loss due to predation in Michigan maize fields', *Agric. Ecosyst. Environ.* 66(3), 189–96.

Masuda, T. and Goldsmith, P. D. (2009) 'World soybean production: area harvested, yield, and long-term projections', *Int. Food Agribus. Manage. Rev.* 12(4), 143–62.

McFadyen, R. E. C. (1998) 'Biological control of weeds', *Annu. Rev. Entomol.* 43, 369–93.

Mirsky, S. B., Ryan, M. R., Teasdale, J. R., Curran, W. S., Reberg-Horton, C. S., Spargo, J. T., Wells, M. S., Keene, C. L. and Moyer, J. W. (2013) 'Overcoming weed management challenges in cover crop-based organic rotational no-till soybean production in the eastern United States', *Weed Technol.* 27(1), 193–203.

Müller-Schärer, H., Scheepens, P. C. and Greaves, M. P. (2000) 'Biological control of weeds in European crops: Recent achievements and future work', *Weed Res.* 40(1), 83–98.

Neeser, C., Dille, J. A., Krishnan, G., Mortensen, D. A., Rawlinson, J. T., Martin, A. R., and Bills, L. B. (2004) 'WeedSoft®: A weed management decision support system', *Weed Res.* 52(1), 115–22.

Neve, P. (2007) 'Challenges for herbicide resistance evolution and management: 50 years after Harper', *Weed Res.* 47(5), 365–9.

Norris, R. F. (2005) 'Ecological bases of interactions between weeds and organisms in other pest categories', *Weed Sci.* 53(6), 909–13.

Norsworthy, J., Griffith, G., Griffin, T., Bagavathiannan, M. and Gbur, E. E. (2014) 'In-field movement of glyphosate-resistant Palmer amaranth (*Amaranthus palmeri*) and its impact on cotton lint yield: Evidence supporting a zero-threshold strategy', *Weed Sci.* 62(2), 237–49.

Norsworthy, J., Ward, S., Shaw, D., Llewellyn, R. S., Nichols, R. L., Webster, T. M., Bradley, K. W., Frisvold, G., Powles, S. B., Burgos, N. R. Witt, W. W. and Barrett, M. (2012) 'Reducing the risks of herbicide resistance: best management practices and recommendations', *Weed Sci.* 60(sp1), 31–62.

Olson, K. D. and Eidman, V. R. (1996) 'A farmer's choice of weed control method and the impacts of policy and risk', *Rev. Agric. Econ.* 14(1), 125–37.

Owen, M. D. K. (2016) 'Diverse approaches to herbicide-resistant weed management', *Weed Sci.* 64(sp1), 570–84.

Owen, M. D. K., Beckie, H. J., Leeson, J. Y., Norsworthy, J. K. and Steck, L. E. (2015) 'Integrated pest management and weed management in the United States and Canada', *Pest Manage. Sci.* 71(3), 357–76.

Owen, M. and Zelaya, I. (2005) 'Herbicide-resistant crops and weed resistance to herbicide', *Pest Manage. Sci.* 61(3), 301–11.

Page, E. R., Tollenaar, M., Lee, E. A., Lukens, L. and Swanton, C. J. (2010) 'Shade avoidance: An integral component of crop-weed competition', *Weed Res.* 50(4), 281–8.

Paolini, R., Principi, M., Froud-Williams, R. J., Del Puglia, S. and Biancardi, E. (1999) 'Competition between sugarbeet and *Sinapsis arvensis* and *Chenopodium album* as affected by timing of nitrogen fertilization', *Weed Res.* 39(6), 425–40.

Pareja, M. R., Staniforth, D. W. and Pareja, G. P. (1985) 'Distribution of weed seed among soil structural units', *Weed Sci.* 33(2), 182–9.

Peña, J. M., Torres-Sánchez, J., de Castro, A. I., Kelly, M. and López-Granados, F. (2013) 'Weed mapping in early-season maize fields using object-based analysis of unmanned aerial vehicle (UAV) images', *PLoS ONE*, 8(10), e77151.

Pike, D. R., McGlamery, M. D. and Knake, E. L. (1991) 'A case study of herbicide use', *Weed Technol.* 5(3), 639–46.

Place, G. T., Reberg-Horton, S. C., Dunphy, J. E. and Smith, A. N. (2009) 'Seeding rate effects on weed control and yield for organic soybean production', *Weed Technol.* 23(4), 497–502.

Price, A. J., Balkcom, K. S., Culpepper, S. A., Kelton, J. A., Nichols, R. L. and Schomberg, H. (2011) 'Glyphosate-resistant Palmer amaranth: A threat to conservation tillage', *J. Soil Water Conserv.* 66(4), 265–75.

Rasmussen, J., Nielsen, J., Garcia-Ruiz, F., Christensen, S. and Streibig, J. C. (2013) Potential uses of small unmanned aircraft systems (UAS) in weed research. *Weed Res.* 53:242–8.

Rasmussen, K. (2002) 'Influence of liquid manure application method on weed control in spring cereals', *Weed Res.* 42(4), 287–98.

Sandermann, H. (2006) 'Plant biotechnology: Ecological case studies on herbicide resistance', *Trends Plant Sci.* 11(7), 324–8.

Schipanski, M. E., Barbercheck, M., Douglas, M. R., Finney, D. M., Haider, K., Kaye, J. P., Kemanian, A. R., Mortensen, D. A., Ryan, M. R., Tooker, J. and White, C. (2014) 'A framework for evaluating ecosystem services provided by cover crops in agroecosystems', *Agric. Sys.* 125(1), 12–22.

Shaner, D. L. (2014) 'Lessons learned from the history of herbicide resistance', *Weed Sci.* 62(2), 427–31.

Shaner, D. L. and Beckie, H. (2014) 'The future for weed control and technology', *Pest Manage. Sci.* 70(9), 1329–39.

Sharon, A., Amsellem, Z. and Gressel, J. (1992) 'Glyphosate suppression of an elicited defense response: increased susceptibility of *Cassia obtusifolia* to a mycoherbicide', *Plant Physiol.* 98(2), 654–9.

Slaugher, D. C., Giles, D. K. and Downey, D. (2008) 'Autonomous robotic weed control systems: A review', *Comput. Electron. Agric.* 61(1), 63–78.

Smith, R. G., Atwood, L. W., Morris, M. B. and Mortensen, D. A. (2016) 'Evidence for indirect effects of pesticide seed treatments on weed seed banks in maize and soybean', *Agric. Ecosyst. Environ.* 216(2), 216–73.

Smith, R. G., Mortensen, D. A. and Ryan, M. R. (2009) 'A new hypothesis for the functional role of diversity in mediating resource pools and weed-crop competition in agroecosystems', *Weed Res.* 50(1), 37–48.

Stanislaus, M. and Cheng, C. (2002) 'Genetically engineered self-destruction: an alternative to herbicides for cover crop systems', *Weed Sci.* 50(6), 794–801.

Teasdale, J. R. (1996) 'Contribution of cover crops to weed management in sustainable agricultural systems', *J. Prod. Agric.* 9(4), 475–9.

Templeton, G. E. (1986) 'Mycoherbicide research at the University of Arkansas: past, present, and future', *Weed Sci.* 34(Suppl. 1), 35–7.

Thill, D. C. and Mallory-Smith, C. A. (1997) 'The nature and consequence of weed spread in cropping systems', *Weed Sci.* 45(3), 337–42.

Ulloa, S. M., Datta, A. and Knezevic, S. Z. (2010) 'Tolerance of selected weed species to broadcast flaming at different growth stages', *Crop Prot.* 29(12), 1381–8.

Van Acker, R. C., Swanton, C. J. and Weise, S. F. (1993) 'The critical period of weed control in soybean [*Glycine max* (L.) Merr.]', *Weed Sci.* 41(2), 194–200.

Van Der Weide, R. Y., Bleeker, P. O., Achten, V. T. J. M., Lotz, L. A., Fogelberg, F. and Belander, B. (2008) 'Innovation in mechanical weed control in crop rows', *Weed Res.* 48(3), 215–24.

van Heemst, H. D. J. (1985) 'The influence of weed competition on crop yield', *Agric. Sys.* 18(2), 81–93.

Venkatesh, R., Harrison, S. K. and Riedel, R. M. (2000) 'Weed hosts of soybean cyst nematode', *Weed Technol.* 14(1), 156–60.

Walsh, M. J., Harrington, R. B. and Powles, S. B. (2012) 'Harrington seed destructor: a new nonchemical weed control tool for global grain crops', *Crop Sci.* 52(3), 1343–7.

Walsh, M. J. and Powles, S. B. (2007) 'Management strategies for herbicide-resistant weed populations in Australian dryland crop production systems', *Weed Technol.* 21(2), 332–8.

Walsh, M. J. and Powles, S. B. (2014) 'High seed retention at maturity of annual weeds infesting crop fields highlights the potential for harvest weed seed control', *Weed Technol.* 28(3), 486–93.

Weber, C. R., Shibles, R. M. and Blyth, D. E. (1966) 'Effect of plant population and row spacing on soybean development and production', *Agron. J.* 58(1), 99–102.

Weber, C. R. and Staniforth, D. (1957) 'Competitive relationships in variable weed and soybean stands', *Agron. J.* 49(8), 440–4.

Webster, T. M., Simmons, D. B., Culpepper, A. S., Grey, T. L., Bridges, D. C. and Scully, B. T. (2016) 'Factors affecting potential for Palmer amaranth (*Amaranthus palmeri*) suppression by winter rye in Georgia, USA', *Field Crops Res.* 192(1), 103–9.

Westerman, P. R., Liebman, M., Menalled, F. D., Heggenstaller, A. H., Hartzler, R. G. and Dixon, P. M. (2005) 'Are many little hammers effective? Velvetleaf (*Abutilon theophrasti*) population dynamics in two- and four-year crop rotation systems', *Weed Sci.* 53(3), 382–92.

Wisler, G. C. and Norris, R. F. (2005) 'Interactions between weeds and cultivated plants as related to management of plant pathogens', *Weed Sci.* 53(6), 914–17.

Woodhouse, P. (2010) 'Beyond industrial agriculture? Some questions about farm size, productivity and sustainability', *J. Agrar. Change* 10(3), 437–53.

Wortman, S. E., Francis, C. A., Bernards, M. A., Blankenship, E. E. and Lindquist, J. L. (2013) 'Mechanical termination of diverse cover crop mixtures for improved weed suppression in organic cropping systems', *Weed Sci.* 61(1), 162–70.

Wortmann, C. (1993) 'Contribution of bean morphological characteristics to weed suppression', *Agron. J.* 85(4), 840–3.

Wyse, D. L. (1994) 'New technologies and approaches for weed management in sustainable agriculture systems', *Weed Technol.* 8(2), 403–7.

Yelverton, F. H. and Coble, H. D. (1991) 'Narrow row spacing and canopy formation reduces weed resurgence in soybeans (*Glycine max*)', *Weed Technol.* 5(1), 169–74.

Yu, Q. and Powles, S. (2014) 'Metabolism-based herbicide resistance and cross-resistance in crop weeds: A threat to herbicide sustainability and global crop production', *Plant Physiol.* 166(3), 1106–18.

Part 2

Food and other uses

Chapter 8

Nutritional, nutraceutical and functional properties of soybeans

Suzanne Hendrich, Iowa State University, USA

1 Introduction
2 Nutritional effects of soybean oils
3 Nutritional and nutraceutical functions of soybean proteins in cardiovascular disease
4 Health effects of soybean proteins and associated constituents
5 Other health or adverse effects of soybean foods and ingredients
6 Conclusion and future trends
7 References

1 Introduction

Soybeans, foods derived from soybeans (e.g. tofu, soymilk, soy infant formula and tempeh), and food and dietary supplement ingredients derived from soybeans (e.g. soybean oil, soybean proteins and isoflavones) have been under intensive research for their health effects especially over the past 25 years.

This intensive research derives from the recognition of soybeans as having desirable nutritional properties, containing about twice the protein of other legumes/serving, with good protein quality, such that some soybean protein ingredients have protein-digestibility-corrected amino acid scores commensurate with proteins thought to be optimal to meet human protein needs (Messina 1999). Soybean oil is used as the nutritional standard fat source for the AIN-93G diet for growing rodents (Reeves 1997); its high content of polyunsaturated fats (PUFAs), including a high ratio of α-linolenic:linoleic acids (~1:7.5) (Messina 1999), make it well aligned to meet human as well as rodent requirements for n-3 and n-6 essential fatty acids.

Soybeans are also a nearly unique source of isoflavones in the human diet. The phyto-oestrogens associated with soybean proteins may mimic human oestrogens in useful ways, such as in alleviating menopausal symptoms, and perhaps protecting from heart disease and some types of cancer (Messina 1999). In 1999, a US Food and Drug Administration health claim was approved stating that soybean foods and foods with ingredients,

http://dx.doi.org/10.19103/AS.2017.0034.23

which provide 25 g of soybean protein per day as part of a diet low in saturated fat and cholesterol, may reduce the risk of heart disease (Stein 2000).

In South Korea, several soybean ingredients are endorsed by the government's functional food code, permitted to carry claims of health benefits at specified doses. Soy protein, soy isoflavones and soy fibre as well as lecithin are listed in South Korea's functional food code (Lupton et al. 2014). This review will examine the recent human clinical studies related to the efficacy and safety of soybean foods and food or dietary supplement ingredients including soybean oils, proteins, isoflavones and selected other compounds for which soybeans are a significant source.

2 Nutritional effects of soybean oils

A recent systematic literature review of 15 randomized controlled trials of nearly 60 000 participants linked decreased intake of saturated fats in favour of increased intake of PUFAs with modest decrease of cardiovascular disease risk (Hooper et al. 2015), supporting a positive role for soybean oil in human health. However, partial hydrogenation of soybean and other vegetable oils to yield cheap, functional food fats suitable for production of pastries and other baked goods was also determined to increase the dietary content of trans-fatty acids.

A 35-day randomized crossover study was performed with 30 mildly hyper-cholesterolaemic post-menopausal women (low-density lipoprotein (LDL) cholesterol ≥120 mg/dL) who consumed diets with 25% of energy from fat, 2/3 of which was from either partially hydrogenated soybean oil (PHSO) or corn oil. The corn oil diet lowered LDL cholesterol by 10% compared with the PHSO diet; trans-fatty acid content was 4.3% of energy in the PHSO diet vs 0.3% in the corn oil diet; PUFA content of the corn oil diet was twice that of the PHSO diet (Vega-Lopez et al. 2009).

Various strategies to substitute other fats and oils for trans-fatty acid sources reduced cardiovascular disease risk factors according to a meta-analysis of controlled human trials (Mozaffarian and Clarke 2009). Some recent research has developed ways to remove trans-fatty acids from PHSOs. An example is the hydrolysis of fatty acids from PHSO with a commercially available *Candida antarctica* lipase A favouring hydrolysis of trans- over cis-unsaturated fatty acids by ~4-fold. This processing was capable of removing ~3/4 of the trans-fatty acids from PHSO (Jala et al. 2013). It seems feasible to improve PHSO products by removing their trans-fatty acids. But retaining functionality in terms of good eating quality of baked goods will also require attention.

A recent 4-week study of 34 participants who ate 20 g of either palm oil mayonnaise or soybean oil mayonnaise showed significantly lower total cholesterol (185 mg/dL vs 196 mg/dL) and LDL cholesterol (118 mg/dL vs 125 mg/dL) after soybean oil intake compared with palm oil intake (Karupaiah et al. 2016). This specifically demonstrated a modest benefit of soybean oil compared with another common food fat. Information on the connection between dietary fatty acid composition and heart disease continues to be refined. Soybean oil is among the healthier choices in dietary fats, but even with the removal of trans-fats from PHSO, the lesser PUFA content of PHSO compared with soybean oil warrants consumer advice to choose diet not containing PHSO, as one among many strategies to diminish cardiovascular disease risk.

3 Nutritional and nutraceutical functions of soybean proteins in cardiovascular disease

Studies continue to emerge on the efficacy of soybean protein to benefit human health. This review will attempt to distinguish among human clinical studies that focused on providing soybean proteins per se vs studies providing soybean proteins within foods from whole soybeans. In human clinical trials, few studies have distinguished the effects of soybean proteins from those of the isoflavones or of other constituents that are present in the diet containing soybean proteins.

A study of groups of 30–33 moderately hypercholesterolaemic women showed that after 12 weeks of supplementing their diets with 42 g of protein from milk, soy with 80 mg isoflavones or soy with only trace amounts of isoflavones, LDL cholesterol was lowered to a greater extent by soy protein containing isoflavones than by soy protein without isoflavones. Neither treatment differed from milk protein in the final LDL cholesterol (Gardner et al. 2001).

An abstract from a trial published at the same time, of a study of groups of 16–18 moderately hypercholesterolaemic women given 28 g supplemental protein for 4 weeks, showed that soy beta-conglycinin with 60 mg isoflavones lowered LDL and total cholesterol at 4 weeks compared with baseline but beta-conglycinin without isoflavones did not. Women given a casein control treatment also showed lower LDL cholesterol after 4 weeks compared with baseline. The three treatments did not differ in effects on LDL cholesterol or total cholesterol after 4 weeks (Lee et al. 2001).

A group of 90 post-menopausal, prehypertensive, moderately hypercholesterolaemic women given 40 g whole soy flour containing 50 mg isoflavones showed lower LDL cholesterol and C-reactive protein (CRP) after 6 months of treatment compared with groups of 90 women given 60 mg daidzein in milk powder or milk powder alone over the same time period. All 270 women were shown to produce equol, a daidzein metabolite of interest to various health endpoints (Liu et al. 2015b).

Groups of 80–82 hypercholesterolaemic adults were given 25 g soy protein, 12.5 g soy protein + 12.5 g whey protein, or 25 g whey protein; none of the three treatments lowered LDL cholesterol after 6 weeks of treatment (Padhi et al. 2015).

A meta-analysis of five clinical trials of a total of 104 hypercholesterolaemic participants given soy isoflavones (15 mg/d or more) in soy protein or as a supplement for 3–6 months compared with control participants given soy protein with <5 mg isoflavones/d or a placebo supplement showed no benefit of isoflavones to lower LDL cholesterol (Qin et al. 2013). A meta-analysis that included 35 randomized controlled trials of soy foods, soy ingredients and/or isoflavones examined the effects on LDL cholesterol in 2670 participants given these products for 1–12 months, with a mean soy protein intake of 30 g (range of 14–50 g/d). This analysis showed a mean benefit of soy products of a 5 mg/dL decrease in total and LDL cholesterol. Isoflavones per se showed no benefit on serum lipids (Tokede et al. 2015).

These recent studies support a modest benefit of foods or ingredients providing soy protein and its associated isoflavones on serum lipids associated with cardiovascular disease risk. Extreme variability among participants in isoflavone bioavailability as observed, for example, by Liu et al. (2015b) may be interfering with the ability to discern an action of isoflavones. The importance of isoflavone bioavailability to lessening cholesterol was illustrated in a study of 38 Golden Syrian hamsters fed soy protein for 4 weeks. Compared with casein-fed controls, the soy-protein-fed hamsters had significantly lower LDL cholesterol. But all of the cholesterol-lowering effect of soy protein could be attributed to nine of the 38 individuals who excreted about four times greater amounts of isoflavones than did the majority of the hamsters (Ye et al. 2006). A similar proportion of women (~25%) fed soy foods in single meals showed more than 2-fold greater urinary excretion of isoflavones than did the majority of participants (Xu et al. 1995). This phenomenon was apparently determined by isoflavone-degrading microbial species in the gut microbiome (Renouf et al. 2011). It remains to be determined if a subset of humans with a 'high isoflavones excreting' phenotype may account for the observed modest benefits of ingestion of soy protein or soy foods on serum lipids. It seems that a combination of soy protein and isoflavones but not isoflavones per se is required for the cholesterol-lowering effects of soy protein.

4 Health effects of soybean proteins and associated constituents

4.1 Effects on blood pressure, kidney and arterial function

Research has recently been done on various other effects of soybean foods and ingredients on the cardiovascular system, beyond serum lipids. Sixty post-menopausal women replaced non-soy protein in their diet with 25 g of protein from soy nuts for 8 weeks. After 8 weeks on diets with and without the soy nuts, 49 women who did not have characteristics of metabolic syndrome showed significantly decreased systolic blood pressure and CRP upon consuming soy nuts.

In 11 women with metabolic syndrome, only four women who were equol producers showed decreased diastolic blood pressure, CRP and serum triglycerides. Out of 49 women, 28 women without metabolic syndrome and who were equol producers showed significantly decreased diastolic blood pressure after consuming soy nuts (Acharjee et al. 2015). This suggests a novel strategy to benefit individuals with metabolic syndrome: introduction of equol-producing bacteria into their guts. This would require significant preliminary work to establish the safety and efficacy of such a manipulation of human gut microbes.

In groups of 90 prehypertensive, post-menopausal women, 40 g soy flour or 60 mg daidzein in milk powder prevented decrease in glomerular filtration rate that occurred in women given 40 g milk powder per day as a placebo after 6 months (Liu et al. 2014). In this study, neither blood pressure nor brachial flow-mediated dilation, an indicator of arterial function, was different between treatment groups (Liu et al. 2015b). On the basis of limited data, healthy post-menopausal women may particularly benefit from whole soy foods to help maintain healthy blood pressure. More work on this potential benefit of soybean foods in both sexes and across the lifespan is indicated.

4.2 Effects on blood glucose and diabetes

Effects of soy on blood glucose are another important public health interest, with burgeoning rates of type 2 diabetes across the globe. Three groups of ~50 women with fasting glucose status of 100–126 mg/dL, 2-h postprandial blood glucose of 140–200 mg/dL or newly diagnosed type 2 diabetes not requiring medication were given 10 g soy protein alone, or soy protein with 50 mg daidzein or with 50 mg genistein for 6 months. Responses to oral glucose tolerance tests and fasting glucose and haemoglobin A1C did not differ between the treatments (Ye et al. 2015).

Adding tofu (25 g protein) to a standard meal of 50 g glucose from rice significantly diminished glucose response over time compared with rice alone, whereas 25 g of protein from chicken, fish or egg white did not significantly decrease the blood glucose response to rice in 15 Chinese men in a randomized crossover study (Quek et al. 2016). This effect of tofu suggests a benefit for glycaemic control in individuals with prediabetes or diabetes. The mechanism for this effect is not yet clear, but potentially suggests a role for isoflavones as the other proteins fed are not associated with these compounds.

In a group of 34 women with gestational diabetes, substituting soy protein for half of the animal protein in their diets for 6 weeks during gestation improved fasting plasma glucose, triglycerides and homeostatic model assessment of insulin resistance (HOMA-IR) status compared with a control group of 34 women, matched for week of gestation and BMI (Jamilian and Asemi 2015). Diets provided 0.8 g total protein/kg body weight. At birth, the women who had eaten soy protein had significantly fewer babies with hyperbilirubinaemia or the need to be hospitalized. Additional studies are needed to confirm this finding, which promises an important benefit for pregnant women and newborn infants. The inclusion of soy protein in human diets, rather than isoflavones per se, seems to confer benefits on blood glucose, but its effects on men have not been studied much. Because type 2 diabetes is strongly associated with obesity, obesity prevention as well as approaches to preventing and mitigating diabetes might well include further study of soybean foods and soy protein.

4.3 Effects on obesity

Type 2 diabetes risk is the disease most strongly associated with obesity (Eckel et al. 2011). It follows that some nutrition research has been directed towards the effects of soybean foods and components on obesity. Over 16 weeks in a randomized design, groups of approximately 60 eighteen-year-old women consumed soy protein or casein beverages containing 20 g protein and 500 kcals (Berger et al. 2014). Neither protein shake prevented slight weight gain (~1 lb) in this population. A high-protein soy snack delayed the onset time of eating dinner and decreased appetite in 31 adolescents (age 17 years) compared with a high-fat snack or no snack (Leidy et al. 2015). In 21 adults fed lunch meals matched for macronutrient content either with beef or soy as the protein source (24 g of each), onset of eating dinner and overall appetite were similar for both treatments (Douglas et al. 2015). These studies suggest that soy protein may not especially benefit satiety or prevent obesity. Soy protein snacks might be useful in controlling appetite, but probably not more so than other high-protein foods of good amino acid quality.

4.4 Effects on cancer risk

Human epidemiological studies have tended to support the association of decreased cancer risk with increased intake of soybean foods (Messina et al. 1994). This association remains the subject of considerable research. Genistein has been investigated in animal models and in a small number of clinical trials of prostate cancer patients (Perabo et al.

2008). The animal models indicate a potential for genistein to prevent prostate cancer. The human trials have studied prostate cancer patients for varying lengths of time, but for not longer than six months, with no obvious benefits across a range of doses (60–600 mg genistein/day) and across a range of cancer stages (Perabo et al. 2008).

A more recent meta-analysis of 2 prostate cancer prevention trials and 6 prostate cancer treatment trials indicated a significant reduction of prostate cancer risk, especially in men older than 65 years taking 60 mg isoflavones/day for 12 months or soy protein isolate containing isoflavones for 6 months compared with placebo treatments (van Die et al. 2014). The prostate cancer treatment trials of soy/isoflavones showed no clear benefits of soy components on prostate-specific antigen or sex hormones (van Die et al. 2014).

In a review of breast cancer epidemiological studies of the relation between soy foods and/or isoflavone contents of the body, only one of the six large prospective cohort studies showed a positive association between miso soup and dietary isoflavone intake and decreased breast cancer risk (179 cases of breast cancer in 21 852 Japanese women) (Enderlin et al. 2009). The other five such studies showed no association between dietary soy intake and breast cancer risk, in studies totalling more than 200 000 women in Great Britain, France, Japan and the USA. It may be that the range of soy food intakes studied was too narrow to see a benefit of soy. Of the 28 case–control studies associating soy/isoflavone intake with breast cancer risk, more than half of the studies showed decreased risk of breast cancer with increased soy/isoflavone intake, but two studies showed increased risk of breast cancer with increased isoflavone intake (Enderlin et al. 2009). Groups of 70 women recently diagnosed with early stage breast cancer consumed 26 g of soy protein or milk protein for 7–30 days before breast surgery. In a subset of women clustered as having especially high plasma genistein levels after treatment, their mammary tissue showed more than 2-fold greater expression of a set of genes (e.g. fibroblast growth factor receptor-2 (FGFR2)) related to stimulation of cell proliferation (Shike et al. 2014).

Because isoflavones are oestrogenic, and some mammary neoplasms proliferate in response to oestrogen (Renoir et al. 2013), it is reasonable to be cautious in recommending soy foods to prevent breast cancer as the research cited above supports the concept that some women with breast cancer might be harmed by soy foods or isoflavones supplements. A clinical trial of women at high risk for breast cancer or who had previously had breast cancer provided 50 mg isoflavones or a placebo to groups of 46 or 49 women, respectively, for 12 months. Mammogram density, an early indicator of breast cancer, did not differ between the two groups (Wu et al. 2015). Thus recent work does not support soy or isoflavone interventions for women with breast cancer, breast cancer survivors, or women at high risk for breast cancer. Soy or isoflavones may not be harmful to most of those women, but more progress in individualizing cancer prevention strategies is needed. Soy foods remain somewhat promising in their ability to prevent breast and prostate cancer, and identification of characteristics that can better guide decisions on cancer preventive strategies may be on the near horizon.

4.5 Effects on menopause

Along with increased risk of chronic diseases such as cancer, ageing also involves women undergoing menopause. Soy protein, presumably due to oestrogenic effects of isoflavones, seems to benefit the symptoms of menopause and post-menopause. A randomized double-blind trial of groups of 20 post-menopausal women given soy powder with 90 mg isoflavones or low-dose hormone therapy for 16 weeks showed reduced vaginal dryness with both of these treatments compared with placebo (Carmignani et al. 2015). Only the hormone therapy improved vaginal maturation index.

In a group of 30 post-menopausal women in Iran given 33 g soy protein containing 54 mg isoflavones/day for eight weeks, menopausal symptoms according to the Kupperman index abated significantly compared with a group of 31 placebo-treated women (Husain et al. 2015). Post-menopausal women given isoflavone supplements ranging from 50 to 220 mg/d for 50 days in a randomized crossover design (n = 14) showed improved bone calcium retention compared with baseline, but did not experience as much bone calcium retention as when given risedronate (Pawlowski et al. 2015). The participants in this study experienced this benefit of isoflavones regardless of their ability to produce equol.

Recently, an analysis of numerous studies on isoflavones and vasomotor symptoms (hot flushes) during menopause resulted in a consensus that doses of isoflavones up to 150 mg/day may serve as a first choice option for relief from these symptoms that is both effective and safe (Schmidt et al. 2016). Thus, with respect to several aspects of health during and after menopause, soy foods containing isoflavones and isoflavones as dietary supplements may be recommended.

4.6 Effects on reproductive function

The oestrogenicity of soybean foods, ingredients and supplements, while apparently beneficial during menopause, has also put these dietary constituents under scrutiny for adverse effects on reproductive function (Maqbool et al. 2016). Very high doses of phyto-oestrogenic isoflavones from subterranean clover in livestock forage caused infertility in cattle and sheep (Lindner 1976). Claims that the long history of use of soybean foods in human diets has not been associated with adverse reproductive effects have led to research.

A long-term project on infant formula quality that examined the nutritional effects of soy and cow's milk formulas presented the chance to examine the health effects of early dietary exposure to soybean isoflavones. Adults aged 20–34 years who had ingested soy (n = 248) or cow's milk formula (n = 563) as infants showed no differences in reproductive history or function except slightly more menstrual discomfort and slightly longer menstruation in women fed soy formula compared with those fed cow's milk formula as babies (Strom et al. 2001).

A recent analysis of reproductive organ development in children showed that groups of ~30 infants fed breast milk or formulas based on soy milk or cow's milk for the first year of life did not differ at five years of age in reproductive organ size (breast bud, ovaries, uteri, testes) assessed by sonogram (Andres et al. 2015). The intent is to follow-up these children into adolescence, which will be important. Although limited in scope, the available research supports a long history of use of soy infant formula without apparent long-term adverse reproductive effects.

In groups of ~70 healthy post-menopausal women given 80 or 120 mg isoflavones/day or placebo for 36 months, endometrial thickness declined across this time period but did not differ among treatments. Circulating hormone levels (oestradiol, estrone sulphate and thyroxine) and adverse events did not differ between treatments (Alekel et al. 2015), supporting the long-term safety of isoflavone supplements after menopause. To date, soybean isoflavones, either in soy infant formula or as dietary supplements, showed no ill effects on human reproductive system development or function.

4.7 Effects on cognition

Because intake of isoflavones or soybean foods has been much studied at least in older women, effects of soy/isoflavones on cognition, which tends to decline with ageing, have also been studied. A meta-analysis of ten randomized placebo-controlled trials of soy isoflavones with a total of more than 1000 post-menopausal participants showed that mean summary cognitive function and visual memory test scores were significantly improved in women taking isoflavone supplements for 6 weeks to 30 months (Cheng et al. 2015). Six of the 10 trials included showed such a benefit of isoflavones; none of the included trials showed a detriment of isoflavones to cognitive function.

In 300 women given 25 g milk protein powder placebo or 25 g soy protein with 90 mg isoflavones for a mean of 2.7 years, composite global cognition score was not significantly associated with change in urinary isoflavone excretion over the trial period. General intelligence score was slightly but significantly inversely associated with urinary isoflavone excretion (St John et al. 2014). This needs further investigation. In groups of 65 individuals with Alzheimer's disease, 100 mg isoflavones/day for six months showed no benefit versus a placebo across several cognitive and mood tests (Gleason et al. 2015).

These results together support more study, especially of the ability of isoflavone supplements to prevent cognitive decline before the onset of overt cognitive disease. Soy lecithin may be of benefit as well. A 3-month trial of 300 mg phosphatidylserine and 240 mg phosphatidic acid/day (PS + PA, derived from soy lecithin) in 31 healthy elders (60–80 years old) showed improved memory and less seasonal depression compared with 25 participants on a placebo (More et al. 2014). A 2-month study of the same treatments in individuals with Alzheimer's disease showed maintenance of daily functioning and stabilization of mood in 53 participants on PS + PA, whereas daily functioning and positive mood declined in 39 participants given a placebo (More et al. 2014). These trials indicate that research on soy foods and their components needs to be continued in this field, and not just be focused on isoflavones.

5 Other health or adverse effects of soybean foods and ingredients

Soybean-based infant formulas have a long history of use as an alternative to cow's milk formulas when breastfeeding is not feasible. The addition of fructooligosaccharides (FOS) has been explored as a prebiotic strategy to increase infant gut contents of probiotic (beneficial) bacteria, mainly of the genera *Bifidobacterium* and *Lactobacillus*, which may protect from infectious illnesses. But cow's milk formulas containing FOS caused softer, watery stools, raising concerns about FOS as a formula additive. A recent clinical trial provided groups of ~50 newborn full-term infants with soy formula or one of two soy formulas + FOS (2.5 g/L) for the first month of life (Lasekan et al. 2015). No differences were found in stool consistency, body weight gain, formula intake, vomiting or hydration status (urine specific gravity) among the three treatment groups. This suggests that FOS supplementation of soy formula is feasible, and might provide a useful alternative to cow's milk formulas containing FOS.

Beneficial interactions between foods or food ingredients and specific nutrients are an interesting and unstudied area. A recent example involving soybean saponins perhaps suggests that this topic deserves further attention. ZIP4 proteins in the small intestine are essential for zinc uptake. Enhancing ZIP4 expression and/or function enhances zinc absorption. Poor zinc status is a significant problem in low-income world regions and in elders. Soybean extracts were tested for effects on ZIP4 in cell models simulating the human small intestine, including Caco-2 cells. Soyasaponin-rich soy extract and soyasaponin Bb, a major saponin glucoside in soybeans, in similar concentrations in purified form as present in the soy extract significantly increased ZIP4 expression by blocking ZIP4 degradation in AsPC1 cells (human pancreatic tumour cells that express ZIP4 similar to the human intestine) (Ayako et al. 2015). Soyasaponins are not appreciably absorbed by humans, as shown in a small short-term feeding trial of eight women given a soyasaponin extract that showed the absence of saponins or their deglucosylated metabolites in urine (Hu et al. 2004). It is feasible that the effect of soyasaponin Bb on ZIP4 derives from cell surface interactions between the saponin and ZIP4 protein domains. Effects of soybean foods or soyasaponin Bb on zinc status will need to be confirmed in vivo; such research is warranted based on the prospect of a novel benefit of soybean foods.

Soy protein might adversely influence haem iron uptake in comparison with pea or lentil proteins, according to a study of 15 women given single doses of each protein with haem iron. Soy protein permitted 10–20% less haem iron absorption than did the other legume proteins (Weinborn et al. 2015). The implications of this finding for iron uptake from intact foods remain to be seen, and would seem to be of practical significance only in extreme dietary circumstances such as when someone ingests very limited protein sources containing a source of haem iron (meat or seafood) for fairly long periods of time. Important interactions between soybean foods and nutrient uptake seemingly would have been discovered by now, given how commonly soybeans are included in human diets, at least in some major world regions. These studies on iron (Weinborn et al. 2015) and zinc (Ayako et al. 2015) suggest some need for such investigation.

Another interaction between soybean food intake and a nutritionally important endogenous substance, uric acid, was studied in 450 post-menopausal women with prehypertension or prediabetes in China (Liu et al. 2015a); data from two trials were combined in which soy protein or soy flour, or an analogous amount of isoflavone supplement, was provided and compared with milk powder for 6 months. Isoflavones or milk powder treatment groups showed 10% decrease in serum uric acid from baseline, and soybean protein/flour caused less decrease in uric acid (5% decrease). Participants were asked to limit their intake of soy foods to two servings per week during the trial, which might have caused the decrease in uric acid from baseline. It might be worth determining the mechanisms for decreased uric acid by isoflavones or milk powder and what constituents in soy foods might interfere with that seeming effect of isoflavones.

A recent systematic review of botanical dietary supplements including soybean isoflavone supplements has been provided by Di Lorenzo et al. (2015). Of the 95 scientific papers reviewed in this study, 30 highlighted adverse effects of soybean supplements (Di Lorenzo et al. 2015). These findings reflect the well-known allergenicity of soybean proteins which are among the eight major food allergens (Sathe et al. 2016). The ingestion of a brand of soy milk containing seaweed disrupted thyroid function in eight patients, but this seemed very likely to be due to an exceptionally high level of iodine from the seaweed in this soy milk (~25 mg/L) (Crawford et al. 2010), and not due to the soy milk. Causal links between ingestion of soybean isoflavones and reproductive organ functional changes reported as adverse effects in this systematic review remain to be established. In summary, recent studies indicate no reason to worry about the potential harm from intake of soy infant formulas. The human food supply contains an increasingly complex array of ingredients and mixtures, including dietary supplements. 'Buyer beware' remains worthy advice in the use of food and supplements not bearing scientifically validated health claims. Food producers should also be vigilant to assure that potentially harmful amounts of constituents, which may even be essential nutrients (e.g. iodine), are avoided.

6 Conclusion and future trends

Soybean foods, soy proteins and/or isoflavones, which are uniquely associated with soy protein, may benefit human health by lessening some cardiovascular disease risk factors, diminishing cancer risk and cognitive decline with ageing. Isoflavones have been recently recommended by gynaecological researchers as a first line of remedy for menopausal vasomotor symptoms. Considerable inter-individual variability in uptake of isoflavones due to their apparent degradation by gut microbes remains an impediment to ascertaining the health efficacy of these soybean constituents; more attention to this aspect of these compounds is consistent with advances in individualizing chronic disease prevention approaches. Human diets may benefit from increased inclusion of soybean foods.

7 References

Acharjee, S., Zhou, J. R., Elajami, T. K. and Welty, F. K. 2015. Effect of soy nuts and equol status on blood pressure, lipids and inflammation in postmenopausal women stratified by metabolic syndrome status. *Metabolism* 64(2):236–43.

Alekel, D. L., Genschel, U., Koehler, K. J., Hofmann, H., Van Loan, M. D., Beer, B. S., Hanson, L. N., Peterson, C. T. and Kurzer, M. S. 2015. Soy isoflavones for reducing bone loss study: Effects of a 3-year trial on hormones, adverse events, and endometrial thickness in postmenopausal women. *Menopause* 22(2):185–97.

Andres, A., Moore, M. B., Linam, L. E., Casey, P. H., Cleves, M. A. and Badger, T. M. 2015. Compared with feeding infants breast milk or cow-milk formula, soy formula feeding does not affect subsequent reproductive organ size at 5 years of age. *J. Nutr.* 145(5):871–5.

Ayako, H., Katsuma, O., Masakazu, T., Kumiko, K., Hiroshi, N., Shuichi, E., Yusaku, M., Seiji, M., Masaya, N., Kazuhiro, I., Hajime, O., Andrews, G. K. and Taiho, K. 2015. Soybean extracts increase cell surface ZIP4 abundance and cellular zinc levels: a potential novel strategy to enhance zinc absorption by ZIP4 targeting. *Biochem. J.* 472(2):183–93.

Berger, P. K., Principe, J. L., Laing, E. M., Henley, E. C., Pollock, N. K., Taylor, R. G., Blair, R. M., Baile, C. A., Hall, D. B. and Lewis, R. D. 2014. Weight gain in college females is not prevented by isoflavone-rich soy protein: a randomized controlled trial. *Nutr. Res.* 34(1):66–73.

Carmignani, L. O., Pedro, A. O., Montemor, E. B., Arias, V. A., Costa-Paiva, L. H. and Pinto-Neto, A. M. 2015. Effects of a soy-based dietary supplement compared with low-dose hormone therapy on the urogenital system: a randomized, double-blind, controlled clinical trial. *Menopause* 22(7):741–9.

Cheng, P. F., Chen, J. J., Zhou, X. Y., Ren, Y. F., Huang, W., Zhou, J. J. and Xie, P. 2015. Do soy isoflavones improve cognitive function in postmenopausal women? A meta-analysis. *Menopause* 22(2):198–206.

Crawford, B. A., Cowell, C. T., Emder, P. J., Learoyd, D. L., Chua, E. L., Sinn, J. and Jack, M. M. 2010. Iodine toxicity from soy milk and seaweed ingestion is associated with serious thyroid dysfunction. *Med. J. Aust.* 193(7):413–5.

Di Lorenzo, C., Ceschi, A., Kupferschmidt, H., Lüde, S., De Souza Nascimento, E., Dos Santos, A., Colombo, F., Frigerio, G., Nørby, K., Plumb, J., Finglas, P. and Restani, P. 2015. Adverse effects of plant food supplements and botanical preparations: a systematic review with critical evaluation of causality. *Br. J. Clin. Pharmacol.* 79(4):578–92.

Douglas, S. M., Lasley, T. R. and Leidy, H. J. 2015. Consuming beef vs. soy protein has little effect on appetite, satiety, and food intake in healthy adults. *J. Nutr.* 145(5):1010–6.

Eckel, R. H., Kahn, S. E., Ferrannini, E., Goldfine, A. B., Nathan, D. M., Schwartz, M. W., Smith, R. J. and Smith, S. R. 2011. Obesity and type 2 diabetes: what can be unified and what needs to be individualized? *J. Clin. Endocrinol. Metab.* 96(6):1654–63.

Enderlin, C. A., Ann Coleman, E., Stewart, C. B. and Hakkak, R. 2009. Dietary soy intake and breast cancer risk. *Oncol. Nurs. Forum* 36(5):531–9.

Gardner, C. D., Newell, K. A., Cherin, R. and Haskell, W. L. 2001. The effect of soy protein with or without isoflavones relative to milk protein on plasma lipids in hypercholesterolemic postmenopausal women. *Am. J. Clin. Nutr.* 73(4):728–35.

Gleason, C. E., Fischer, B. L., Dowling, N. M., Setchell, K. D., Atwood, C. S., Carlsson, C. M. and Asthana, S. 2015. Cognitive effects of soy isoflavones in patients with Alzheimer's disease. *J. Alzheimers Dis.* 47(4):1009–19.

Hooper, L., Martin, N., Abdelhamid, A. and Davey Smith, G. 2015. Reduction in saturated fat intake for cardiovascular disease. *Cochrane Database Syst. Rev.* (6):Cd011737.

Hu, J., Reddy, M. B., Hendrich, S. and Murphy, P. A. 2004. Soyasaponin I and Sapongel B have limited absorption by Caco-2 intestinal cells and limited bioavailability in women. *J. Nutr.* 134(8):1867–73.

Husain, D., Khanna, K., Puri, S. and Haghighizadeh, M. 2015. Supplementation of soy isoflavones improved sex hormones, blood pressure, and postmenopausal symptoms. *J. Am. Coll. Nutr.* 34(1):42–8.

Jala, R. C. R., Xu, X. and Guo, Z. 2013. Enzymatic production of trans fatty acid free fat from partially hydrogenated soybean oil (PHSO) – Theory, strategy and practicability. *Food Chem.* 141(3):1934–40.

Jamilian, M. and Asemi, Z. 2015. The effect of soy intake on metabolic profiles of women with gestational diabetes mellitus. *J. Clin. Endocrinol. Metab.* 100(12):4654–61.

Karupaiah, T., Chuah, K.-A., Chinna, K., Ryosuke, M., Yasunobu, M., Sundram, K. and Michihiro, S. 2016. Comparing effects of soybean oil- and palm olein-based mayonnaise consumption on the plasma lipid and lipoprotein profiles in human subjects: a double-blind randomized controlled trial with cross-over design. *Lipids Health Dis.* 15:1–11.

Lasekan, J., Baggs, G., Acosta, S. and Mackey, A. 2015. Soy protein-based infant formulas with supplemental fructooligosaccharides: gastrointestinal tolerance and hydration status in newborn infants. *Nutrients* 7(4):3022–37.

Lee, S.-O., Smith, V., Rickert, D., Hu, J., Reuber, M., Hanson, K., Zheng, Y., Lebepe, S., Johnson, L., Alekel, D. L., Lewis, D., Hendrich, S. and Murphy, P. 2001. Beta-conglycinin deficient in isoflavones and saponins did not lower plasma cholesterol over time in moderately hypercholesterolemic women. *FASEB J.* 15:A274.

Leidy, H. J., Todd, C. B., Zino, A. Z., Immel, J. E., Mukherjea, R., Shafer, R. S., Ortinau, L. C. and Braun, M. 2015. Consuming high-protein soy snacks affects appetite control, satiety, and diet quality in young people and influences select aspects of mood and cognition. *J. Nutr.* 145(7):1614–22.

Lindner, H. R. 1976. Occurrence of anabolic agents in plants and their importance. *Environ. Qual. Saf. Suppl.* (5):151–8.

Liu, Z. M., Ho, C. S., Chen, Y. M. and Woo, J. 2015a. Can soy intake affect serum uric acid level? Pooled analysis from two 6-month randomized controlled trials among Chinese postmenopausal women with prediabetes or prehypertension. *Eur. J. Nutr.* 54(1):51–8.

Liu, Z. M., Ho, S. C., Chen, Y. M., Tang, N. and Woo, J. 2014. Effect of whole soy and purified isoflavone daidzein on renal function–a 6-month randomized controlled trial in equol-producing postmenopausal women with prehypertension. *Clin. Biochem.* 47(13–14):1250–6.

Liu, Z. M., Ho, S. C., Chen, Y. M., Tomlinson, B., Ho, S., To, K. and Woo, J. 2015b. Effect of whole soy and purified daidzein on ambulatory blood pressure and endothelial function–a 6-month double-blind, randomized controlled trial among Chinese postmenopausal women with prehypertension. *Eur. J. Clin. Nutr.* 69(10):1161–8.

Lupton, J., Atkinson, S., Chang, N., Fraga, C., Levy, J., Messina, M., Richardson, D., Ommen, B., Yang, Y., Griffiths, J. and Hathcock, J. 2014. Exploring the benefits and challenges of establishing a DRI-like process for bioactives. *Eur. J. Nutr.* 53 Suppl 1:1–9. Springer Science & Business Media BV.

Maqbool, F., Mostafalou, S., Bahadar, H. and Abdollahi, M. 2016. Review of endocrine disorders associated with environmental toxicants and possible involved mechanisms. *Life Sci.* 145:265–73.

Messina, M. J. 1999. Legumes and soybeans: overview of their nutritional profiles and health effects. *Am. J. Clin. Nutr.* 70 (3 Suppl):439s–50s.

Messina, M. J., Persky, V., Setchell, K. D. and Barnes, S. 1994. Soy intake and cancer risk: a review of the in vitro and in vivo data. *Nutr. Cancer* 21(2):113–31.

More, M. I., Freitas, U. and Rutenberg, D. 2014. Positive effects of soy lecithin-derived phosphatidylserine plus phosphatidic acid on memory, cognition, daily functioning, and mood in elderly patients with Alzheimer's disease and dementia. *Adv. Ther.* 31(12):1247–62.

Mozaffarian, D. and Clarke, R. 2009. Quantitative effects on cardiovascular risk factors and coronary heart disease risk of replacing partially hydrogenated vegetable oils with other fats and oils. *Eur. J. Clin. Nutr.* 63:S22–S33.

Padhi, E. M., Blewett, H. J., Duncan, A. M., Guzman, R. P., Hawke, A., Seetharaman, K., Tsao, R., Wolever, T. M. and Ramdath, D. D. 2015. Whole soy flour incorporated into a muffin and consumed at 2 doses of soy protein does not lower LDL cholesterol in a randomized, double-blind controlled trial of hypercholesterolemic adults. *J. Nutr.* 145(12):2665–74.

Pawlowski, J. W., Martin, B. R., McCabe, G. P., McCabe, L., Jackson, G. S., Peacock, M., Barnes, S. and Weaver, C. M. 2015. Impact of equol-producing capacity and soy-isoflavone profiles of supplements on bone calcium retention in postmenopausal women: a randomized crossover trial. *Am. J. Clin. Nutr.* 102(3):695–703.

Perabo, F. G., Von Low, E. C., Ellinger, J., von Rucker, A., Muller, S. C. and Bastian, P. J. 2008. Soy isoflavone genistein in prevention and treatment of prostate cancer. *Prostate Cancer Prostatic Dis.* 11(1):6–12.

Qin, Y., Niu, K., Zeng, Y., Liu, P., Yi, L., Zhang, T., Zhang, Q. Y., Zhu, J. D. and Mi, M. T. 2013. Isoflavones for hypercholesterolaemia in adults. *Cochrane Database Syst. Rev.* (6):Cd009518.

Quek, R., Bi, X. and Henry, C. J. 2016. Impact of protein-rich meals on glycaemic response of rice. *Br. J. Nutr.* 115(7):1194–201.

Reeves, P. G. 1997. Components of the AIN-93 diets as improvements in the AIN-76A diet. *J. Nutr.* 127(5 Suppl):838s–41s.

Renoir, J. M., Marsaud, V. and Lazennec, G. 2013. Estrogen receptor signaling as a target for novel breast cancer therapeutics. *Biochem. Pharmacol.* 85(4):449–65.

Renouf, M. and Hendrich, S. 2011. Bacteroides uniformis is a putative bacterial species associated with the degradation of the isoflavone genistein in human feces. *J. Nutr.* 141(6):1120–6.

Sathe, S. K., Liu, C. and Zaffran, V. D. 2016. Food allergy. *Annu. Rev. Food Sci. Technol.* 7:191–220.

Schmidt, M., Arjomand-Wolkart, K., Birkhauser, M. H., Genazzani, A. R., Gruber, D. M., Huber, J., Kolbl, H., Kreft, S., Leodolter, S., Linsberger, D., Metka, M., Simoncini, T. and Vrabic Dezman, L. 2016. Consensus: soy isoflavones as a first-line approach to the treatment of menopausal vasomotor complaints. *Gynecol. Endocrinol.* 32(6):427–30.

Shike, M., Doane, A. S., Russo, L., Cabal, R., Reis-Filho, J. S., Gerald, W., Cody, H., Khanin, R., Bromberg, J. and Norton, L. 2014. The effects of soy supplementation on gene expression in breast cancer: a randomized placebo-controlled study. *J. Natl. Cancer. Inst.* 106(9).

St John, J. A., Henderson, V. W., Hodis, H. N., Kono, N., McCleary, C. A., Franke, A. A. and Mack, W. J. 2014. Associations between urine excretion of isoflavonoids and cognition in postmenopausal women in the Women's Isoflavone Soy Health clinical trial. *J. Am. Geriatr. Soc.* 62(4):629–35.

Stein, K. 2000. FDA approves health claim labeling for foods containing soy protein. *J. Am. Diet. Assoc.* 100(3):292.

Strom, B. L., Schinnar, R., Ziegler, E. E., Barnhart, K. T., Sammel, M. D., Macones, G. A., Stallings, V. A., Drulis, J. M., Nelson, S. E. and Hanson, S. A. 2001. Exposure to soy-based formula in infancy and endocrinological and reproductive outcomes in young adulthood. *JAMA* 286(7):807–14.

Tokede, O. A., Onabanjo, T. A., Yansane, A., Gaziano, J. M. and Djousse, L. 2015. Soya products and serum lipids: a meta-analysis of randomised controlled trials. *Br. J. Nutr.* 114(6):831–43.

van Die, M. D., Bone, K. M., Williams, S. G. and Pirotta, M. V. 2014. Soy and soy isoflavones in prostate cancer: a systematic review and meta-analysis of randomized controlled trials. *BJU Int.* 113(5b):E119–30.

Vega-Lopez, S., Matthan, N. R., Ausman, L. M., Ai, M., Otokozawa, S., Schaefer, E. J. and Lichtenstein, A. H. 2009. Substitution of vegetable oil for a partially-hydrogenated fat favorably alters cardiovascular disease risk factors in moderately hypercholesterolemic postmenopausal women. *Atherosclerosis* 207(1):208–12.

Weinborn, V., Pizarro, F., Olivares, M., Brito, A., Arredondo, M., Flores, S. and Valenzuela, C. 2015. The effect of plant proteins derived from cereals and legumes on heme iron absorption. *Nutrients* 7(11):8977–86.

Wu, A. H., Spicer, D., Garcia, A., Tseng, C. C., Hovanessian-Larsen, L., Sheth, P., Martin, S. E., Hawes, D., Russell, C., MacDonald, H., Tripathy, D., Su, M. Y., Ursin, G. and Pike, M. C. 2015. Double-blind randomized 12-month soy intervention had no effects on breast MRI fibroglandular tissue density or mammographic density. *Cancer Prev. Res. (Phila)* 8(10):942–51.

Xu, X. I. A., Harris, K. S., Wang, H.-J., Murphy, P. A. and Hendrich, S. 1995. Bioavailability of soybean isoflavones depends upon gut microflora in women. *J. Nutr.* 125(9):2307–15.

Ye, Y. B., Chen, A. L., Lu, W., Zhuo, S. Y., Liu, J., Guan, J. H., Deng, W. P., Fang, S., Li, Y. B. and Chen, Y. M. 2015. Daidzein and genistein fail to improve glycemic control and insulin sensitivity in Chinese women with impaired glucose regulation: A double-blind, randomized, placebo-controlled trial. *Mol. Nutr. Food Res.* 59(2):240–9.

Ye, Z., Renouf, M., Lee, S. O., Hauck, C. C., Murphy, P. A. and Hendrich, S. 2006. High urinary isoflavone excretion phenotype decreases plasma cholesterol in golden Syrian hamsters fed soy protein. *J. Nutr.* 136(11):2773–8.

Chapter 9

Improving the nutritional value of soybean

Istvan Rajcan and Jocelyne Letarte, University of Guelph, Canada

1 Introduction
2 Soybean protein content and profile
3 Soybean oil profile
4 Tocopherols
5 Isoflavones
6 Saponins
7 Future trends and conclusion
8 References

1 Introduction

The importance of soybean in many aspects of our society is now well established. It is a major source of protein and oil, and is therefore a prominent crop in the food industry, be it for animals or humans. Moreover, soybean is also being more recognized as a source of fuel and of biologically active compounds. In fact, eating soy products on a regular basis has been associated with a reduction of coronary diseases, diabetes, obesity and risk of cancer (Anderson et al. 1995; Zhang et al. 2003; Zhan and Ho 2005; Cena and Steinberg 2011). In addition to oil production and feed for animals, the worldwide consumer demand for soybean products has increased tremendously in the last twenty-five years and is likely to increase even more with the pressure to resolve the problem of malnutrition still pervasive in the developing countries (Hartman et al. 2011).

Even though it is a legume, soybean is not considered a pulse crop as most of it is produced as an oil crop. The rest of the production is for soy foods such as tofu, soymilk, miso and tempeh. Depending on the end product, the soybean variety chosen is extremely important and has a major impact on the quality of the desired end product. For example in the livestock industry, the need to increase the content of sulphur-containing amino acids is critical to the well-being of animals, while a lower content of anti-nutrients would help animals to thrive and grow healthier. On the other hand, for the oil industry, the protein content of a soybean line might not be as important as the fatty acids profile, where a desire to lower the linolenic acid content is significant to avoid a rapid oxidation of soybean oil that shortens the shelf life of soy products. In Asian cuisine, the texture and taste of tofu are intimately linked to the ratio of two major proteins present in soybean seeds, which are affected in turn by the genotype and environment. These are just a few

http://dx.doi.org/10.19103/AS.2017.0034.24

examples of the challenges researchers have to face to meet the demands of consumers or the livestock and health industry.

In this chapter we will limit ourselves to the genetics and breeding of soybean to improve the nutritional value of the seed as the end product. We have divided the chapter into five sections according to the specific areas of current research in breeding. The first one is on the protein content and profile of soybean seeds. The second section covers breeding for modified fatty acids profile that could extend soybean oil's shelf life or address various health or nutritional issues. The last three sections will explore how breeding could improve tocopherol, isoflavone and saponin contents of soybean seeds, which are known to have important health benefits for humans.

2 Soybean protein content and profile

Until recently, breeding programmes had mostly focused on increasing crop yield or improving pests or disease resistance under specific regional conditions and cultural practices. With selective breeding, hundreds of high-yielding cultivars have become available and nowadays the amount of soybean harvested year after year contributes to around 70% of the world protein meal consumption (http://soystats.com/international-world-protein-meal-consumption/). After vast achievements in these areas of soybean cultivation, improving protein content and quality is now at the forefront of breeders' efforts. Protein content or composition is cultivar dependent and affected by environmental conditions (Paek et al. 1997). The main objectives behind the genetic improvement of soybean proteins have been to increase total seed protein content, improve the amino acid composition, eliminate food allergens and design specialty soybean lines for the food or health industry that contain proteins with specific properties (Clarke and Wiseman 2000; Medic et al. 2014).

Protein content in soybean is a complex quantitatively inherited trait controlled by multiple genes, the environment and the interaction between the two. Depending on the line and growing conditions, soybean commercial seeds can have a protein content that vary from 35 to 40% on a dry weight basis (Krishnan 2005). However, there is often an inverse correlation between protein content and yield when soybean lines reach a 50% protein content or higher causing the yield to decrease (Brim and Burton 1979; Carter et al. 1982; Burton 1987; Wilcox and Zhang 1997; Wilcox and Shibles 2001). From a different perspective, continual selection for greater yield has in fact decreased protein content and elevated oil content (Rincker et al. 2014). These are important issues breeders have to take into account when releasing new varieties.

Recent efforts have been made on the molecular and genetics understanding of protein content in soybean. Several quantitative trait loci (QTLs) related to protein content have been identified. In the USDA Soybean Genetics Database (https://www.soybase.org/) 166 QTLs have been listed as being associated with seed protein content as of January, 2017. These QTLs were mapped in segregation populations, recombinant inbred lines or backcross populations, and some of them were linked to chromosomes regions mapped repetitively by different studies.

The first major QTL associated with protein and oil contents was found on chromosome (Chr.) 20 or linkage group I (Diers et al. 1992) and was later confirmed in other studies (Chung et al. 2003; Nichols et al. 2006). Subsequently, other chromosome regions were also identified as being highly linked to soybean seed protein content: Chr. 4: 43.6–47.7 Mb; Chr. 5: 39.7–41.4 Mb; Chr. 7: 4.2–9.6 Mb; Chr. 8: 5.8–10.2 Mb; Chr. 14: 4.8–9.6 Mb; Chr. 15: 0.0–7.5 Mb; Chr. 18: 47.9–54.0 Mb; Chr. 19: 35.5–42.1 Mb and again Chr. 20: 2.1–34.2 Mb (Qi et al. 2016; Zhang et al. 2015; Qi et al. 2011). However, most of the protein QTLs reported to date have not been confirmed or ultimately cloned (Phansak et al. 2016). According to Phansak et al. (2016) the additive effect values for these QTLs might have been amplified by the small size of the mapping populations used, which resulted in an intrinsic QTL detection problem known as selection bias (Kearsey and Farquhar 1998; Xu 2003; Broman and Sen 2009).

To assess the allelic diversity of QTLs associated with soybean seed protein content in germplasm collections (Germplasm Resources Information Network (GRIN) https://www.ars-grin.gov/npgs), genome-wide association studies (GWASs) were also conducted (Thornsberry et al. 2001; Hwang et al. 2014; Vaughn et al. 2014). Recently, Bandillo et al. (2015) have also done an analysis using GWAS involving the USDA Soybean Germplasm collection. They used the collection of 14,000 soybean accessions [*Glycine max* (L.) Merr. and *G. soja* Siebold & Zucc.] using a 50K-SNP chip and screened about 12 000 accessions related to protein and oil phenotypic data. Their study is one of the largest reported in plants and identified single nucleotide polymorphisms (SNPs) with strong signals on chromosomes 20 and 15 and weaker signals on chromosomes 13, 6 and 5. The chromosome 20 region mentioned previously, known to be important for protein and oil contents (Diers et al. 1992; Hwang et al. 2014; Vaughn et al. 2014), was further narrowed down to include only three plausible candidate genes (Glyma20 g21030, Glyma20 g21040 and Glyma20 g21080) (Bandillo et al. 2015). Glyma20 g21040 has no known function, while Glyma20 g21030 is annotated as a putative ammonium transporter, AMT1 (Sohlenkamp et al. 2002). Glyma20 g21080 is another candidate gene within the block region, which is annotated as an ATP synthase D chain (Bandillo et al. 2015). Further research is required to confirm if one of these genes plays an important role in the synthesis and accumulation of protein and oil contents as part of the QTL on chromosome 20 of soybean.

In addition to a need for increasing soybean protein content, considerable efforts are also devoted on the profile of the proteins stored in the seeds. Proteins stored in the seeds are divided into four major groups (2S, 7S, 11S and 15S) according to their coefficient of sedimentation. Among these groups, 7S and 11S, called β-conglycinin and glycinin respectively, are the most prevalent storage proteins, forming 70% of the bulk of proteins present in the seeds (Yaklich et al. 1999). Research in this important aspect of protein composition in soybean is driven by the soybean meal industry. Soybean meal, a by-product of the oil extraction industry, has a critical role in the animal feed industry. In fact, 97% of feed formulation used for livestock in the USA (and 63% worldwide) comes from soybean meal (Fu et al. 2007).

Even with an overall rich amino acids profile in comparison to cereals, soybean is however poor in some essential amino acids for monogastric animals (e.g. swine and poultry) such as lysine (Lys), threonine (Thr), methionine (Met) and cysteine (Cys) (Kwanyuen et al. 1997; Van Warrington et al. 2014). Increasing these amino acids in soybean seeds would benefit the feed industry. Although this is a complex economical issue, a conservative estimate in saving up to 5% could be achieved by the livestock industry by just improving the amino acid profile of soybean meal (Chung and Pettigrew 1998). Since the implementation of the Better Bean Initiative in 2002 in the USA, research efforts have been spent on finding ways to increase sulphur-containing amino acids like Met and Cys in soybean seeds, which are the two lowest amino acids, in terms of concentration, present in this legume.

Different undertakings have been explored to increase the Met/Cys concentration in soybean. Mutations or genetic engineering methods were performed in an attempt to create soybean phenotypes high in these amino acids (Townsend and Thomas 1994; Dinkins et al. 2001; Imsande 2001; Takahashi et al. 2003; Kim and Krishnan 2004; Krishnan 2005). Most of these studies have been successful at implementing a new gene or at selecting the desired mutant, but the final outcome – that is, producing soybean seeds with a higher level of sulphur-containing amino acids – has so far been elusive or in the case of the brazil nut gene (Townsend and Thomas 1994) not marketable due to its allergenic effect (Nordlee et al. 1996). More recently, Zhang et al. (2014) have introduced a synthetic gene MB-16 in the soybean genome to boost seed Met content. MB-16, an 11-kDa de novo protein enriched with the amino acids methionine, threonine, lysine, and leucine was developed initially for expression in rumen bacteria (Beauregard et al. 1995; Simons et al. 2001). With β-conglycinin or cruciferin seed-specific protein storage promoters attached to MB-16, soybean homozygous lines with single locus integrations were identified and analysed. MB-16 protein levels were the highest in full-sized green seed but the protein was found absent in mature seed. However, amino acid analysis of mature seed, in their best transgenic line, showed a significant increase of 16.2 and 65.9% in methionine and cysteine, respectively, as compared to the parent line. Although much research is still required to understand protein stability in transgenic plants overall, this study gives encouraging results on the feasibility of using foreign proteins in soybean to improve the specific nutritional needs requested by the feed industry.

Alternative approaches to the use of foreign genes introgression in soybean are the identification of cultivars that accumulate proteins rich in sulphur-containing amino acids or rebalance the proteome by decreasing proteins low in Met in favour for those with a higher content. Glycinin (11S) is a hexamer of 360 kDa composed of five major subunits (acidic and basic), namely G1, G2, G3, G4 and G5. It contains 3–4 times more Met and Cys than β-conglycinin (7S), a trimer (α,α′ and β subunits) of approximately of 180 kDa (Rajcan et al. 2005). By promoting, for example, the accumulation of the 11S globulin while reducing the expression of 7S, soybean with a higher level of sulphur-containing amino acids could be achieved. Therefore, mapping QTLs for storage proteins is an important step for breeders to get information about the genetics behind the control of proteins expressed in the seeds.

By using 101 F6-derived recombinant inbred lines from a cross between N87-984-16 and TN93-99, Panthee et al. (2004) were able to identify three QTLs for glycinin on chromosomes 17, 19 and 20 and two QTLs on chromosomes 16 and 17 were found for β-conglycinin. Moreover, a single dominant locus qBSC-1 from the β-subunit content of β-conglycinin was reported by Wang et al. (2014) and mapped to a region of 11.9 cM on chromosome 20 in a F2 segregation population of 85 individuals and subsequently narrowed down to 2.5 cM between BARCSOYSSR_20_0997 and BARCSOYSSR_20_0910 in a residual heterozygous line of 246 individuals. Recently Ma et al. (2016) also analysed the genetic mechanism regulating the glycinin and β-conglycinin contents of soybean by using a recombinant inbred line (RIL) population from a cross between Kefeng No.1 and Nannong 1138-2. They used 221 markers to map QTLs for glycinin and β-conglycinin contents, and to find the ratio of glycinin to β-conglycinin (RGC) and the sum of glycinin and β-conglycinin. A total of 35 QTLs (25 significant) were detected – among them 14 were consistent with previous reports (Csanadi et al. 2001; Chapman et al. 2003; Kabelka et al. 2004; Panthee et al. 2004; Panthee et al. 2006a, 2006b; Palomeque et al. 2009; Liang et al. 2010; Yoshikawa et al. 2010; Lu et al. 2013;

Pandurangan et al. 2012; Rossi et al. 2013), indicating that some causal genes might be present in these regions, which would be useful for molecular-assisted breeding to improve soybean quality.

Nonetheless few studies have been conducted on the genetics related to the amino acid composition of soybean seed. A landmark study by Panthee et al. (2006a) used 101 F6-derived recombinant inbred lines to identify soybean linkage groups associated with sulphur-containing amino acid concentrations. They found QTLs for Met on chromosomes 7, 13 and 18; Cys on chromosomes 1, 13 and 18; and Met/Cys on chromosomes 7 and 13. In a more recent study, Fallen et al. (2013) used the Universal Soy Linkage Panel 1.0 (USLP 1.0) (Hyten et al. 2010) with 1,536 SNPs to detect QTLs associated with amino acids in a F5:9-derived RIL population developed from a cross – Essex x Williams 82. They found QTLs associated with Cys on Chr. 20, Met on Chr. 13 and Thr on Chr. 13. Warrington et al. (2015) have also pursued this exploration of the genetic basis of amino acids composition in soybean seeds by identifying QTLs associated with specific amino acids in a 'Benning' x 'Danbaekkong' population. Four QTLs were detected on chromosomes 14, 15, 17 and 20. The QTL on chromosome 20 explained 55% of the phenotypic variation since this chromosomal region was associated to the contents of Lys, Thr, Met, Cys and Met/Cys. They also found two additional QTLs for Lys on Chr. 8 and 20; three QTLs for Thr on Chr. 1, 9 and 17; three QTLs for Met on Chr. 6, 9 and 10; and one QTL for Cys on Chr. 10. Another recent study by Wang et al. (2015) also looked at the identification and validation of QTLs associated with Met and Cys contents in two recombinant inbred lines. From the eight QTLs initially identified in the mapping population for Met and Cys contents, three QTLs were subsequently validated in a second population on chromosomes 4, 17 and 20. These stable QTLs could be good candidates for marker-assisted selection in breeding. The desired alleles of traits of interest can be easily introgressed into a commercial soybean line and then expressed under these new conditions if the QTLs selected originally were independent of environments and genetic backgrounds (Palomeque et al. 2010).

In addition to improving amino acid content and profile in soybean, there is also the important aspect of eliminating or decreasing anti-nutritional proteins present in soybean meal. In aquaculture, this is a crucial problem. For sustainable purpose, the fish industry is now routinely substituting a portion of fishmeal for plant protein at a low concentration level. However, most types of fish will have a limited growth rate or be affected negatively if this proportion is increased over a certain threshold (Herman and Schmit 2016). The breeding of low anti-nutritional bioactivity soybean could help achieve this goal. The USDA soybean collection has null lines for anti-nutritional bioactive seed proteins. These include the Kunitz trypsin inhibitor (KTI) null (Orf and Hymowitz 1979; Hymowitz 1986), soybean agglutinin (SBA) null (Orf et al. 1978; Goldberg et al. 1983) and immuno-dominant soybean allergen P34 protein null (Josephs et al. 2006). Each of these null lines can partially address the anti-nutritional effects following soybean meal consumption. However, a Triple Null stack of the recessive nulls of KTI, SBA and P34 in the standard cultivar Williams has been recently produced (Schmidt et al. 2015; Herman and Schmidt 2016). Proteomic analysis of the Triple Null line shows that it lacks three bioactive proteins detrimental to nutrition while at the same time preserving the full complement of the other protein constituents with no further collateral bioactive protein alterations. Therefore, this Triple Null line could be used as a potential platform to design commercial soybean line that retains high yield and resistance to pests while enhancing their capacity to optimize the feed performance of this important legume.

3 Soybean oil profile

Based on its fatty acid profile, soybean oil is considered one of the healthiest vegetable oils. It is also the most widely used edible oil in the United States (Gillman et al. 2014b) at 66% of total edible fats (http://soystats.com/archives/2012/non-frames.htm; as compiled from the USDA statistics). Wild-type soybean oil consists mainly of triacylglycerols, which contain five principal fatty acids: palmitic acid (C16:0, ~100 g kg^{-1}), stearic acid (C18:0, ~40 g kg^{-1}), oleic acid (C18:1, ~220 g kg^{-1}), linoleic acid (C18:2, ~540 g kg^{-1}) and linolenic acid (C18:3, ~100 g kg^{-1}) (Wilson 2004). The majority of soybean oil produced in oil crushing plants is used as edible oil for salad or as cooking oil for frying or baking, representing approximately 53 and 21% of soybean oil consumption, respectively (Gillman et al. 2014b as per http://soystats.com/archives/2012/page_30.htm, compiled from USDA statistics; verified online 02 October 2017).

Whereas the conventional fatty acid is generally considered healthy, it is not perfect due to a relatively high level of palmitic acid and linolenic acid. The former has been associated with LDL cholesterol (Primomo et al. 2002) that has been associated with heart disease (Khosla and Sundrum 1996) and the latter with the need for partial hydrogenation leading to trans-fatty acids as a by-product (Bilyeau et al. 2011). A number of research labs and breeding programmes around the world have attempted and succeeded in modification of the soybean oil's fatty acid profile in different ways as discussed below.

Although the level of palmitic acid is considered to be moderate (at approx. 110 g kg^{-1}), reduction (Primomo et al. 2002) of this fatty acid has been accomplished by searching for natural allelic variation or mutagenesis. Early work has resulted in the development of low-palmitic lines with as little as 40 g kg-1 palmitic acid in the oil, including A18 (Fehr et al. 1991), C1943 and N94-2575 (Burton et al. 1998; Primomo et al. 2002). The alleles in these and other soybean lines have been given different names as follows: The alleles conferring low palmitic acid in these lines were given different names – fap1 (in C1726), fap3 (in A22), fapnc (in N79-2077), fap*or fapx (in ELLP2), and sop1 (in J3) (Cardinal et al. 2007). Upon isolating full-length cDNAs of three of the four unique FATB genes expressed in soybean, Cardinal et al. (2007) reported that the isoform-designated GmFATB1a represented the specific gene deleted in lines possessing the fapnc locus. The same group was able to develop allele-specific primers corresponding to GmFATB1a mutant allele, which accounted for 62–70% of the genotypic variation in palmitate content and could be used in marker-assisted breeding for low palmitate content. Fillman et al. (2014) identified a new null allele of the fap3-ug/GmFATB1A (derived from line ELLP2), which was present in line RG3 (Primomo et al. 2002). The splicing defect mutation in a beta-ketoacyl-[acyl-carrier-protein] synthase III candidate gene, which was located in the fap1 genomic region, derived originally from ethyl methane sulphonate mutant line C1726 (Cardinal et al. 2014), was also present in line RG3. More recently, Thapa et al. (2016) identified two lines with reduced levels of seed palmitic acid, each of which having a 30% reduction in palmitic acid. Both lines carried distinct mutations in the FATB1A (Glyma.05G012300) with one line carrying a single base change that disrupted the splicing of the FATB1A mRNA and the other line containing a missense mutation (aspartic acid substituted for glycine) in a conserved domain of the FATB1 enzyme – both imparting low palmitic acid phenotype (Thapa et al. 2016). Goettel et al. (2016), using transcriptome sequencing, identified and characterized a 254-kb genomic deletion on a duplicated chromosome segment that resulted in a low level of palmitic acid in soybean seeds. The functional conservation of homoeologous genes may have contributed to the gene dosage effects, explaining why the large deletion did not cause lethal effects or completely eliminate palmitic acid in the low-palmitic line, N0304-303-3 (Goettel et al. 2016). Besides the reduction of palmitic acid, an increase in the same has also been studied to address a number of industrial applications; it is also served as a source of edible oils for human consumption (Head et al. 2012). The 3-ketoacyl-ACP synthase II (KASII) enzyme represents a branch point in the pathway of oil biosynthesis in soybean, which leads to the elongation of palmitic acid to stearic acid (Head et al. 2012). Candidate gene sequencing has been used to determine the genetic cause of elevated levels in three mutant soybean lines in these lines ranging from 12.9 to 15.6% of palmitic acid. It has been reported that all three lines contained nucleotide changes in the coding sequence for the KASIIa gene (Head et al. 2012).

A reduction in linolenic acid has been the subject of much research and breeding over the past several decades, primarily due to the low linolenic oil's advantage in that it does not require partial hydrogenation, which leads to the production of detrimental trans-fatty acids as a by-product. It was shown that single recessive alleles could reduce linolenic acid content to about the 4% level (Rahman et al. 1998; Ross et al. 2000; Stojsin et al. 1998; Takagi et al. 1999; Wilcox and Cavins 1985). Reinprecht et al. (2009) reported that a mutation in the Fad3A gene introduced a stop codon in exon 6 that would prematurely terminate translation and a second mutation in the 5′ splice site of intron 5 of the Fad3B gene may result in abnormal mRNA splicing products in the soybean line RG10. Both mutations would result in a non-functional enzyme, and molecular markers were developed for these mutations to simplify and accelerate introgression of the low linolenic (<2.5%) trait from RG10 into elite soybean cultivars. To further reduce linolenic acid levels below 3% and ultimately to approximately 1%, additional mutations were necessary (Fehr et al. 1992; Bilyeu et al. 2011). Bilyeu et al. (2011) suggested that 1% linolenic acid soybean germplasm is the result of combinations of mutations in three independent members of soybean omega-3 fatty acid desaturase (FAD3) genes that corresponded to fan1, fan2 and fan3 loci. The same group was able to show that different combinations of mutant FAD3 alleles from CX1512-44 and an RG-10–derived soybean line were able to produce less than 3% linolenic acid, producing only 1% linolenic acid in the seed oil. In a later study, Shi et al. (2015) successfully developed TaqMan or Kompetitive Allele Specific PCR assays to detect and discriminate mutant alleles of FAD3A (CX1512-44 and C1640 sources) and FAD3C (CX1512-44 source) for low linolenic acid content soybean. These assayed markers have been validated in multiple populations and against a panel of diverse soybean lines, showing a perfect association between the SNP alleles and their corresponding fatty acid phenotypes (Shi et al. 2015).

In terms of combining the low linolenic and palmitic acids, Cherrak et al. (2003) showed that modified fatty acid soybean lines developed with approximately 4% 16:0 and 18:3, respectively, showed very weak positive correlations between 16:0 concentration and seed yield (r = 0.12) and between 16:0 and seed oil concentration (r = 0.13). With no correlation found between 18:3 levels and seed yield, the results indicated that breeding for reduced 16:0 and 18:3 could not have a negative impact on seed yield or oil concentration (Cherrak et al. 2003).

Increasing oleic acid from the wild-type level of 18–20% in the oil has been a goal of several breeding programmes and research groups around the world but primarily in the USA and Japan. The main reason lies in that consumption of high oleic acid oils is desirable because this monounsaturated fatty acid not only improves shelf life but also reduces the need for hydrogenation, similar to low linolenic acid, which adds cost and increases the cost of the oil and generates unwanted trans-fat (Ascherio and Willett 1997).

In the de novo lipid biosynthetic pathway, conversion of oleic acid (18:1) to linoleic acid (18:2) is catalysed by the delta-twelve fatty acid desaturase 2 enzyme (FAD2) (Okuley et al. 1994). In contrast to *Arabidopsis* and maize, soybean has multiple copies of the FAD2 desaturase genes consisting of FAD2-2A (Glyma19g32930), FAD2-2B (Glyma19g32940) and FAD2-2C (Glyma03g30070), which were reported as being widely expressed in the vegetative tissues of the soybean plant (Schlueter et al. 2007). Upchurch and Ramirez (2010) compared the steady-state transcript abundance during seed development of oleate-ACP thioesterase (GmFATB1a), delta-9 stearoyl acyl carrier protein desaturase (GmSACPD) and omega-6 fatty acid desaturase (GmFAD2) genes in five natural mid-oleic varieties and mutant M23 to gene expression data for the conventional non-mid-oleic cultivar. They found that, relative to Dare, there were instances where lower expression of GmFATB1a, GmFAD2-1A, GmFAD2-1B, GmFAD2-2 and GmFAD2-3 and higher expression of the GmSACPD-C might be associated with the mid-oleic seed phenotype M23 (Upchurch and Ramirez 2010). Pham et al. (2010) reported the development of conventional soybean lines with 80% oleic acid in the oil in two different ways, each requiring the contribution of only two genes. The high oleic acid soybean germplasm with the desirable oleic acid profile was also stable in two production environments (Pham et al. 2010), which was an issue with some of the previously reported mutations (Lee et al. 2009; Oliva et al. 2006). Highly efficient molecular markers have been developed using the presumed causative sequence polymorphisms in the FAD2-1B alleles for tracking the mutant alleles (Pham et al. 2010). The high oleic acid trait has not been easy to develop concurrently with high yield. La et al. (2014) reported that when averaged across all locations and several populations, seed yield of the high-oleic lines derived from 17D were within 2% of the normal oleic lines but yield between high and normal oleic lines varied among populations. The high-oleic lines averaged significantly higher in protein than normal oleic lines in all six populations. Oil was also higher in the high-oleic lines than normal-oleic lines from the 17D mutation populations but was lower in the high-oleic lines derived from M23. The high-oleic soybeans derived by combining the mutant alleles S117N FAD2-1A and P137R FAD2-1B gave comparable yields with the potential to generate more oil with greater functionality and a higher-protein meal than soybeans with normal oleic acid content (La et al. 2014). Furthermore, the same group from the University of Missouri has been issued a patent, U.S. Pat. No. 9,035,129, which describes a soybean that contains three to four times more oleic acid than conventional soybean varieties (Missouri Soybeans 2017). The total amount of oleic acid in the soybean oil of these seeds is between 80 and 85%, which significantly exceeds the industry's 70% minimum standard to be classified as high oleic. 'The Curators of the University of Missouri and the USDA Agricultural Research Service (ARS) are joint owners of the patent, and the Missouri Soybean Merchandising Council is the exclusive licensee' (Missouri Soybeans 2017). Besides the natural and induced mutations that have been developed by different research programmes, transgenic-based high-oleic soybeans with approximately 80% of oleic acid in the oil have been grown for a few years in 12 different states (United Soybean Board 2017). These soybeans have been developed and marketed mostly by Monsanto and Dupont. The area planted achieved 450 000 acres in 2016 and the current premium of 40–60 cents per bushel is expected to double in 2017 (United Soybean Board 2017), representing a fast-growing market that might replace commodity oilseed soybean in the future.

Stearic acid (C18:0) is considered to be health-neutral saturated fatty acid, which unlike the palmitic has not been associated with LDL cholesterol or heart disease. Soybean oil generally contains about 3–4% stearic acid; therefore, to make the oil more amenable to margarine production, attempts have been made to increase it to a target level of approximately 20% in the oil. Besides this application, the automotive industry stearic acid has a major use in the production of rubber for tires and other rubber parts (Luyt 1993). Increased levels of stearic acid, in particular mutant soybean lines, have been associated with molecular changes that negatively impact the activity of the delta-9-stearoylacyl carrier protein desaturase (SACPD) enzyme (Wilson 2004), which inserts the first double bond into the 18-C fatty acid chain at C9 position. Spencer et al. (2003) have localized the gene encoding this enzyme to the Fas locus on soybean chromosome 14. At least six additional allelic mutations have been reported for the locus (Pantalone et al. 2002). Boersma et al. (2012) reported a mutation in RG 7 soybean line's SACPD-C gene and an independent, presumably allelic, mutation within the SACPD-C gene in line RG8, which had about 106 g kg^{-1} stearic acid. The progeny of a cross between the high-stearic RG7 and low-palmitic RG2 lines had further increased seed stearic acid of almost 180 g kg^{-1} (Boersma et al. 2012). The same group reported development of molecular markers diagnostic for the RG7 and RG8 SACPD-C mutations that could aid in high-stearic soybean cultivar development (Boersma et al. 2012). It has been reported that, although no yield penalty was found for moderate elevated seed stearic acid content in soybean seeds, genetic alteration SACPD activity in particular has a role in the establishment and maintenance of symbiotic nitrogen fixation (Gillman et al. 2014a) and plant defence (Krishnan et al. 2016).

4 Tocopherols

Vitamin E represents a group of powerful antioxidant molecules known as α-, β-, γ- and δ-tocopherols and their four corresponding unsaturated derivatives: α-, β-, γ- and δ-tocotrienols (Hussain et al. 2013). The tocopherols are composed of a polar chromanol group with a lipophilic tail attached to it, whereas the tocotrienols consist only of a lipophilic tail (Bramley 2000; Munne-Bosch and Alegre 2002). The tocopherol isomers differ from one another with respect to the number and position of methyl groups present on the chromanol group. Alpha-tocopherol has three methyl groups followed by β- and γ-tocopherols with two on each and finally δ-tocopherol with only one methyl. A higher capacity to release hydrogen atoms correlates with the number of methyl substituents on the chromanol ring. Consequently, α-tocopherol is the most active form of vitamin E (DellaPenna 2005) and the potential activity for β-, γ- and δ-tocopherols is 50, 10 and 3% of the one measured in α-tocopherol (Kamal-Eldin and Appelqvist 1996; Shintani and DellaPenna 1998).

In plants, tocopherols act as antioxidant molecules that scavenge free radicals including the protection of chloroplasts from photooxidative damage (Munne-Bosch and Alegre 2002), the support of seed longevity, and the preservation of lipids from oxidation during germination and early seedling growth (Sattler et al. 2004). In humans, vitamin E plays an important role in supporting a strong immune system and also in the prevention or retardation of chronic diseases such as cardiovascular diseases (Stampfer and Rimm 1995; Pryor 2000), cancer (Constantinou et al. 2008), arthritis (Li et al. 2016b) or Alzheimer's disease (Grimm et al. 2016). The human liver absorbs all tocopherols equally (Traber 1996) but only α-tocopherol is selectively retained by the hepatic α-tocopherol transfer protein (TTP) to facilitate the excretion of α-tocopherol from hepatic cells to circulating lipoproteins (Traber 1996; Thakur et al. 2010; Hussain et al. 2013). As a result of its high activity in humans, α-tocopherol has aroused the greatest interest in research.

Vitamin E used for human or animal nutrition, or in the pharmaceutical and cosmetic industry, comes mainly from chemically synthesized α-tocopherol, which contributes to about 85% of the world market (Carrera and Seguin 2016). The remaining comes from natural sources (canola, oil palm, safflower, soybean and sunflower). However, natural vitamin E is known to be more biopotent than the synthetic version (Jensen and Lauridsen 2007); for this reason it would be highly advantageous to increase the market share of natural vitamin E in the health and food industry. Moreover, with fossil fuel resources declining, the production cost of natural vitamin E is expected to become much lower than the chemically derived tocopherols (Valentin and Qi 2005).

Tocopherols are important components of oil seeds; as soybean supplies more than 60% of the oil extracted worldwide (http://soystats.com/international-world-oilseed-production/), it could be used as a greater source of vitamin E by different industries. In addition, the current oil extraction process in soybean already enriches the tocopherol fraction, where it is collected with the deodorizer distillate. Then, following additional purification steps, a tocopherol-rich oil is obtained (Valentin and Qi 2005). This oil contains several tocopherol isomers, consisting of about 10% α-tocopherol, 2–5% β-tocopherol, 60–70% γ-tocopherol and 20–30% δ-tocopherol. A constant and uniform supply of vitamin E is important for the industries and being able to rely on a stable source is also critical; accordingly, breeders are looking at ways to increase tocopherols in soy. Carrera and Seguin (2016) have just recently published an excellent review covering the different aspects surrounding tocopherols in soybean, from their synthesis and biological functions in plants to genetic factors affecting their concentrations, followed by the role that environmental and agronomic factors play as well on tocopherols concentration. Consequently, we will only cover the most salient studies related to the genetics of tocopherols.

From several reports (Carrao-Panizzi and Erhan 2007; Britz et al. 2008; Seguin et al. 2010; Carrera et al. 2014; Lee et al. 2015; Shaw et al. 2016), a clear consensus has emerged that genotype, environment and the interaction between the two play an important role in the determination of tocopherol concentration in soybean seed. Nonetheless, genotype variation has the most effect on tocopherol content or composition and has therefore been the main focus of research efforts. In addition, heritability of tocopherols in plant seeds also seems to be relatively high (Dwiyanti et al. 2007; Yang et al. 2010; Del Moral et al. 2011; Shaw et al. 2016). For example, in 79 recombinant in bred lines grown in six environments, Shaw et al. (2016) estimated broad-sense heritabilities for α-, γ- and δ-tocopherols to be 38, 47 and 35%, respectively. Previously, Dwiyanti et al. (2007) had also found a strong broad-sense heritability value of 65% for α-tocopherol concentration in 140 F2 soybean seeds grown in a greenhouse. These results in soybean and other crops (Yang et al. 2010; Del Moral et al. 2011) give additional support on the potential success of eventually breeding high-tocopherol soybean lines. A running start to achieve this goal would be the screening of the soybean germplasm collections to identify potential genotypes with higher tocopherol content than the commercial soybean cultivars.

Therefore, Ujiie et al. (2005) performed a screening of 909 soybean (*Glycine max* (L.) Merr.) and 200 wild soybean (*Glycine soja* Sieb. and Zucc.) accessions from a worldwide germplasm collection and were able to identify from HPLC analysis three accessions with high α-tocopherol content (>20%): Dobrogeance, Dobruoza 14 Pancevo and Keszthelyi Aproszemu Sarga (KAS). One of them, KAS, was further analysed by QTL mapping using an F2 segregating population derived from a cross between Japanese cultivars,

Ichihime (α-tocopherol ratio <10%) and KAS (Dwiyanti et al. 2007). The result has led to the identification of a major QTL located on chromosome 9 and the potential gene, γ-tocopherol methyltransferase-3 (γ-TMT3), responsible for higher α-tocopherol content in this genotype (Dwiyanti et al. 2011). Moreover, in the Williams 82 soybean reference genome (Schmutz et al. 2010) three copies of γ-TMT have also been identified: γ-TMT3 (Glyma.09G222800) located on Chr. 9, then γ-TMT1 (Glyma.12G014200) and γ-TMT2 (Glyma.12G014300), both located in tandem on chromosome 12. Studies on the expression of γ-TMT2 in maize and *Arabidopsis* were also done and they showed an increase of α-tocopherol content, which means this gene is also functional (Zhang et al. 2013).

Gamma-tocopherol methyltransferase is a key enzyme in the synthesis of α-tocopherol and β-tocopherol. The first step in the formation of tocopherols is the synthesis of homogentisic acid (HGA), the tocopherol aromatic headgroup, from p-hydroxyphenylpyruvic (HPP), catalysed by the enzyme, HPP acid dioxygenase. The next enzyme in the pathway is homogentisate phytyl transferase that uses two precursors, HGA and phytyl diphosphate, to catalyse the formation of the common precursor of the tocopherols biosynthetic pathway: 2-methyl-6-phytyl-1,4-benzoquinone (MPBQ). A methyl alkyl is then added to MPBQ by the methyltransferase (MPBQ-MT). This results in the production of 2,3-dimethyl-6-phytyl-plastoquinol (DMPBQ). MPBQ and DMPBQ are then cyclized by tocopherol cyclase to form δ-tocopherol and γ-tocopherol, respectively. An additional methylation step of δ-tocopherol and γ-tocopherol produces β-tocopherol and α-tocopherol, respectively. The last methylation step, to form α- and β-tocopherol, is performed by the key enzyme, γ-tocopherol methyltransferase (γ-TMT). Genes encoding γ-TMT have been isolated from *Arabidopsis thaliana* (Shintani and DellaPenna 1998), *Perilla frutescens* (Tavva et al. 2007) and *Brassica napus* (Chen et al. 2012)

During seed maturation, γ-TMT3 expression has been found to be higher in soybean seeds of RILs with the KAS allele in comparison to RIL seeds with the Ichihime allele (Dwiyanti et al. 2011). After further analysis, Dwiyanti et al. (2011) also observed SNPs in MYBCORE and CAAT box motifs in the promoter region of the gene, and this difference could explain the elevated α-tocopherol content measured in the seeds of this genotype.

In addition to screening for genotype variation, molecular engineering is another avenue being explored to increase α-tocopherol content. Indeed, the very potent activity of α-tocopherol in humans has prompted researchers to focus on other molecular ways of increasing α-tocopherol in plants. In *Arabidopsis*, all the key enzymes involved in the tocopherol biosynthetic pathway have been cloned and characterized (Sattler et al. 2004; Farré et al. 2014). Since γ-tocopherol is converted into α-tocopherol by the enzyme γ-TMT, an expected increase in α-tocopherol content could be achieved in plant seeds with an overexpression of γ-TMT. Shintani and DellaPenna (1998) tried this approach when the γ-TMT gene was cloned and overexpressed using a seed-specific promoter in *Arabidopsis*. This resulted in transgenic seeds with a higher α-tocopherol content (>80-fold increase) at the expense of γ-tocopherol, without however altering the total tocopherol content in the seeds. More recently, this was also done with γ-tocopherol methyltransferase from *B. napus* (BnTMT) and under the control of an *Arabidopsis* promoter and inserted in soybean (Chen et al. 2012). In this case, increases of about 11-fold and 18.0-fold were measured for α-tocopherol and β-tocopherol, respectively. The overexpression of γ-TMT attached to different promoters in *Arabidopsis* (Shintani and DellPenna 1998), soybean seeds (Chen et al. 2012), perilla seeds (Lee et al. 2008) or lettuce leaves (Li et al. 2011) has given so far ambivalent results since total tocopherols tend to not change significantly and the observed increase in α-tocopherol is often done to the detriment of the other isomers. Van Eenennaam et al. (2003) have also shown the complexity of increasing tocopherols by molecular engineering. They transformed soy with a construct made of two *Arabidopsis* genes involved in the tocopherol biosynthetic pathway (At-VTE3 corresponding to MPBQ-MT and At-VTE4 corresponding to γ-TMT). When At-VTE3 was co-expressed with At-VTE4 in soybean, the seed accumulated up to >95% α-tocopherol, about an eightfold increase of α-tocopherol in comparison to the normal 10% measured in a regular cultivar. However, like the studies mentioned above, seed-specific expression of At-VTE3 and At-VTE4 as single genes or in combination did not cause a significant effect on total tocopherol levels in the soybean seed (Van Eenennaam et al. 2003).

Although the tocopherol biosynthetic pathway seems relatively simple, the molecular mechanisms regulating it are not fully known or understood. Moreover, the transformation of soy by molecular engineering is not widely accepted by the public in some countries and this might be an obstacle in the future (Carrera and Seguin 2016). Conventional breeding with marker-assisted selection still remains an efficient method to improve the nutritional value of this worldwide crop. However, there are yet too few QTLs associated with individual tocopherol or total vitamin E (Li et al. 2010; Li et al. 2016a) that are stable and can be detected in multiple environments. Markers closely associated with QTLs related to vitamin E content would facilitate the identification of soybean lines high in vitamin E and make the selection process more effective (Li et al. 2016a).

5 Isoflavones

Isoflavones are a group of secondary metabolites synthesized almost exclusively in legumes, especially soybean seeds (Primomo et al. 2005a). They have received considerable attention because of their reported effects as nutraceuticals in that clinical studies showed them to be associated with prevention and treatment of a number of chronic diseases such as certain types of cancer, osteoporosis, heart disease and relief from menopause symptoms (as reviewed in Messina 1999; Munro et al. 2003). Soybean seeds contain three major isoflavone aglycones (i.e. daidzein, genistein and glycitein) as well as their corresponding glucosides, acetyl- and malonyl-ester glucosides (Primomo et al. 2005a). Breeding for high isoflavone content has been challenging, because the trait is heavily influenced by the environment (Primomo et al. 2005b). Therefore, identification of molecular markers linked to QTL for isoflavones has been used as an approach towards understanding the genetic basis of this trait and facilitating molecular marker development that can be used in MAS. The first reports on molecular markers for isoflavones in soybean have been published by Kassem et al. (2001) and Meksem et al. (2001). In the cross Essex x Forrest, they found significant associations of glycitein content with SSR markers Satt237 (LG N) and Satt251 (LG B1), and daidzein content with Satt080 (LG N) and Satt276 (LG A1) (Meksem et al. 2001). Although the reported QTLs were associated with daidzein, genistein and glycitein, only QTLs with major effects were found and no total isoflavone QTLs were found perhaps due to the small population size (Primomo et al. 2005b). The first report on soybean isoflavone QTL identified using northern soybean germplasm was published by Primomo et al. (2005a,b). In that study, a population of 207 F6 lines derived from the cross AC756 x RCAT Angora was used to identify 32 QTLs across environments by single-factor ANOVA with individual locus explaining up to 10.5% ($P < 0.0001$) of the phenotypic variation (Primomo et al. 2005b). In addition, interval mapping and composite interval mapping identified nine genomic regions (LGs A1, C2, D1a, F, G, H, J, K and M) associated with isoflavone content of which some were also associated with agronomic traits or protein content (LGs M, J or K) and significant epistasis was also found between markers for each isoflavone (Primomo et al. 2005b). In a related study, using two different RIL populations, it was shown that 'high' isoflavone RIL group produced higher yield, matured later, and were taller than the 'low' isoflavone RIL group and that there was a significant negative association between isoflavone and protein content (Primomo et al. 2005a). This leads to the conclusion that soybean cultivars could be developed with desirable seed isoflavone and superior agronomic traits. Using an RIL population derived from the cross of two Chinese cultivars, Jindou 23 and Huibuzhi, six QTLs for isoflavone content were mapped to linkage groups J, N, D2 and G (Liang et al. 2010). A major QTL found on Gm05 (Glycine max chromosome 5; formerly linkage group A1), which had a significant effect on each individual isoflavone plus total seed isoflavone accumulation (named qGEN5, qDAI5, qGLY5 and qTOT5, respectively), was validated in F7-RIL population derived from the cross Magellan 9 PI 437654 (Gutierrez-Gonzales et al. 2011). Smallwood et al. (2014) identified 21 QTLs for isoflavone accumulation using the cross, Williams 82 x Essex, of which 12 had been reported in the same genomic regions in independent studies on chromosomes 6 (Primomo et al. 2005b; Gutierrez-Gonzalez et al. 2011), 13 (Zeng et al. 2009) and 19 (Gutierrez-Gonzalez et al. 2009, 2010), which represented a positional confirmation.

Gutierrez-Gonzales et al. (2010) studied transcripts of the most important genes for isoflavone biosynthesis and found that the expression of CHS7, CHS8 and IFS2 correlated with isoflavone accumulation under well-watered conditions. However, two isoflavone synthase genes in soybean (IFS1 and IFS2) showed different patterns of expression by which IFS2 seemed to be the main contributor to decreasing isoflavone accumulation under drought conditions (Gutierrez-Gozales et al. 2010). The study by Chen et al. (2011) found that at the R5 stage the expression of 14 key genes and their homologs involved in isoflavone synthesis occurred in all plant parts with the minimum being in pods and seeds and the maximum being in leaves and roots. The gene expression data coincided with isoflavone concentrations in the same plant parts.

Morrison et al. (2008) studied changes in isoflavone concentration over decades of breeding using a collection of 14 historical cultivars released from 1934 to 1992, which were grown in Ottawa for 12 years under identical cultural conditions. During that period, daidzein, genistein and total isoflavones increased by 1.04, 1.47 and 0.98% per year, respectively. In addition, genotype main effects + genotype x environment interaction biplots showed that recent cultivars with high isoflavone concentration were more prone to environmental influence than older cultivars (Morrison et al. 2008). Besides the genetic characterization, understanding the environmental factors has received much attention by soybean physiologists as well. Isoflavone profile was determined in a set of commercial cultivars that were grown in 76 environments in Argentina (29–38°S) (Carrera and Dardanelli 2016). Multiple linear regressions were developed including total isoflavones (TI), aglycones (AGL), glucosides (GLC), malonyl glucosides (MAL) and acetyl glucosides (ACE), for the following parameters: 1) temperature during seed fill (TmR5R7) and 2) precipitation minus potential evapotranspiration during the reproductive period (pp-PETR1R7) and the combination of 1) and 2) (Carrera and Dardanelli 2016). It was reported that cumulative solar radiation predicted isoflavone content but was less robust than TmR5R7 and pp-PETR1R7, and when pp-PETR1R7 was below 70 mm (indicating drought), TI, AGL, GLC, MAL and ACE decreased linearly with rising temperatures and water deficit (decreasing values of pp-PETR1R7) with an additive effect on isoflavones.

6 Saponins

Another important biologically active group of molecules in soybean that have also sparked more interest in research in recent years are the saponins. Saponins are steroid or triterpenoid glycosides found in many plants with soybean containing the highest concentration. Raw soybean can have a content of 0.6 to 6.2% dry weight of saponins (or soyasaponins) (Shiraiwa et al. 1991) depending on the cultivar, the age or part of the plant, growing conditions, location, etc. (Price et al. 1987; Shimoyamada et al. 1990; Shi et al. 2004). Saponins consist of a polycyclic aglycone attached to one or more polar sugar side chains. The aglycone part, which is also called sapogenin, is either a steroid or a triterpene. Soyasaponins are classified in four major groups according to their aglycone part: groups A, B, E and 2,3-dihydro-2,5-dihydroxy-6-methyl-4H-pyran-4-one (DDMP) (Guang et al. 2014). DDMP saponins are degraded into group B and E saponins during food processing (Kudou et al. 1992, 1993). For the illustrations of the chemical structures of each group please refer to Takada et al. (2013) and the review of Guang et al. (2014). Approximately 36 saponins have been identified in soybean with some of them present in parts of the plant and others present after food processing (Park et al. 2016). With their numerous chemical structures, their physiological effects on plants, animals or humans are varied. Saponins are amphilic molecules and this allows them to act as detergents and form oil-in-water emulsions with foaming capacity (Isanga and Zhang 2008). In plants, they have an important role in biotic or abiotic stresses (Woldemicheal and Wink 2002; Barlie et al. 2007). In humans, soyasaponins have received wide attention not only for their recognized health benefits but also for their adverse sensory effects in soy products. The undesirable bitter and astringent taste that is often present in soy foods is partly due to soyasaponins A (Okubo et al. 1992), whereas the health benefits saponins have on humans are mainly attributed to soyasaponins B. These health benefits are varied and range from anti-cancer activity to antiviral, antihepatotoxic or hypocholesterolaemic effect (Fournier et al. 1998; Kinjo et al. 1998; Gurfinkel and Rao 2003; Shi et al. 2004; Chang et al. 2006).

With a higher interest in these molecules from the food and health industry, more research is coming forth on saponin content or profile in soybeans. A recent study done by Park et al. (2016) looked at group A soyasaponins known to cause a disagreeable aftertaste. Group A soyasaponins are divided into three types: Aa-, Ab- and A0- (Takada et al. 2013). In the Aa- subgroup, the saponins (Aa, Au, Ae, Ax, Ay and Ag) have an acetyl xylose at the terminal of the sugar chain attached at the C-22 position of soyasapogenol A (SS-A). For the Ab- saponins (Ab, Ac, Af, Ad, Az and Ah), there is an acetyl glucose placed at the same position (Krishnamurthy et al. 2014). The third subgroup of soyasaponins A, saponins A0- (A0-αg, A0-βg, A0-γg, A0-αa, A0-βa and A0-γa), have no terminal acetylated sugar at their C-22 sugar chain (Takada et al. 2013). Among all these soyasaponins A: Aa, Ab and A0-αg are the major components for each subgroup. The sugar terminal of the C-22 sugar chain in Aa and Ab subgroups is controlled by Sg-1a and Sg-1b, respectively. The Sg-1a allele encodes a xylosyltransferase and Sg-1b allele encodes a glucosyltransferase, both of which are codominant-allelic alternatives at a single locus, Sg-1, which is mapped on soybean chromosome 7 (Takada et al. 2010; Sayama et al. 2012; Park et al. 2016). Since the bitter aftertaste is associated with saponins of group A, Park et al. (2016) suggested that a genetic removal of these molecules could eliminate this undesirable characteristic in soy products. They therefore investigated a group A acetylsaponin-deficient soybean mutant (PE1515) selected after mutagenesis on the cultivar Pungsannamul with ethyl methanesulphonate. The phenotypic and genotypic variation was analysed for the mutant PE151. The analysis of this mutant line showed that the Ab saponins subgroup was absent and A0-αg saponins accumulated as the major saponins. PE1515 has a single base substitution that results in a premature stop codon in the Sg-1 gene. They intend to use this mutant line in soybean breeding programmes to improve the taste in soybean-derived foods.

Studies have also been conducted on polymorphic variation, geographical distribution or QTLs associated to specific groups of soyasaponins A, B, E or DDMP. Seed saponins were evaluated in 3025 wild soybean (*G. soja* Sieb. and Zucc.) accessions collected from nine regions in Korea and analysed by thin-layer chromatography to determine polymorphic variations and geographical distributions (Panneerselvam et al. 2013). They found that saponin composition in seed hypocotyls was primarily divided into seven phenotypes, designated as Aa, Ab, AaBc, AbBc, Aa+α, AaBc+α and AbBc+α. The predominant phenotypes were AaBc (55%), Aa (33%), AaBc+α (7.5%) and Aa+α (3.3%). They also discovered three accessions as mutants: the accession CWS0115 did not produce saponins Aa and Ax, CWS2133 did not produce saponins Aa and Ab, and CWS5095 did not produce any group A saponins. Although further analysis is required, these newly determined mutants could be used in soy breeding to produce new varieties with good taste.

In a similar analysis, Krishnamurthy et al. (2014) extended this study to other regions, namely China, Korea, Japan and Russia Far East and found eight common distinguishable saponin phenotypes (Aa, Ab, AaBc, AbBc, Aa+α, Ab+α, AaBc+α and AbBc+α). Six new triterpene saponins were detected in +α type and collectively named as Sg-6 saponins. The frequency of this Sg-6 allele was 17.6% in Chinese, 10.0% in Korean and 1.0% in Japanese wild soybean.

The most recent study to our knowledge related specifically to QTLs was done by Huang et al. (2012) and looked at the identification of the QTLs associated with total soyasaponin content through a F2 population, derived from a cross between Ha 91016 (higher soyasaponin content cultivar, 16.8 mg g^{-1}) and N98-9445A (lower soyasaponin content, only 5.7 mg g^{-1}). Two QTLs were identified: qSAP_1 (located between sat_044-satt102 on chromosome 9) could explain 12.6% of phenotypic variance and qSAP_2 (located between satt368 and sat_413 on chromosome 1) could explain 15.8% of phenotypic variance. After a validation analysis, these two QTLs could be used in marker-assisted breeding.

7 Future trends and conclusion

As the largest oilseed crop in the world, soybean is grown primarily for its high oil and protein concentration in the seed. Much of the extracted oil (on average 20% of the seed volume) is used for the production of edible oil but the rest is used for biodiesel and industrial applications as well. Soybean oil is considered one of the healthiest vegetable oil but further improvements have been made possible as a result of breeding efforts. Such improvements include oil stability, increased shelf life and lower saturated fat, especially palmitic acid and higher stearic acid for margarine production.

Whereas the average concentration of protein is about 40%, breeding efforts have concentrated on increasing it to as much as 50% to meet the demand of the soy food industry such as tofu and soymilk. Additional genetic improvement has been made to change the ratio of protein subunits and amino acid composition for various food applications such as tofu quality.

Numerous QTL studies have been identified to be associated with protein and oil concentration and composition, and molecular markers have been developed that are either PCR-based or gene-specific in nature. These markers provide useful tools to plant breeders working on improving the nutritional quality of soybean seed for various applications.

A large body of research in soybean has been done mainly on the bioactive components isoflavones or tocopherols. However, there are many more phytochemicals (Kang et al. 2010) present in soy (e.g. phytic acid, phytosterols, saponins, lecithins, lignans and ester glycosides) that need to be explored since their actions, and their synergic or antagonistic effects in soy following food processing are not fully understood. Molecular work and breeding are required to gain more knowledge on the mechanisms affecting their concentrations, compositions and expressions in different parts of the plants or their stability after food processing. By understanding their governing factors, it would be possible to alter their amount or profile in soybean according to the demands of the food and nutraceutical industries.

While improvements continue to be made, a challenge for the breeders has been the sometimes conflicting health reports and market development intelligence that influence the final breeding targets for nutritional quality traits. Another challenge lies in the placement of soybean cultivars with added nutritional value, such as high isoflavone ones, in the marketplace with appropriate compensations for the health benefit and breeding efforts involved.

8 References

Anderson, J. W., Johnstone, B. M. and Cook-Newell, M. E. (1995). Meta-analysis of the effects of soy protein intake on serum lipids. *N. Engl. J. Med.* 333:276–82.

Ascherio, A. and Willett, W. C. (1997). Health effects of trans fatty acids. *Am. J. Clin. Nutr.* 66 (Suppl 4): 1006–10.

Bandillo, N., Jarquin, D., Song, Q., Nelson, R., Cregan, P., Specht, J. and Lorenz, A. (2015). A population structure and genome-wide association analysis on the USDA Soybean Germplasm Collection. *Plant Genome* 8: 1–13.

Barlie, E., Bonanomi, G., Antignani, V., Zolfaghari, B., Sajjadi, S. E., Scala, F. and Lanzotti, V. (2007). Saponins from *Allium minutiflorum* with antifungal activity. *Phytochem* 68: 596–603.

Beauregard, M., Dupont, C., Teather, R. M. and Hefford, M. A. (1995). Design, expression, and initial characterization of MB1, a *de novo* protein enriched in essential amino acids. *Bio. Technol.* 13: 974–81.

Bilyeu, K., Gillman, J. D. and LeRoy, A. R. (2011). Novel FAD3 mutant allele combinations produce soybeans containing 1% linolenic acid in the seed oil. *Crop. Sci.* 51: 259–64.

Boersma, J. G., Ablett, G. R., Grainger, C., Gillman, J. D., Bilyeu, K. D. and Rajcan, I. (2012). New mutations in a *delta-9-stearoyl-ACP desaturase* gene associated with enhanced stearic acid levels in soybean seed. *Crop Sci.* 52: 1736–42.

Bramley, P. M., Elmadfa, I., Kafatos, A., Kelly, F. J., Manios, Y., Roxborough, H. E., Schuch, W., Sheehy, P. J. A. and Wagner, K. H. (2000). Vitamin E. *Sci. Food Agric.* 80: 913–38.

Brim, C. A. and Burton, J. W. (1979). Recurrent selection in soybeans. 2. Selection for increased percent protein in seeds. *Crop Sci.* 19: 494–8.

Britz, S. J., Kremer, D. F. and Kenworthy, W. J. (2008). Tocopherols in soybean seeds: genetic variation and environmental effects in fieldgrown crops. *J. Am. Oil Chem. Soc.* 85: 931–6.

Broman, K. W. and Sen, S. (2009). *A Guide to QTL Mapping with R/qtl*, Springer, New York.

Burton, J. W. (1987). Quantitative genetics: Results relevant to soybean breeding, pp. 211–42 in *Soybeans: Improvement, Production, and Uses*, edited by J. R. Wilcox, ASA, CSSA, and SSSA, Madison, WI.

Burton, J. W., Wilcox, J. R., Wilson, R. F., Novitzky, W. P. and Rebetzke, G. J. (1998). Registration of low palmitic acid soybean germplasm lines N94-2575 and C1943, *Crop Sci.* 38: 1407.

Cardinal, A. J., Whetten, R., Wang, S., Auclair, J., Hyten, D., Cregan, P., Bachlava, E., Gillman, J., Dewey, R., Upchurch, G., Miranda, L. and Burton, J. W. (2014). Mapping the low palmitate fap1 mutation and validation of its effects in soybean oil and agronomic traits in three soybean populations. *Theor. Appl. Genet.* 127: 97–111.

Cardinal, A. J., Burton, J. W., Camacho-Roger, A. M., Yang, J. H., Wilson, R. F. and Dewey, R. E. (2007). Molecular analysis of soybean lines with low palmitic acid content in the seed oil. *Crop Sci.* 47:304–10.

Carrao–Panizzi, M. C. and Erhan, S. Z. (2007). Environmental and genetic variation of soybean tocopherol content under Brazilian growing conditions. *J. Am. Oil Chem. Soc.* 84: 921–8.

Carrera, C. S., Dardanelli, J. L. and Soldini, D. O. (2014). Genotypic and environmental variation in seed nutraceutical and industrial composition of non-transgenic soybean (*Glycine max*) genotypes. *Crop Pasture Sci.* 65: 1311–22.

Carrera, C. S. and Dardanelli, J. L. (2016). Changes in the relationship between temperature during the seed-filling period and soyabean seed isoflavones under water-deficit conditions. *J. Agron. Crop Sci.* 202: 421–32.

Carrera, C. S. and Seguin, P. (2016). Factors affecting tocopherol concentrations in soybean seeds. *J. Agric. Food Chem.* 64: 9465–74.

Carter, T. E., Burton, J. W. and Brim, C. A. (1982). Recurrent selection for percent protein in soybean seed – Indirect effects on plant-N accumulation and distribution. *Crop Sci.* 22: 513–19 .

Cena, E. R. and Steinberg, F. M. (2011). Soy may help protect against cardiovascular disease. *California Agric.* 65: 118–23.

Chang, W. W., Yu, C. Y., Lin, T. W., Wang, P. H. and Tsai, Y. C. (2006). Soyasaponin I decreases the expression of alpha-2,3-linked sialic acid on the cell surface and suppresses the metastatic potential of B16F10 melanoma cells. *Biochem, Biophys. Res. Commun.* 341: 614–19.

Chapman, A., Pantalone, V. R., Usten, A., Allen, F. L., Landau-Ellis, D., Trigiano, R. N. and Gresshoff, P. M. (2003). Quantitative trait loci for agronomic and seed quality traits in an F_2 and $F_{4:6}$ soybean population. *Euphytica* 129: 387–93.

Chen, D. F., Zhang, M., Wang, Y. Q. and Chen, X. W. (2012). Expression of γ-tocopherol methyltransferase gene from *Brassica napus* increased α-tocopherol content in soybean seed. *Biol. Plantarum.* 56: 131–4.

Chen, H., Seguin, P., Jabaji, S. and Liu, W. (2011). Spatial distribution of isoflavones and isoflavone-related gene expression in high and low isoflavone soybean cultivars. *Can. J. Plant Sci.* 91: 697–705.

Cherrak, C. M., Pantalone, V. R., Meyer, E. J. et al. (2003). Low-palmitic, low-linolenic soybean development. *J. Amer. Oil Chem. Soc.* 80: 539–43.

Chung, C. and Pettigrew, J. E. (1998). Economics of soybean biotechnology in the livestock industry. *Int. Food Agribus. Man.* 1: 373–85.

Chung, J., Babka, H. L., Graef, G. L., Staswick, P. E., Lee, D. J., Cregan, P. B., Shoemaker, R. C. and Specht, J. E. (2003). The seed protein, oil, and yield QTL on soybean linkage group I. *Crop Sci.* 43:1053–67.

Clarke, E. J. and Wiseman, J. (2000). Developments in plant breeding for improved nutritional quality of soya beans I. Protein and amino acid content. *J. Agric. Sci.* 134: 111–24.

Constantinou, C., Papas, A. and Constantinou, A. I. (2008). Vitamin E and cancer: An insight into the anticancer activities of vitamin E isomers and analogs. *Int. J. Cancer* 123: 739–52.
Csanádi, G., Vollmann, J., Stift, G.and Lelley, T. (2001). Seed quality QTLs identified in a molecular map of early maturing soybean. *Theor. Appl. Genet.* 103: 912–19.
Del Moral, L., Pérez-Vich, B., Fernandez-Martinez, M. and Velasco, L. (2011). Inheritance of increased seed tocopherol content in sunflower line IAST-413. *Plant Breed.* 130: 540–3.
DellaPenna, D. (2005). Progress in the dissection and manipulation of vitamin E synthesis. *Trends Plant Sci.* 10: 574–9.
Diers, B. W., Keim, P., Fehr, W. R. and Shoemaker, R. C. (1992). RFPL analysis of soybean seed protein and oil content. *Theor. Appl. Genet.* 83: 608–12.
Dinkins, R. D., Reddy, R., Meurer, C. A., Yan, B., Trick, H., Thibaud-Nissen, F. et al. (2001) Increased sulfur amino acids in soybean plants overexpressing the maize 15 kD zein protein. *Vitro Cell. Dev. Biol. Plant* 37: 742–7.
Dwiyanti, M. S., Ujiie, A., Thuy, L. T. B., Yamada, T. and Kitamura, K. (2007). Genetic analysis of high α-tocopherol content in soybean seeds. *Breed. Sci.* 57: 23–8.
Dwiyanti, M. S., Yamada, T., Sato, M., Abe, J. and Kitamura, K. (2011). Genetic variation of γ-tocopherol methyltransferase gene contributes to elevated α-tocopherol content in soybean seeds. *BMC Plant Biol.* 11: 1.
Fallen, B. D., Hatcher, C. N., Allen, F. L., Kopsell, D. A., Saxton, A. M., Chen, P., Kantartzi, S. K., Cregan, P. B., Hyten, D. L. and Pantalone, V. R. (2013). Soybean seed amino acid content QTL detected using the Universal Soy Linkage Panel 1.0 with 1,536 SNPs. *J. Plant Genome Sci.* 3: 68–79.
Farré, G., Blancquaert, D., Capell, T., Van Der Straeten, D., Christou, P. and Zhu, C. (2014). Engineering complex metabolic pathways in plants. *Annu. Rev. Plant Biol.* 65: 187–23.
Fehr, W. R., Welke, G. A., Hammond, E. G., Duvick, D. N. and Cianzio, S. R. (1991). Inheritance of reduced palmitic acid content in seed oil of soybean. *Crop Sci.* 31: 88–9.
Fehr, W. R., Welke, G. A., Hammond, E. G., Duvick, D. N. and Cianzio, S. R. (1992). Inheritance of reduced linolenic acid content in soybean genotypes A16 and A17. *Crop Sci.* 32: 903–6.
Fournier, D. B., Erdman, J. W. and Gordon, G. B. (1998). Soy, its components and cancer prevention: a review of *in vitro*, animal and human data. *Cancer Epidemiol. Biomarkers Prev.* 7: 1055–65.
Fu, C. J., Jez, J. M., Kerley, M. S., Allee, G. L. and Krishnan, H. B. (2007). Identification, characterization, epitope mapping, and three dimensional modeling of the alpha-subunit of beta-conglycinin of soybean, a potential allergen for young pigs. *J. Agric. Food Chem.* 55: 4014–20.
Gillman, J. D., Stacey, M. G., Cui, Y., Berg, H. R. and Stacey, G. (2014a). Deletions of the SACPD-C locus elevate seed stearic acid levels but also result in fatty acid and morphological alterations in nitrogen fixing nodules. *BMC Plant Biol.* 14: 143.
Gillman, J. D., Tetlow, A., Hagely, K., Boersma, J. G., Cardinal, A., Rajcan, I. and Bilyeu, K. (2014b). Identification of the molecular genetic basis of the low palmitic acid seed oil trait in soybean mutant line RG3 and association analysis of molecular markers with elevated seed stearic acid and reduced seed palmitic acid. *Mol. Breed.* 34: 447–55.
Goettel, W., Ramirez, M., Upchurch, R. G. and An, Y. C. (2016). Identification and characterization of large DNA deletions affecting oil quality traits in soybean seeds through transcriptome sequencing analysis. *Theor. Appl. Genet.* 129: 1577–593.
Goldberg, R. B., Hoschek, G. and Vodkin, L. O. (1983). An insertion sequence blocks the expression of a soybean lectin gene. *Cell* 33: 465–75.
Grimm, M. O. W., Mett, J. and Hartmann, T. (2016). The impact of vitamin E and other fat-soluble vitamins on Alzheimer's disease. *Int. J. Mol. Sci.* 17: 1785.
Guang, C., Chen, J., Sang, S. and Cheng, S. (2014). Biological functionality of soyasaponins and soyasapogenols. *J. Agric. Food Chem.* 62: 8247–55.
Gurfinkel, D. M. and Rao, A. V. (2003). Soyasaponins: the relationship between chemical structure and colon anticarcinogenic activity. *Nutr. Cancer* 47: 24–33.
Gutierrez-Gonzalez, J. J., Guttikonda, S. K., Tran, L. S. P., Aldrich, D. L., Zhong, R., Yu, O., Nguyen, H. T. and Sleper, D. A. (2010). Differential expression of isoflavone biosynthetic genes in soybean during water deficits. *Plant Cell Physiol.* 51: 936–48.

Gutierrez-Gonzalez, J. J., Vuong, T. D., Zhong, R., Yu, O., Lee, J.-D., Shannon, G., Ellersieck, M., Nguyen, H. T. and Sleper, D. A. (2011). Major locus and other novel additive and epistatic loci involved in modulation of isoflavone concentration in soybean seeds. *Theor. Appl. Genet.* 123: 1375–85.

Gutierrez-Gonzalez, J. J., Wu, X. L., Zhang, J., Lee, J. D., Ellersieck, M., Shannon, J. G., Yu, O., Nguyen, H. T. and Sleper, D. A. (2009). Genetic control of soybean seed isoflavone content: Importance of statistical model and epistasis in complex traits. *Theor. Appl. Genet.* 119: 1069–83.

Hartman, G. L., West, E. D. and Herman, T. K. (2011). Crops that feed the world 2. Soybean-worldwide production, use, and constraints caused by pathogens and pests. *Food Secur.* 3: 5–17.

Head, K., Galos, T., Fang, Y. and Hudson, K. (2012). Mutations in the soybean 3-ketoacyl-ACP synthase gene are correlated with high levels of seed palmitic acid. *Mol. Breed.* 30: 1519–23.

Herman, E. M. and Schmidt, M. A. (2016). The potential for engineering enhanced functional-feed soybeans for sustainable aquaculture feed. *Front. Plant Sci.* 7: 440.

Huang, S. S., Han, Y. P., Li, C. S., Tian, J., Li, W. B. and Wang, J. A. (2012). Identification of QTLs associated with total soyasaponin content in soybean (*Glycine max* (L.) Merr.). *J. Int. Agric.* 11: 1976–84.

Hussain, N., Irshad, F., Jabeen, Z., Shamsi, I. H., Li, Z. and Jiang, L. (2013). Biosynthesis, structural, and functional attributes of tocopherols in planta; past, present, and future perspectives. *J. Agric. Food Chem.* 61: 6137–49.

Hwang, E.-Y., Song, Q., Jia, G., Specht, J. E., Hyten, D. L. et al. (2014) A genome-wide association study of seed protein and oil content in soybean. *PLoS Genet.* 15: 1.

Hymowitz, T. (1986). Genetics and breeding of soybeans lacking the Kunitz trypsin inhibitor. *Adv. Exp. Med. Biol.* 199: 291–8.

Hyten, D. L., Choi, I. Y., Song, Q., Specht, J. E., Carter Jr, T. E., Shoemaker, R. C., Matukumalli, L. K. and Cregan, P. B. (2010). A high density integrated genetic linkage map of soybean and the development of a 1536 universal soy linkage panel for quantitative trait locus mapping. *Crop Sci.* 50: 960–8.

Imsande, J. (2001). Selection of soybean mutants with increased concentrations of seed methionine and cysteine. *Crop Sci.* 41: 510–15.

Isanga, J. and Zhang, G. N. (2008). Soybean bioactive components and their implications to health – A review. *Food Rev. Int.* 24: 252–76.

Jensen, S. K. and Lauridsen, C. (2007). Alpha-tocopherol stereoisomers. *Vitam. Horm.* 76: 281–308.

Josephs, L. M., Hymowitz, T., Schmidt, M. A. and Herman, E. M. (2006). Evaluation of glycine germplasm for nulls of the immunodominant allergen P34/Gly mBd30k. *Crop Sci.* 46: 1755–63.

Kabelka, E. A., Diers, B. W., Fehr, W. R., LeRoy, A. R., Baianu, I. C., You, T., Neece, D. J. and Nelson, R. L. (2004). Putative alleles for increased yield from soybean plant introductions. *Crop Sci.* 44: 784–91.

Kamal-Eldin, A. and Appelqvist, L. A. (1996). The chemistry and antioxidant properties of tocopherols and tocotrienols. *Lipids* 31: 671–701.

Kang, J., Badger, T. M., Ronis, M. J. J. and Wu, X. (2010). Non-isoflavone phytochemicals in soy and their health effects. *J. Agric. Food Chem.* 58: 8119–33.

Kassem, M. A., Meksem, K., Iqbal, M. J., Njiti, V. N., Banz, W. J. and Winters, T. A. (2001). Definition of soybean genomic regions that control seed phytoestrogen amounts. *J. Biomed. Biotech.* 1: 52–60.

Kearsey, M. J. and Farquhar, A. G. L. (1998). QTL analysis in plants; where are we now? *Heredity* 80: 137–42.

Khosla, P. and Sundram, K. (1996). Effects of dietary fatty acid composition on plasma cholesterol. *Prog. Lipid. Res.* 35: 93–132.

Kim, W. S. and Krishnan, H. B. (2004). Expression of an 11 kDa methionine rich delta-zein in transgenic soybean results in the formation of two types of novel protein bodies in transitional cells situated between the vascular tissue and storage parenchyma cells. *Plant Biotech. J.* 2: 199–210.

Kinjo, J., Imagire, M., Udayama, M., Arao, T. and Nohara, T. (1998). Structure-hepatoprotective relationships study of soyasaponins I-IV having soyasapogenol B as aglycone. *Planta Med.* 64: 233–6.

Krishnamurthy, P., Tsukamoto, C., Singh, R. J., Lee, J. D., Kim, H. S., Yang, S. H. and Chung, G. (2014). The Sg-6 saponins, new components in wild soybean (*Glycine soja* Sieb. and Zucc.): Polymorphism, geographical distribution and inheritance. *Euphytica* 198: 413–24.

Krishnan, H. B. (2005). Engineering soybean for enhanced sulfur amino acid content. *Crop Sci.* 45: 454–61.

Krishnan, H. B., Alaswad, A. A., Oehrle, N. W. and Gillman, J. D. (2016). Deletion of the *SACPD-C* locus alters the symbiotic relationship between *Bradyrhizobium japonicum* USDA110 and soybean, resulting in elicitation of plant defense response and nodulation defects. *Mol. Plant Microbe Interact.* 29: 862–77.

Kudou, S,, Tonomura, M,, Tsukamoto, C,, Shimoyamada, M,, Uchida, T, and Okubo, K. (1992). Isolation and structural elucidation of the major genuine soybean saponin. *Biosci. Biotech. Biochem.* 56: 142–3.

Kudou, S., Tonomura, M., Tsukamoto, C., Uchida, T., Sakabe, T., Tamura, N. and Okubo, K. (1993). Isolation and structural elucidation of DDMP-conjugated soyasaponins as genuine saponins from soybean seeds. *Biosci. Biotech. Biochem.* 57: 546–50.

Kwanyuen, P., Pantaloneb, V. R., Burton, J. W. and Wilson, R. F. (1997). A new approach to genetic alteration of soybean protein composition and quality. *J. Am. Oil Chem. Soc.* 74: 983–7.

La, T. C., Pathan, S. M., Vuong, T., Lee, J.-D., Scaboo, A. M., Smith, J. R., Gillen, A. M., Gillman, J., Ellersieck, M. R., Nguyen, H. T. and Shannon, J. G. (2014). Effect of high-oleic acid soybean on seed oil, protein concentration, and yield. *Crop Sci.* 54: 2054–62.

Lee, B. K., Kim, S. L., Kim, K. H., Yu, S. H., Lee, S. C. et al. (2008). Seed specific expression of perilla γ-tocopherol methyltransferase gene increases α-tocopherol content in transgenic perilla (*Perilla frutescens*). *Plant Cell Tiss. Organ Cult.* 92: 47–54.

Lee, J. D., Woolard, M., Sleper, D. A., Smith, J. R., Pantalone, V. R., Nyinyi, C. N., Cardinal, A. and Shannon, J. G. (2009). Environmental effects on oleic acid in soybean seed oil of Plant Introductions with elevated oleic concentration. *Crop Sci.* 49: 1762–8.

Lee, Y. Y., Park, H. M., Hwang, T. Y., Kim, S. L., Kim, M. J., Lee, S. K., Seo, M. J., Kim, K. J., Kwon, Y. U., Leeb, S. C. and Kima, Y. H. (2015). A correlation between tocopherol content and antioxidant activity in seeds and germinating seeds of soybean cultivars. *J. Sci. Food Agric.* 95: 819–27.

Li, H., Liu, H., Han, Y., Wu, X., Teng, W., Liu, G. and Li, W. (2010). Identification of QTL underlying vitamin E contents in soybean seed among multiple environments. *Theor. Appl. Genet.* 120: 1405–13.

Li, H., Wang, Y., Han, Y., Teng, W., Zhao, X., Li, Y. and Li, W. (2016a). Mapping quantitative trait loci (QTLs) underlying seed vitamin E content in soybean with main, epistatic and QTL x environment effects. *Plant Breed.* 135: 208–14.

Li, X., Dong, Z., Zhang, F., Dong, J. and Zhang, Y. (2016b). Vitamin E slows the progression of osteoarthritis. *Exp. Ther. Med.* 12: 18–22.

Li, Y., Wang, G., Hou, R., Zhou. Y., Gong, R., et al. (2011). Engineering tocopherol biosynthetic pathway in lettuce. *Biol. Plant* 55: 453–60.

Liang, H. Z., Yu, Y. L., Wang, S. F., Lian, Y., Wang, T. F., Wei, Y. L., Gong, P. T., Liu, X. Y., Fang, X. J. and Zhang, M. C. (2010). QTL mapping of isoflavone, oil and protein contents in soybean (*Glycine max* L. Merr.). *Agric. Sci. China* 9: 1108–16.

Lu, W., Wen, Z., Li, H., Yuan, D., Li, J., Zhang, H., Huang, Z., Cui, S. and Du, W. (2013). Identification of the quantitative trait loci (QTL) underlying water soluble protein content in soybean. *Theor. Appl. Genet.* 126: 425–33.

Luyt, A. S. (1993). A differential scanning calorimeter study of the interaction of zinc benzothiazole-2-thiolate, sulfur, stearic acid, and zinc stearate in the absence and presence of polyisoprene. *J. Appl. Polym. Sci.* 47: 2177–85.

Ma, Y., Kan, G., Zhang, X., Wang, Y., Zhang, W., Du, H. and Yu, D. (2016). Quantitative trait loci (QTL) mapping for glycinin and β-conglycinin contents is soybean (*Glycine max* L. Merr) *J. Agric. Chem.* 64: 3473–83.

Medic, J., Atkinson, C. and Hurburgh Jr, C. R. (2014). Current knowledge in soybean composition. *J. Am. Oil Soc.* 91: 363–84.

Meksem, K., Njiti, V. N., Banz, W. J., Iqbal, M. J., Kassem, M. M., Hyten, D. L., Yuang, J., Winters, T. A. and Lightfoot, D. A. (2001). Genomic regions that underlie soybean seed isoflavone content. *J. Biomed. Biotechnol.* 1: 38–44.

Messina, M. J. (1999). Legumes and soybeans: Overview of their nutritional profiles and health effects. *Am. J. Clin. Nutr.* 70: 439s–450s.

Missouri Soybeans. 2017. *Missouri Now Holds Patent for High Oleic Soybeans*. Missouri Soybean Merchandising Council. Online resource available at: https://mosoy.org/missouri-now-holds-patent-for-high-oleic-soybeans/; verified on 24 June 2017.

Morrison, M. J., Cober, E. R., Saleem, M. F., McLaughlin, N. B., Fregeau-Reid, J., Ma, B. L., Yan, W. and Woodrow, L. (2008). Changes in isoflavone concentration with 58 years of genetic improvement of short-season soybean cultivars in Canada. *Crop Sci.* 48: 2201–8.

Munne-Bosch, S. and Alegre, L. (2002). The function of tocopherols and tocotrienols in plants. *Crit. Rev. Plant Sci.* 21: 31–57.

Nichols, D. M., Glover, K. D., Carlson, S. R., Specht, J. E. and Diers, B. W. (2006). Fine mapping of a seed protein QTL on soybean linkage group I and its correlated effects on agronomic traits. *Crop Sci.* 46: 834–9.

Nordlee, J. A., Taylor, S. L., Townsend, J. A., Thomas, L. A. and Bush, R. K. (1996). Identification of a Brazil-nut allergen in transgenic soybeans. *N Engl. J. Med.* 334: 688–92.

Okubo, K., Iijima, M., Kobayashi, Y., Yoshikoshi, M., Uchida T. and Kudou, S. (1992). Components responsible for the undesirable taste of soybean seeds. *Biosci. Biotechnol. Biochem.* 56: 99–103.

Okuley, J., Lightner, J., Feldmann, K., Yadav, N., Lark, E. and Browse, J. (1994). Arabidopsis FAD2 gene encodes the enzyme that is essential for polyunsaturated lipid synthesis. *Plant Cell* 6: 147–58.

Oliva, M. L., Shannon, J. G., Sleper, D. A., Ellersieck, M. R., Cardinal, A. J., Paris, R. L. and Lee, J. D. (2006). Stability of fatty acid profile in soybean genotypes with modified seed oil composition. *Crop Sci.* 46: 2069–75.

Orf, J. H. and Hymowitz, T. (1979). The inheritance of the absence of the Kunitz trypsin inhibitor in seed protein of soybeans. *Crop Sci.* 19: 107–9.

Orf, J. H., Hymowitz, T., Pull, S. P. and Pueppke, S. G. (1978). Inheritance of soybean seed lectin. *Crop Sci.* 18: 899–900.

Paek, N. C., Imsande, J., Shoemaker, R. C. and Shibles, R. (1997). Nutritional control of soybean seed storage protein. *Crop Sci.* 37: 498–503.

Palomeque, L., Li-Jun, L., Li, W., Hedges, B., Cober, E. R. and Rajcan, I. (2009). QTL in mega-environments: II. Agronomic trait QTL colocalized with seed yield QTL detected in a population derived from a cross of high-yielding adapted x high-yielding exotic soybean lines. *Theor. Appl. Genet.* 119: 429–36.

Palomeque, L., Liu, L. J., Li, W., Hedges, B. R., Cober, E. R., Smid, M. P., Lukens, L. and Rajcan, I. (2010). Validation of mega-environment universal and specific QTL associated with seed yield and agronomic traits in soybeans. *Theor. Appl. Genet.* 120: 997–1003.

Pandurangan, S., Pajak, A., Molnar, S. J., Cober, E. R., Dhaubhadel, S., Hernández-Sebastia, C., Kaiser, W. M., Nelson, R. L., Huber, S. C. and Marsolais, F. (2012). Relationship between asparagine metabolism and protein concentration in soybean seed. *J. Exp. Bot.* 63: 3173–84.

Panneerselvam, K., Tsukamoto, C., Honda, N., Kikuchi, A., Lee, J., Yang, S. and Chung, G. (2013). Saponin polymorphism in the Korean wild soybean (*Glycine soja* Sieb. and Zucc.). *Plant Breed.* 132: 121–6.

Pantalone, V. R., Wilson, R. F., Novitzky, W. P. and Burton, J. W. (2002). Genetic regulation of elevated stearic acid concentration in soybean oil. *J. Am. Oil Chem. Soc.* 79: 543–53.

Panthee, D., Pantalone, V., Sams, C., Saxton, A., West, D., Orf, J. and Killam, A. (2006a). Quantitative trait loci controlling sulfur containing amino acids, methionine and cysteine, in soybean seeds. *Theor. Appl. Genet.* 112: 546–53.

Panthee, D., Pantalone, V., Saxton, A., West, D. and Sams, C. (2006b). Genomic regions associated with amino acid composition in soybean. *Mol. Breed.* 17: 79–89.

Panthee, D. R., Kwanyuen, P., Sams, C. E., West, D. R., Saxton, A. M. and Pantalone, V. R. (2004). Quantitative trait loci for β-conglycinin (7S) and glycinin (11S) fractions of soybean storage protein. *J. Am. Oil Chem. Soc.* 81: 1005–12.

Park, J., Kim, J. H., Krishnamurthy, P., Tsukamoto, C., Song, J. T., Chung, G., Shannon, J. G. and Lee, J. D. (2016). Characterization of a new allele of the saponin-synthesizing gene *Sg-1* in soybean. *Crop Sci.* 56: 385–91.

Pham, A. T., Lee, J. D., Shannon, J. G. and Bilyeu, K. D. (2010). Mutant alleles of FAD2-1A and FAD2-1B combine to produce soybeans with the high oleic acid seed oil trait. *BMC Plant Biol.* 10: 195.

Phansak, P., Soonsuwon, W., Hyten, D. L., Song, O., Cregan, P. B., Graef, G. L. and Specht, J. E. (2016). Multi-population selective genotyping to identify soybean [*Glycine max* (L.) Merr.] seed protein and oil QTLs. G3 (Bethseda) 6: 1635–48.

Price, K. R., Johnson, I. T. and Fenwick, G. R. (1987). The chemistry and biological significance of saponins in food and feeding stuffs. *CRC Crit. Rev. Food Sci. Nutr.* 26: 27–135.

Primomo, V. S., Falk, D. E., Tanner, J. W. and Rajcan, I. (2002). Inheritance and interaction of low palmitic and low linolenic soybean. *Crop Sci.* 42: 31–6.

Primomo, V. S., Poysa, V., Ablett, G. R., Jackson, C. J. and Rajcan, I. (2005a). Agronomic performance of recombinant inbred line populations segregating for isoflavone content in soybean seeds. *Crop Sci.* 45: 2203–11.

Primomo, V. S., Poysa, V., Ablett, G. R., Jackson, C. J. and Rajcan, I. (2005b). Mapping QTL for individual and total isoflavone content in soybean. *Crop Sci.* 45: 2454–64.

Pryor, W. A. (2000). Vitamin E and heart disease: Basic science to clinical intervention trials. *Free Radic. Biol. Med.* 28: 141–64.

Qi, Z., Pan, J., Han, X., Qi, H., Xin, D., Li, W., Mao, X., Wang, Z., Jiang, H., Liu, C., Hu, Z., Hu, G., Zhu, R. and Chen, Q. (2016). Identification of major QTLs and epistatic interactions for seed protein concentration in soybean under multiple environments based on a high-density map. *Mol. Breed.* 36: 55.

Qi, Z. M., Sun, Y. N., Wu, Q., Liu, C. Y., Hu, G. H. and Chen, Q. S. (2011). A meta-analysis of seed protein concentration QTL in soybean. *Can. J. Plant Sci.* 91: 221–30.

Rahman, S., Kinoshita, T., Anai, T., Arima S. and Takagi Y. (1998). Genetic relationships of soybean mutants for different linolenic acid contents. *Crop Sci.* 38: 702–6.

Rajcan, I., Hou, G. and Weir, A. (2005). Advances in breeding of seed-quality traits in soybean. *J. Crop Improv.* 14: 145–74.

Reinprecht, Y., Luk-Labey, S. Y., Larsen R. J., Poysa, V. W., Yu, K., Rajcan, I., Ablett, G. R. and Pauls, K. P. (2009). Molecular basis of the low linolenic acid trait in soybean EMS mutant line RG10. *Plant Breed.* 128: 253–258.

Rincker, K., Nelson, R., Specht, J., Sleper, D., Cary, T., Cianzio, S. R., Casteel, S. et al. (2014). Genetic improvement of U.S. soybean in maturity groups II, III, and IV. *Crop Sci.* 54:1419–32.

Ross, A. J., Fehr, W. R., Welke, G. A. and Cianzio, S. R. (2000). Agronomic and seed traits of 1%-linolenate soybean genotypes. *Crop Sci.* 40: 383–6.

Rossi, M. E., Orf, J. H., Liu, L. J., Dong, Z. and Rajcan, I. (2013). Genetic basis of soybean adaptation to North American vs. Asian megaenvironments in two independent populations from Canadian x Chinese crosses. *Theor. Appl. Genet.* 126: 1809–23.

Sattler, S. E., Gilliland, L. U., Magallanes-Lundback, M., Pollard, M. and DellaPenna, D. (2004). Vitamin E is essential for seed longevity and for preventing lipid peroxidation during germination. *Plant Cell* 16: 1419–32.

Sayama, T., Ono, E., Takagi, K., Takada, Y., Horikawa, M., Naka-moto, Y., Hirose, A., Sasama, H., Ohashi, M. and Hasegawa, H. (2012). The *sg-1* glycosyltransferase locus regulates structural diversity of triterpenoid saponins of soybean. *Plant Cell* 24: 2123–8.

Schlueter, J. A., Vasylenko-Sanders, I. F., Deshpande, S., Yi, J., Siegfried, M., Roe, B. A., Schlueter, S. D., Scheffler, B. E. and Shoemaker, R. C. (2007). The FAD2 gene family of soybean: Insights into the structural and functional divergence of a paleopolyploid genome. *Crop Sci.* 47: S14–26.

Schmidt, M. A., Hymowitz, T. and Herman, E. M. (2015). Breeding and characterization of soybean *Triple Null*; a stack of recessive alleles of kunitz trypsin inhibitor, soybean agglutinin, and P34 allergen nulls. *Plant Breed.* 134: 310–15.

Schmutz, J., Cannon, S. B., Schlueter, J., Ma, J., Mitros, T., Nelson, W., Hyten, D. L., Song, Q., Thelen, J. J., Cheng, J. et al. (2010). Genome sequence of the palaeopolyploid soybean. *Nature* 463: 78–183.

Seguin, P., Tremblay, G., Pageau, D. and Liu, W. (2010). Soybean tocopherol concentrations are affected by crop management. *J. Agric. Food Chem.* 58: 5495–5501.

Shaw, E. J., Kakuda, Y. and Rajcan, I. (2016). Effect of genotype, environment, and genotype x environment interaction on tocopherol accumulation in soybean seed. *Crop Sci.* 56: 40–50.

Shi, J., Arunasalam, K., Yeung, D., Kakuda, Y., Mittal, G. and Jiang, Y. (2004). Saponins from edible legumes: Chemistry, processing, and health benefits. *J. Med. Food* 7: 67–78.

Shi, Z., Bachleda, N., Pham, A. T., Bilyeu, K., Shannon, G., Nguyen, H. and Li, Z. (2015). High-throughput and functional SNP detection assays for oleic and linolenic acids in soybean. *Mol. Breed.* 35: 176.

Shimoyamada, M., Kudo, S., Okubo, K., Yamauchi, F. and Harada, K. (1990). Distributions of saponin constituents in some varieties of soybean plant. *Agric. Biol. Chem.* 54: 77–81.

Shintani, D. and DellaPenna, D. (1998). Elevating the vitamin E content of plants through metabolic engineering. *Science* 282 (5396): 2098–2100.

Shiraiwa, M., Harad, K. and Okubo, K. (1991). Composition and content of saponins in soybean seed according to variety, cultivation year and maturity (analytical chemistry). *Agric. Biol. Chem.* 55: 323–31.

Simons, B., Scholl, D., Cyr, T. and Alice, H. M. (2001). Effects of increased loop flexibility on the structure and stability of a *de novo* designed helical protein. *Protein Pept. Lett.* 8: 89–96.

Smallwood, C. J., Nyinyi, C. N., Kopsell, D. A., Sams, C. E., West, D. R., Chen, P., Kantartzi, S. K., Cregan, P. B., Hyten, D. L. and Pantalone, V. R. (2014). Detection and confirmation of quantitative trait loci for soybean seed isoflavones. *Crop Sci.* 54: 595–606.

Sohlenkamp, C., Wood, C. C., Roeb, G. W. and Udvardi, M. K. (2002). Characterization of *Arabidopsis* AtAMT2, a high-affinity ammonium transporter of the plasma membrane. *Plant Physiol.* 130: 1788–96.

Spencer, M. M., Pantalone, V. R., Meyer, E. J., Landau-Ellis, D. and Hyten, D. L. (2003). Mapping the *Fas* locus controlling stearic content in soybean. *Theor. Appl. Genet.* 106: 615–19.

Stampfer, M. J. and Rimm. E. B. (1995). Epidemiologic evidence for vitamin E in prevention of cardiovascular disease. *Am. J. Clin. Nutr.* 62: 1365S–9S.

Stojšin, D., Ablett, G. R., Luzzi, B. M. and Tanner, J. W. (1998). Use of gene substitution values to quantify partial dominance in low palmitic acid soybean. *Crop Sci.* 38: 1437–41.

Takada, Y., Sasama, H., Sayama, T., Kikuchi, A., Kato, S., Ishimoto, M. and Tsukamoto, C. (2013). Genetic and chemical analysis of a key biosynthetic step for soyasapogenol A, an aglycone of group A saponins that influence soymilk flavor. *Theor. Appl. Genet.* 126: 721–31.

Takada, Y., Sayama, T., Kikuchi, A., Kato, S., Tatsuzaki, N., Naka-moto, Y., Suzuki A., Tsukamoto, C. and Ishimoto, M. (2010). Genetic analysis of variation in sugar chain composition at the C-22 position of group A saponins in soybean, *Glycine max* (L.) Merrill. *Breed. Sci.* 60: 3–8.

Takagi, Y., Rahman, S. M., Anai, T., Wasala, S. K., Kinoshita, T. and Khalekuzzaman, M. (1999) Development of a reduced linolenate soy mutant by re-irradiation and its genetic analysis. *Breed. Sci.* 49: 1–5.

Takahashi, M., Uematsu, Y., Kashiwaba, K., Yagasaki, K., Hajika, M., Matsunaga, R., Komatsu, K. and Ishimoto, M. (2003). Accumulation of high levels of free amino acids in soybean seeds through integration of mutations conferring seed protein deficiency. *Planta* 217: 577–86.

Tavva, V. S., Kim, Y. H., Kagan, I. A., Dinkins, R. D., Kim, K. H. and Collins, G. B. (2007). Increased α-tocopherol content in soybean seed overexpressing the *Perilla frutescens* γ-tocopherol methyltransferase gene. *Plant Cell Rep.* 26: 61–70.

Thakur, V., Morley, S. and Manor, D. (2010). Hepatic α-tocopherol transfer protein: ligand-induced protection from proteasomal degradation. *Biochemistry* 49: 9339–44.

Thapa, R., Carrero-Colón, M. and Hudson, K. A. (2016). New alleles of FATB1A to reduce palmitic acid levels in soybean. *Crop Sci.* 56: 1076–80.

Thornsberry, J. M., Goodman, M. M., Doebley, J., Kresovich, S., Nelson, D. et al. (2001) Dwarf8 polymorphisms associate with variation in flowering time. *Nat. Genet.* 28: 286–9.

Townsend, J. A. and Thomas, L. A. (1994). Factors which influence the *Agrobacterium*-mediated transformation of soybean. *J. Cell Biochem. Suppl.* 18A:X1–X014.

Traber, M. G. (1996). Vitamin E in humans: Demand and delivery. *Annu. Rev. Nutr.* 16: 321–47.

Ujiie, A., Yamada, T., Fujimoto, K., Endo, Y. and Kitamura, K. (2005). Identification of soybean varieties with high α-tocopherol content. *Breed. Sci.* 55: 123–5.

United Soybean Board. 2017. High oleic soybean. Online resource available at: https://unitedsoybean.org/media-center/issue-briefs/high-oleic-soybeans/; verified 24 June 2017.

Upchurch, R. G. and Ramirez, M. E. (2010). Gene expression profiles of soybeans with mid-oleic acid seed phenotype. *J. Am. Oil Chem. Soc.* 87: 857–64.

Valentin, H. E. and Qi, Q. (2005). Biotechnological production and application of vitamin E: Current state and prospects. *Appl. Microbiol. Biotechnol.* 68: 436–44.

Van Eenennaam, A. L., Lincoln, K., Durrett, T. P., Valentin, H. E., Shewmaker, C. K., Thorne, G. M., Jiang, J., Baszis, S. R., Levering, C. K., Aasen, E. D., Hao, M., Stein, J. C., Norris, S. R. and Last, R. L. (2003). Engineering vitamin E content: From *Arabidopsis* mutant to soy oil. *Plant Cell* 15: 3007–19.

Van Warrington, C., Abdel-Haleem, H., Orf, J. H., Killam, A. S., Bajjalieh, N., Li, Z. and Boerma, H. R. (2014). Resource allocation for selection of seed protein and amino acids in Soybean. *Crop Sci.* 54: 963–70.

Vaughn, J. N., Nelson, R. L., Song, Q., Cregan, P. B. and Li, Z. (2014). The genetic architecture of seed composition in soybean is refined by genome-wide association scans across multiple populations. G3 (Bethesda) 4: 2283–94.

Wang, J., Liu, L., Guo, Y., Wang, Y. H., Zhang, L., Jin, L. G., Guan, R. X., Liu, Z. X., Wang, L. L., Chang, R. Z. and Qiu, L. J. (2014). A dominant locus, qBSC-1, controls β subunit content of seed storage protein in soybean (*Glycine max* (L.) Merr.). *J. Integr. Agric.* 13: 1854–64.

Wang, X., Jiang, G.-L., Song, Q., Cregan, P. B., Scott, R. A., Zhang, J., Yen, Y. and Brown, M. (2015). Quantitative trait locus analysis of seed sulfur-containing amino acids in two recombinant inbred line populations of soybean. *Euphitica* 201: 293–305.

Warrington, C. V., Abdel-Haleem, H., Hyten, D. L., Cregan, P. B., Orf, J. H., Killam, A. S., Bajjalieh, N., Li, Z. and Boerma, H. R. (2015). QTL for seed protein and amino acids in the Benning x Danbaekkong soybean population. *Theor. Appl. Genet.* 128: 839–50.

Wilcox, J. R. and Cavins, J. F. (1985). Inheritance of low linolenic acid content of the seed oil of a mutant in *Glycine max. Theor. Appl. Genet.* 71: 74–8.

Wilcox, J. R. and Shibles, R. M. (2001). Interrelationships among seed quality attributes in soybean. *Crop Sci.* 41: 11–14.

Wilcox, J. R. and Zhang, G. D. (1997). Relationships between seed yield and seed protein in determinate and indeterminate soybean populations. *Crop Sci.* 37: 361–4.

Wilson, R. F. (2004). Seed composition. In: Boerma, H. R. and Specht, J. (Eds) *Soybeans: Improvement, Production, and uses*, vol. 16. American Society of Agronomy, Madison, pp. 621–77.

Woldemicheal, G. M. and Wink, M. (2002). Triterpene glycosides of *Lupinus angustifolius. Phytochem* 60: 323–7.

Xu, S. (2003). Theoretical basis of the Beavis effect. *Genetics* 165: 2259–68.

Yaklich, R., Helm, R., Cockrell, G. and Herman, E. (1999). Analysis of the distribution of the major soybean seed allergens in a core collection of *Glycine max* accessions. *Crop Sci.* 39: 1444–7.

Yang, X., Yan, J., Shah, T., Warburton, M. L., Li, Q., Li, L., Gao, Y., Chai, Y., Fu, Z. and Zhou, Y. (2010). Genetic analysis and characterization of a new maize association mapping panel for quantitative trait loci dissection. *Theor. Appl. Genet.* 121: 417–31.

Yoshikawa, T., Okumoto, Y., Ogata, D., Sayama, T., Teraishi, M., Terai, M., Toda, T., Yamada, K., Yagasaki, K., Yamada, N., Tsukiyama, T., Yamada, T. and Tanisaka, T. (2010). Transgressive segregation of isoflavone contents under the control of four QTLs in a cross between distantly related soybean varieties. *Breed, Sci.* 60: 243–54.

Zeng, G. L., Li, D. M., Han, Y. P., Teng, W. L., Wang, J., Qiu, L. Q. and Li, W. B. (2009). Identification of QTL underlying isoflavone contents in soybean seeds among multiple environments. *Theor. Appl. Genet.* 118: 1455–63.

Zhan, S. and Ho, S. C. (2005). Meta-analysis of the effects of soy protein containing isoflavones on the lipid profile. *Am. J. Clin. Nutr.* 81: 397–408.

Zhang, L., Luo, Y., Zhu, Y., Zhang, L., Zhang, W., Chen, R., Xu, M., Fan, Y. and Wang, L. (2013). GmTMT2a from soybean elevates the α-tocopherol content in corn and Arabidopsis. *Transgenic Res.* 22: 1021–8.

Zhang, X., Shu, X. O., Gao, Y. T., Yang, G., Li, Q., Li, H., Jin, F. and Zheng, W. (2003). Soy food consumption is associated with lower risk of coronary heart disease in Chinese women. *J. Nutr.* 133: 2874–8.

Zhang, Y., Schernthaner, J., Labbe, N., Hefford, M. A., Zhao, J. and Simmonds, D. H. (2014). Improved protein quality in transgenic soybean expressing a de novo synthetic protein, MB-16. *Transgenic Res.* 23: 455–67.

Zhang, Y. H., Liu, M. F., He, J. B., Wang, Y. F., Xing, G. N., Li, Y., Yang, S. P., Zhao, T. J. and Gai, J. Y. (2015). Marker-assisted breeding for transgressive seed protein content in soybean [*Glycine max* (L.) Merr]. *Theor. Appl. Genet.* 128: 1061–72.

Chapter 10

Allergens in soybean

Eliot M. Herman, University of Arizona, USA

1 Introduction
2 Clinical allergies to soybean
3 Infant food allergies and intolerance
4 Soybean IgE-binding proteins
5 Mitigating soybean allergens
6 Soybean intolerance in aquaculture
7 Food allergy, biotechnology and the risk of extrinsic introduced allergens
8 Summary
9 Where to look for further information
10 Acknowledgements and dedication
11 References

1 Introduction

Food allergies and food intolerance including celiac disease and infantile food protein-induced enterocolitis syndrome (FPIES) are a growing concern in the industrialized countries where the percentage of the population that exhibit clinical food allergies has increased rapidly over the past few decades (Branum and Lukacs 2008). Food allergies are often manifested by an escalating series of responses to allergen challenge, beginning with mild atopic reaction, then after repeated exposure the patient may escalate to intense reactions culminating in a risk of fatal anaphylaxis.

Most clinically relevant aspects of food intolerance are initiated with gut exposure. The gut contains most of the human body's immune system and is the major site where hypersensitive reactions are provoked and it is a primary site of adverse response. Gut injury and damage are reflected as gut health and may manifest as related systemic health issues. Among the major plant-sourced foods, wheat and soybean are most often cited as sources of food-related intolerance due to their broad inclusion in processed foods. Other plant sources of allergens include seeds such as peanuts, tree nuts, sesame and sunflower as well as fruit and vegetables including apple, tomato, kiwi, papaya and carrots that can provoke severe and sometimes life-threatening anaphylactic responses. Food allergy and intolerance develops from sensitization by proteins that could otherwise be safely consumed.

http://dx.doi.org/10.19103/AS.2017.0034.25

Preventing potentially life-threatening allergic responses is a societal medical problem and a critical question of liability and regulation for the food industry. There is no simple solution to the food allergy problem. Indeed, the food allergy problem has become progressively more difficult to manage as the rate of food-allergic people continues to grow (Burks 2003; Sicherer et al. 1998; Sicherer and Sampson 2007, 2010). The current primary treatment protocol for food allergies is physical avoidance, that is, to identify food risk and for the affected individual to take positive steps to avoid the allergen's consumption. To aid in avoidance the United States has regulations of the Food Allergen Labeling and Consumer Protection Act of 2004 that requires plain language labelling of the 'Big Eight' allergens comprising the inclusion of wheat, soybean, peanut, tree nuts, fish, shellfish, dairy and eggs that are the most common and often hidden ingredients of processed foods. Although labelling appears to be a simple solution, its implementation in the real world is far more complex as labels are often separated from food, especially in the growing tendency of industrialized countries to have a substantial fraction of meals away from the home.

Many of the most at-risk individuals, infants and children, either cannot read labels or will not reliably take the warning seriously. Soybean allergens are especially difficult to avoid due to its wide inclusion in processed foods. Soybean ingestion allergies are manifested not only from consumption of obvious soybean foods such as tofu and soy milk, but also from a wide range of processed foods that contain soybean protein (Bando et al. 1998; Herian et al. 1993; Ogawa et al. 1991, 2000; Tsuji et al. 1995). Other exposures can occur from non-food uses of soybean, for example, in cosmetics (Shaffrali and Gawkrodger 2001). The wide use of soybean-based baby formula exposes babies to soybeans and potential soybean sensitivity (Hepple et al. 1987, 1989).

2 Clinical allergies to soybean

Manifestation of clinical allergies begins early in life and may continue to be a lifelong problem. The basic outline of an immunoglobulin E (IgE)-mediated allergy is that the immune system's T cells direct B cells to produce antibodies with a specificity for individual antigens that bind to the Fcre receptor on mast cells. These mast cells are then sensitized to react to exposure by the allergen and when IgE binds the antigen it provokes the immediate release of histamine that causes a cascade of physiological events leading to the overt clinical allergy response. Within minutes after exposure respiratory distress, skin reactions, gastrointestinal distress and changes in the cardiovascular system are manifested. A systemic allergenic reaction or anaphylaxis that is potentially fatal can result causing about three deaths per 10 000 people per year in the United States. This acute response is treated with prompt administration of epinephrine.

Food allergies are provoked after exposure of an individual predisposed to develop sensitivity. There can be a significant genetic component with similar allergies occurring in related people. For a predisposed individual, exposure that induces the allergy often occurs at an early age. Studies have shown that peanut exposure during the first couple of years of life may induce lifelong peanut sensitivity (see Burks 2003 and Wood 2003 for reviews). Other studies have suggested that exposure may occur *in utero* as the result of the mother's consumption of peanuts. Considered together, the acquisition of peanut sensitivity is a complex process involving both genetic and exposure parameters, some of which may occur prior to birth and may also involve co-sensitization with other allergens especially those of legume origin. Only about 20% of children with peanut sensitivity outgrow the peanut sensitivity by adulthood (Al- Muhsen et al. 2003). In contrast, about 8% of all children have soybean sensitivity but only 2% of adults indicating that 75% of sensitive children will outgrow a soybean allergy (Cordle 2004; Gupta et al. 2011).

Other studies have indicated that perhaps 50% of soybean-sensitive babies and children will outgrow soybean hypersensitivity. Hypersensitivity to food allergens can be transferred from one person to another by tissue transplantation. Bone marrow and liver transplants from peanut-sensitive donors have transferred peanut hypersensitivity to the recipient (Bellou et al. 1997; Legendre et al. 1997). While ingestion food allergies appear to most often develop in children there are also inhalation allergies of aeroallergens that can result from exposure at any age (Gonzalez et al. 1991). Occupational allergies to soybean are most often manifested as asthma in workers involved in agricultural transport and food production industries (Baur et al. 1996, 1998). Soybean inhalation allergy is directed at different proteins than the soybean food allergy. The possible synergy or susceptibility between neonatal and childhood food allergy and the later development of occupational allergy has not been investigated.

Infantile food intolerance

Further complicating the issue of food allergy and intolerance is the manifestation of infantile FPIES (Jarvinen and Nowak-Wegrzyn 2013; Leonard and Nowak-Wegrzyn 2012; Levy and Danon 2003; Levy et al. 1998). This is manifested as an intense non-IgE-mediated gastrointestinal food hypersensitivity most often induced by dairy and soybeans, two the most prevalent constituents of infant foods. Like food allergy the incidence of soybean FPIES is rapidly growing in parallel with the wide use of industrial foods in our society. FPIES is most often resolved by three years of age but the downstream physiological effects of severe gastrointestinal disorder may continue into childhood, possibly producing lifelong effects. Soybean FPIES has the potential to add to the nation's health care costs and medical professional needs. The question of how FPIES intersects with food allergy, and whether its induction is a physiological alternative to developing a conventional food allergy is unknown. Whether FPIES creates the opportunity to develop other sensitivities remains an open question. Clinically FPIES and food allergy, as well as other types of intolerance, are addressed by the same medical specialties and at present addressed by the same type of primary treatment of identifying the offending food for avoidance with a re-assessment later for potentially outgrowing the issue. One of the difficulties in assessing FPIES is that while it has dramatic physiological effects on the gut it is not easily assessed by systemic tests such as skin-wheal hypersensitive responses that are used to screen for IgE-mediated immunological reactions.

3 Infant food allergies and intolerance

Adverse food responses are most intense in immunologically naïve neonates and immunologically immature infants/toddlers. The various adverse responses of allergy and intolerance have the highest frequency in neonates where the maturing immune system and gut change from *de novo* naïve state and establishes the individual's response to its food environment. The increase in food allergy and intolerance growing in the industrialized world shows that naïve guts/immune systems are not reacting well to the food circumstances presented to infants and toddlers. For soybean hypersensitivity there is considerable evidence that formula can elicit cross-sensitivity with milk proteins, primarily casein, that is adversely synergistic for human neonates (Ahn et al. 2003; Candreva et al. 2015, 2016; Curcianrello et al. 2014; Docena et al. 1996; Levy and Danon 2003; Rozenfeld et al. 2002; Smaldini et al. 2012) and may extend to neonatal swine and calves and possibly other animals that are fed milk proteins, such as dogs. How much of this sensitivity is IgE-mediated food allergy and how much of this sensitivity is FPIES remains to be determined but that the two most common sources of FPIES being milk and soybean is likely not coincidental. Until assays for FPIES comparable to the IgE assessments are available, it is difficult to ascertain the natural history of allergy/FPIES interactions or synergy for soybean with other sensitizing sources such as milk casein.

Food sensitization is a gut-mediated process. Soybean provokes adverse responses through gut sensitization and response (Katz et al. 2014; He et al. 2015, Sun et al. 2008). The site where FPIES is provoked remains to be established but is almost certainly the gut. Adverse reactions to food is an inappropriate response to what is otherwise a safe food. How foods elicit adverse responses is complex; soybean seeds accumulate thousands of proteins of which a hundred or more proteins accumulate in sufficient abundance to be at allergenic threshold levels for sensitization or response. The soybean proteins exist in a biological matrix that has many other components including potentially immunogenic complex glycans as well as lipids that could provide an adjuvant effect. The additional substances that may enhance adverse responses are generally termed a 'matrix' and its effect a 'matrix effect'. It is broadly assumed, but not well researched, that the matrix has a key role in initiating the circumstances that leads to food intolerance. Humans and production animals manifest soybean intolerance and hypersensitivity (Dreau et al. 1995; Helm et al. 2003; Li et al. 1990; Schmied et al 2013; Taliercio et al. 2014; Wang et al. 2014a,b; Zheng et al. 2014). Very little soybean-sourced food, or feed, is consumed as the raw natural food but rather almost all of it is processed with heat and is often formulated with and/or consumed with other foods that may also exhibit bioactive characteristics. Food processing by heat, dry or wet conditions, causes a broad range of chemical reactions and protein misfolding. Processing can inactivate bioactivity and conversely induce the formation of new bioactive sites producing physiological response. Co-consumption similarly might lead to synergy where one food's bioactivity causes other foods to be presented to the body in a manner to stimulate a response to a second food. Conversely the activity of one food could also mask the activity of another food. The potential permutations of immunological responses to the complexity of foods is immense. Added to this are the immense and unique immunological variations of all people and animals and how these responses change with developmental/maturation as well as with the constant changes that occur from daily life. Yet it is responding to these complexities that is the normal dynamic process of the gut. From a clinical and research perspective, the gut's response to food can be characterized as a high-order systems response from the input of an immense number of variables.

4 Soybean IgE-binding proteins

Soybeans possess as many as 15 different proteins that show some cross-reactivity to IgEs derived from sensitive people. The major soybean allergens are Gly m Bd 30k that accounts for a large majority of the IgE cross-reactivity (Ogawa et al. 1993) as well as Gly m Bd 28k and Gly m Bd 60k (Ogawa et al. 1995) and glycinin that are secondary but significant allergens (Fig. 1). Gly m Bd 30k is an unusual member of the papain family of cysteine proteases that is accumulated to about 0.5% of the total protein of soybean seeds. This protein, also known as P34, has the structural characteristics of a typical cysteine protease except that it lacks the catalytic site cysteine and is therefore presumptively an inactive protease (Kalinski et al. 1990, 1992). The Gly m Bd 60k allergen is the beta subunit of conglycinin, one of the major storage proteins of soybean (Ogawa et al. 1995; Petruccelli et al. 2005, Wang et al. 2014a,b). Gly m Bd 28k is a conglutin protein with complex glycan side chain that is recognized by IgEs from soybean-sensitive people (Hiemori et al. 2000). Other significant soybean allergens that have been well characterized include glycinin, Gly m 6 (Helm et al. 2000b; Saeed et al. 2016, Wang et al. 2014a,b) and 2S albumin, Gly m 8 (Han et al. 2016).

P34 (Gly m Bd 30k) is the immunodominant allergen of neonates. P34 was recognized as a major soybean allergen by Ogawa et al. (1991, 1993), who surveyed proteins that bound IgEs derived from soybean-sensitive people. When they determined the amino terminal sequence of an approximately 30 kDa IgE-binding protein, they found it corresponded exactly with the cDNA sequence of P34 previously identified and named by Herman and colleagues (Kaliniski et al. 1990, 1992). The human immunoreactivity of P34 is shown in SDS/PAGE and IgE immunoblots shown in Fig. 1 (Yaklich et al. 1999). The upper panel shows a gel of the total protein from a mature soybean seed extract. The middle panel shows the cross-reactivity with an anti-P34 monoclonal antibody resulting in the labelling of a single 34 kDa polypeptide. The bottom panel shows the consequences of probing a

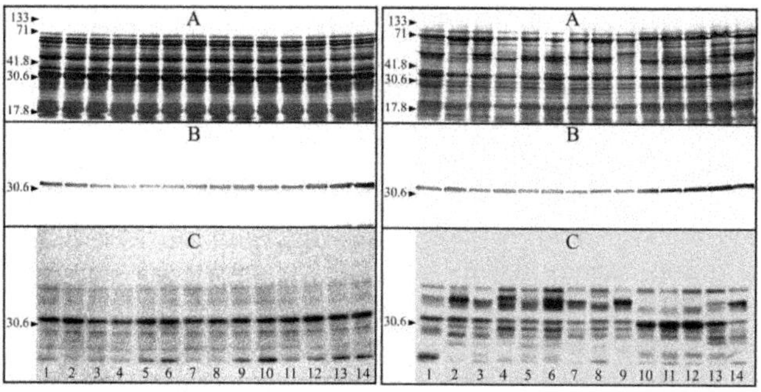

Figure 1 P34 is present in domesticated soybeans and its wild relatives. In cultivated soybeans P34 is the immunodominant protein but in the wild relatives the sera from soy-sensitive infants identifies many other cross-reactive proteins. This figure is derived from Yaklich et al. (1999). Left panel: Immunoblot analysis of the seed extracts of soybean, *Glycine max*. Upper panel (A), amido-black stained blot of the total proteins. Middle panel (B), replicate blot that was treated with anti-P34 serum. Lower panel (C), replicate blot that was treated with human serum. Arrowheads (far left) are the molecular weight rainbow standards. Lanes 1–14 represent the lines D81-8359, CX804-108, NC-2-62, CX797-115, CX797-21, 76-48773, D81-8498, BARC-6, BARC-7, BARC-8, BARC-9, Williams, Miles and Clark, respectively. Right panel: Immunoblot analysis of the seed extracts of *Glycine* species. Upper panel (A), amido-black stained blot of the total proteins. Middle panel (B), replicate blot that was treated with anti-P34 serum. Lower panel (C), replicate blot that was treated with human serum. Arrowheads (far left) are the molecular weight marker standards. Lanes 1–14 represent *G. arenaria* (PI 505204), *G. argyrea* (PI 505151, PI 509451), *G. canescens* (PI 440932, PI 440942, PI 446934), *G. clandestina* (PI 440948, PI 440954, PI 440960), *G. curvata* (PI 505164, PI 505166, PI 505167) and *G. cyrtoloba* (PI 440962, PI 440963), respectively.

replicate blot with sera obtained from a soybean-sensitive person and labelling the bound IgE antibodies with a second antibody. Note that the immunodominant protein indicated by the strong IgE-binding band is the 34-kDa P34 protein.

Surveys of a soybean core collection including wild relatives of soybean showed human IgE and IgG mouse MAb binding in all samples (Yaklich et al. 1999). P34/Gly m Bd 30k is not unique as a papain-superfamily cysteine protease allergen. The dust mite faecal allergen Der1p is a cysteine protease that has been extensively characterized (Yasuhara et al. 2001 for example). The primary cysteine protease of kiwi fruit, actinidin, is a food allergen for sensitive people (Pastorello et al. 1998). Other cysteine proteases induce allergies used to treat food (Nettis et al. 2001 for examples). An allergy to papain (Chambers et al. 1998; Niinimaki et al.1993; Quarre et al. 1995), the archetype cysteine protease, occurs in people who use contact lens treatment solution, used to strip proteins from the lens (Bernstein et al. 1984; Fischer 1985; Santucci et al., 1985), from exposure to fruit juices in throat lozenges (Iliev and Elsner 1997) and from its use as a meat tenderizer (Mansfield and Bowers 1983).

One model for the papain superfamily is the proteolytic activity that has a major role in inducing the immune response to these proteins (Chambers et al. 1998; Deleuran et al. 1998; Yasuhara et al. 2001) that would not apply to P34/Gly m Bd 30k with its apparent lack of intrinsic protease activity. The human immunology of P34/Gly m Bd 30k in soy-sensitive people has been investigated in detail (Helm et al. 1998, 2000; Morita et al. 2011, 2012). The human IgE linear epitopes of P34/Gly m Bd 30k from soy-sensitive people has been determined to consist of 14 epitopes distributed across the mature P34/Gly m Bd 30k sequence. This assay involves using synthetic peptides to create an overlapping array of the entire P34 polypeptide sequence that is then probed with antisera. It is important to note that this procedure identifies only linear-sequence epitopes and cannot identify any of the protein's conformational epitopes. The individual epitopes vary in intensity with some epitopes much 'hotter' than others. The same procedure was used to examine immunological differences among unrelated soy-sensitive people that showed that although P34/Gly m Bd 30k was an immunodominant allergen the individual IgE epitopes varied in intensity among these subjects, although some of the epitopes tended to be stronger in most of the subjects (Helm et al. 1998). The immunodominant epitopes were further examined by producing glycine mutations within the epitope to map the sequence of the critical amino acids (Helm et al. 2000). Among six epitopes examined by this procedure, the antigenicity was entirely abolished by a single glycine substitution in any position within one epitope while in other epitopes there was only a single critical amino acid (Helm et al. 2000).

In the absence of a crystal structure, the P34 epitopes identified by Helm et al. (2000) have been projected onto a three-dimensional model of closely related and archetype papain shown in Fig. 2. The immunological activity of P34/Gly m Bd 30k is unique among seed-source allergens that are otherwise dominated by the 11S, 7S and 2S storage proteins and seed lectins. Papain superfamily allergens are dominant allergens of papaya, pineapple and kiwi; although soybean is unusual with respect to its specific seed allergenic protein, in this context it shares the allergenic characteristics of some fruit.

Soybean and dairy allergy may be synergistic with P34 bridging the sensitivity. Parallel hypersensitive response has been observed in human neonates directed at milk casein and soybean (Candreva et al. 2015). For some infants the joint sensitivity to both milk-based and soybean-based formula is a critical health issue that is only resolved with use of expensive hyposensitive formulas produced from protein hydrolysates. Similar issues

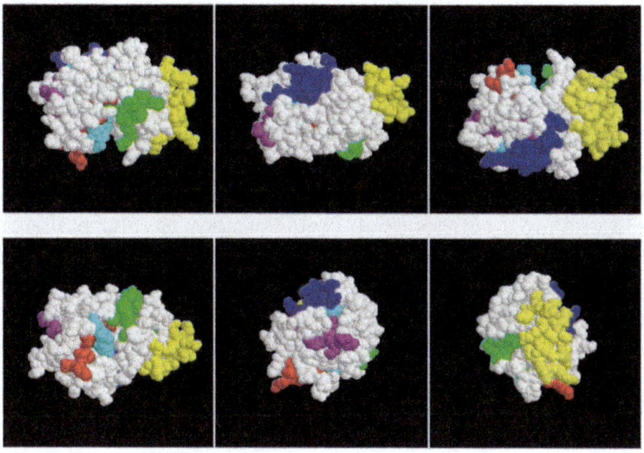

Figure 2 An epitope map of P34 IgE-binding sites described by Helm et al. (1999, 2000) that was projected onto the crystal structure of papain as a structure proxy. Note that the multiple epitopes are widely distributed on the surface of the protein.

may exist for animals, such as dogs, that develop broad food hypersensitivity to multiple potentially allergenic components of dog food. Soybean and dairy sensitivity is linked to FPIES that manifests as a food intolerance rather than an IgE-mediated allergy. The proteins that elicit FPIES still remain to be determined and whether it is casein and P34 and potential synergy is the subject for further research. For reasons yet to be determined P34 appears to have some cross-reactivity with casein (Smaldini et al. 2012) that is hypothesized to be the antigen that induces joint hypersensitivity.

5 Mitigating soybean allergens

The human IgE epitopes of the 11s (glycinin) legumin storage protein have been examined (Xiang et al. 2002). The acidic chain of glycinin shares some epitopes in common with the peanut allergen Ara h 3 (Beardslee et al. 2000). Of these, only one epitope could be converted to a form that did not bind IgE by sequential alanine substitution (Helm et al. 2000). The epitopes of the two other significant soybean allergens Gly m Bd 60k beta conglycinin and Gly m Bd 28k conglutin have not been reported.

Gly m Bd 28k includes IgEs directed at the complex glycan side chain (Hiemori et al. 2000). Studies of soybean allergen epitopes in comparison with other food-source allergens have been useful to understand cross-reactivity and how sensitization to other related foods is acquired once an individual is sensitive to any single allergen source, such as soybean. The close sequence similarity between allergenic proteins such as the 11S, 7S, 2S storage proteins, the seed lectins, as well as widely distributed systemic plant proteins such as profilin and lipid transfer protein creates the circumstances for the immune system to respond to other food sources, thereby broadening the allergenic response. Epitope studies have generally shown that soybean and other plant source allergens have several to many separate linear-sequence IgE-binding sites that vary among sensitized individuals. The multitude of IgE-binding sites (Fig 2) makes the selection of a non-allergenic variant by epitope mutation essential, resulting in the goal to eliminate allergens rather than to identify less allergenic variants.

Attempts have been made to reduce allergens by producing allergen-reduced or allergen-null plants using biotechnology or by selection/breeding. These experiments have demonstrated that it is feasible to silence specific allergens from food plants. From the first attempts to partially silence the rice allergen (Tada et al. 1996, 2003), to completely eliminate a major allergen of soybean (Kinney et al. 2001; Herman et al. 2003) shown in Figure 3, the prospects for using biotechnology to reduce allergen content was established. Subsequent suppression of major peanut (Dodo et al. 2008), tomato (Le et al. 2006a,b; Lorenz et al. 2006) and apple allergens (Gilissen et al. 2005) induced genetic modification has shown this approach is effective at silencing allergens in diverse foods. Parallel approaches of mining germplasm collections of soybean have identified nulls of storage proteins (Takahashi et al. 1994), Gly m Bd 30k/P34 (Josephs et al. 2006), trypsin inhibitor (Hymowitz 1986) and lectin (Goldberg et al. 1983). A P34 null was identified with a systemic search of accessions in the US National Soybean Collection. Two accessions of the same null were identified (Fig. 4) (Jospehs et al. 2006), which resulted from a four-base insertion into the 5′ portion of the P34 gene that altered its translation producing a null or a very inefficient translation shown in Fig. 4 (Bilyeu et al. 2009). Schmidt et al. (2015b) produced a conventionally bred line stacking P34, Kunitz Trypsin inhibitor, and lectin nulls in a Williams 82 background termed 'Triple Null'. It is possible that breeding could inadvertently increase the abundance of intrinsic allergens. Plant breeders increasingly use diverse germplasm to introduce production traits derived from domestic varieties as well as close wild relatives. As shown in Fig. 1 soybean-sensitive human neonates exhibit substantially greater IgE cross-reactivity with wild soybean relatives than with all members of a core collection of domesticated varieties. Based on the core collection survey (Yaklich et al. 1999) the risk of introducing a new dominant allergen by crossing two domesticated soybean varieties appears to be low; however, the risk of introgressing a significant allergen from a wild relative in a wide cross appears to be substantial. With increasing use of genomics-enabled breeding, identifying risk of increased allergens will require attention.

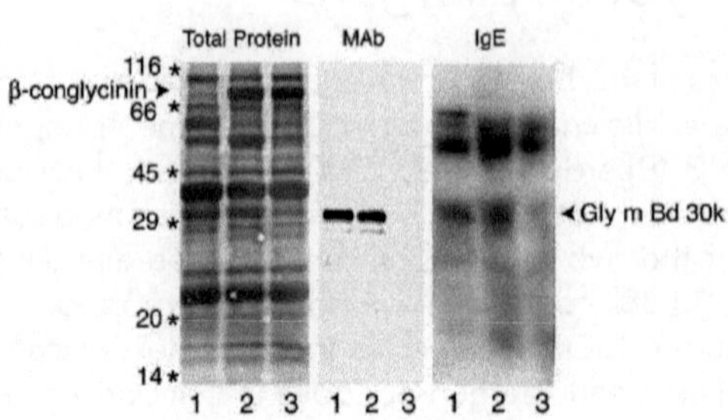

Figure 3 Removal of Gly m Bd 30 K protein by gene silencing. The three panels show replicate samples stained for total protein with amido black and labelled either with anti-Gly m Bd 30 K monoclonal antibody or with a mixture of IgEs from six soybean-sensitive individuals. Lane 1, transgenic control of another soybean suppressing α/α′-conglycinin (arrowhead) that does contain Gly m Bd 30 K. Lane 2, wild-type soybean cv Jack control. Lane 3, R3 generation homozygous transgenic soybean with the Gly m Bd 30 K construct inducing the complete suppression of Gly m Bd 30 K accumulation. The lanes in all the gels were equally loaded with 200 μg of seed protein extracts from the three soybean lines. The Mr indications correspond to Mr markers ran in additional lanes. Figure from Herman et al. (2003).

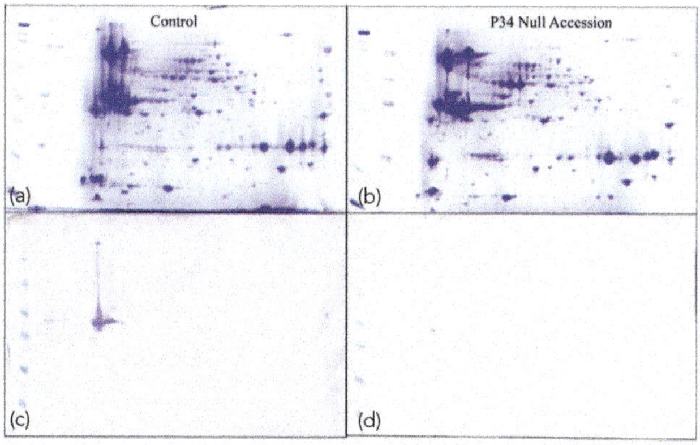

Standard Soybean ttccaccaagttATG GGT TTC CTT GTG TTG CTT CTT TTC TCC.........

MGFLVLLLFS...........................

P34 Null *ttccaaccaagttATG TAT GGG TTT CCT TGT GTT GCT TCT TTT....*

MYGFPCVASFLPLRSLF*..........

Figure 4 A conventional null for P34 was isolated from USDA soybean collection. The upper panel shows two-dimensional IEF/SDS page separation of the seed polypeptides of a standard soybean compared to the P34 null stained for total protein (A & B). The bottom set of panels shows immunoblots labelled with a monoclonal antibody specific for P34. Note the standard soybean with its high content of P34 exhibits a robust label for P34 while in contrast the P34 null exhibits a small amount of cross-reactive P34. The sequence shown is derived from the 5′ genomic and translated N-terminal amino acid sequence of the standard soybean P34 and the null. Analysis of the sequence of the null indicates two potential translational outcomes. The first is due to a frame shift – the nascent P34 polypeptide would be terminated after 17 amino acids when it encounters a stop (*). The second possible translational outcome is to shift the P34 sequence inward to the second ATG shown in red colour resulting in an inefficient translation and accounting for small amount of cross-reactive material labelled in the null's immunoblot. This observation parallels *in vitro* translation of the P34 transcript in rabbit reticulocyte with dog pancreatic membranes for processing that resulted in inefficient translation start within the pro-domain of P34 (Kalinski et al. 1992).

One of the difficulties in using genetic modification or nulls to create low-allergen or hypoallergenic seeds is that for many seeds the allergenic proteins account for the dominant portion of the seed proteome (Fig. 5 and Table 1). This is complicated by different populations being sensitive to different allergenic proteins, for example, soybean Gly m Bd 30k/P34 in a US neonatal population (Cantani and Lucenti 1997; Cordle 2004; Helm et al. 1998, 2000; Yaklich et al. 1999). Different cultivated varieties and their source breeding lines vary greatly in proteins recognized by IgEs from hypersensitive individuals (Yaklich et al. 1999), complicating breeding and varietal development. The individual proteins as compared to generic 'soy protein' that elicits FPIES and whether these proteins overlap the soy allergens are unknown. For the low-allergen lines produced in soybean and other crops there has been little experimentation on naïve models as compared to already sensitized individuals such as IgE assays from human sera samples. One of the problems for food allergy research has been the lack of animal models that faithfully mimic the development of human hypersensitivity to foods (Dearman and Kimber 2009; Helm et al. 2003). Although rodents (Knippels et al. 2005) are often used in soybean allergy

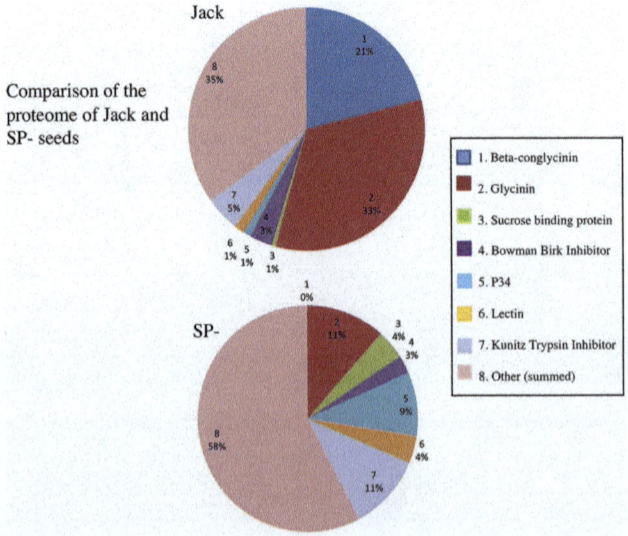

Figure 5 Silencing storage proteins results in rebalancing the seed proteome resulting in hyper-accumulation of the P34 allergen (9-fold increase) and Kunitz Trypsin inhibitor (2.5 fold increase). This experimental result shows that genetic modification of intrinsic protein composition can result in the increased accumulation of intrinsic allergens. This is an experimental line for basic science research that does not have the objective of developing a product. Shown are pie chart representations of the relative abundance of proteins in the proteomes of mature SP− and Jack determined by fractional spot volume. This illustrates the suppression of storage protein accumulation in SP− and its partial replacement by increased lectin, KTI and P34 accumulation. In contrast, BBI, which is a major sulphur sink of soybean seeds, remains unchanged in SP− compared with Jack, indicating the selectivity of proteins that rebalance the shortage of seed storage proteins. Figure reproduced from Schmidt et al. (2011), *Plant Physiology*.

research, their physiology is quite different from humans especially in gut morphology and physiology. Dogs and swine as larger omnivores are closer to the human biology, and both have been used as food allergy models (Helm et al. 2003; Calbrix et al. 2012; Rupa et al. 2008). Swine have been shown to acquire sensitivity to both β-conglycinin and P34 (Fu et al. 2007; Li et al. 1990; Sewekow et al. 2012; Taliercio et al. 2013, 2014). For swine the development of soybean hypersensitivity is a production concern with a significant fraction of neonatal swine developing reactions as the result of formula feeding. There is ongoing research using swine bred to be hypersensitive to soybean allergens (Calbrix et al. 2012). Altering the allergen, or bioactive components of foods, such as soybean, remains an intriguing but a clinically unexplored area of whether a food could be sufficiently altered to impede sensitizing a naïve immune and/or gut system to lower the rate of acquisition of food allergy or FPIES. Reduced allergenic food may be disarmed with respect to the immune system that could build immune tolerance. There is the potential that low-allergen or allergen-null soybeans could be used as therapeutic agents to build immune tolerance but this has not yet been tested. Regulatory agencies have defined the need for labelling allergens in food. It is not clear how these agencies will regulate and approve low-allergen variants of foods such as soybean that are required to

Table 1 Allergen database designation of soybean allergens (extracted from allergenonline.com) illustrating the number and diversity of proteins that are known to elicit a hypersensitive response

Allergen designation	Name	# Amino acids	Example GenBank
Gly m 1	Hull allergen	80	P24337
Gly m 3	Profilin	131	CAA11755
Gly m 4	Pollen-Related Protein	158	CAA42646
Gly m 5.01	α-Conglycinin	605	CAA35691
Gly m 5.02	α′-Conglycinin	621	AB030838.1
Gly m 5.03	β-Conglycinin	439	P25974
Gly m 6.02	Glycinin A2B1a	485	BAA00154
Gly m 6.03	Glycinin 3	562	CAA37044
Gly m 6.04	Glycinin A5B4B3	562	CAA26478
Gly m 6.05	Glycinin A3B4	517	BAB15802
Gly m 8.0	2S Albumin	155	AAD09630
Gly m Bd 28k	28k Allergen	476	BAB21619
Gly m Bd 30k	P34 Allergen	379	P22895
KTI	Kunitz Trypsin Inhibitor	217	CAA45777

be labelled as allergens. Important questions on the threshold response to low-allergen variants compared to standard varieties will need to be established. Much research on the low-allergen variants with sensitized individuals will need to be accomplished before the regulatory agencies will be able to address if this constitutes a safer variant from the perspective of allergens. Animal models such as swine could be used to test this approach and build the database needed to undertake controlled human testing.

6 Soybean intolerance in aquaculture

Soybean intolerance in salmonoids appears to be a parallel process to the mammalian soybean sensitivity that may be addressed by low-allergen variants. Aquaculture is the most rapidly growing of the animal food production sectors, increasing at nearly 10% per year (FAO 2014; Naylor et al. 2009). Aquaculture-sourced fish has surpassed wild-capture fisheries in global annual production. As fish aquaculture displaces wild-capture fisheries, providing aquaculture feed sources imposes an immense challenge. Feed for aquaculture currently over-relies on fishmeal to provide feed protein. Fishmeal is the product of marine forage fish species not used directly for human consumption. Annual fishmeal production has been constant for the last 15 years at approximately 6.5 million metric tons. However, over the past 15 years, the use of fishmeal has increased from less than 10% to about 40% of global fishmeal production and is predicted to increase at about 10% annually for the foreseeable future. The current, or projected, increase of fishmeal production to support aquaculture growth cannot be sustained.

The renewable solution for aquafeed is to use seed-based protein and oil products harvested from terrestrial farms for farmed fish (Barrows et al. 2008; Gatlin et al. 2007; Naylor et al. 2009; Rust et al. 2011). The substitution of plant protein for fishmeal is routine at low levels; increasing the proportion of plant protein in fish feed decreases fish growth rates and feed efficiency, especially in the salmonoids. Atlantic salmon production is a multi-billion dollar global industry that requires alternate feed sources to be sustainable. There is an opportunity to engineer crops as functional feed (Herman 2009; Herman and Schmidt 2016) that specifically addresses the challenge of including a higher proportion of plant-based diet in commercial fish aquaculture systems. Soybean is currently used as a primary source of vegetable protein supplement for aquafeed and is a large US export. Soybean at high-inclusion levels in aquaculture can result in reduced growth rate with soybean's anti-nutritional proteins being a significant factor (Baeverfjord et al. 1996; Bakke-McKellep et al. 2000; Burrells et al. 1999; Lilleeng et al. 2007; Krogdahl et al. 1994, 2010; Sahlmann et al. 2013). The intestinal response to soybean-inclusion feed appears to be similar to the hypersensitive gut responses observed in sensitized mammals. Fish do not produce IgEs so soybean intolerance cannot be characterized as possessing a true allergy, but the response is one of hypersensitivity that results in producing antibodies against the major soybean seed allergens. It is possible that rather being a true allergy the salmonoid response may resemble FPIES. Soybeans such as the conventional line Triple Null (Schmidt et al. 2015b) that stacks conventional nulls of anti-nutritional proteins Kunitz Trypsin Inhibitor (Hymowitz 1986; Jofuku et al. 1989; Orf and Hymowitz 1979), Lectin (Orf et al. 1978; Goldberg et al. 1983) and the immunodominant allergen P34/Gly m Bd 30k (Josephs et al. 2006) are being tested as a potential renewable protein feed source for salmonoids. If successful, the deployment of a low-allergen, low-bioactivity variant of soybean as aquafeed could pave the way for further feed applications and tests necessary to acquire the background data for eventual human use.

7 Food allergy, biotechnology and the risk of extrinsic introduced allergens

Food allergy remains a regulatory and societal concern of induced risk created by soybean biotechnology. An experiment to increase soybean sulphur amino acid content by biotechnology was one of the early key events that raised regulatory and societal concerns of producing potential allergens in transgenic crops. The legacy of one early experiment and its repercussions are still prominent in concerns about the potential for biotechnology-induced food allergy. One approach to enhance sulphur amino acid content of deficient crops is to express the genes of high-sulphur storage proteins, primarily prolamins from cereals and 2S albumins that are proportionately high in methionine.

Among the early experiments, researchers at Pioneer produced a soybean that expressed the Brazil nut 2S albumin (Nordlee et al. 1996). The Brazil nut protein is highly enriched in methionine and appeared to be a good candidate as an ancillary storage protein to increase the content of essential sulphur amino acids in soybean. The development of this product was abandoned when it was determined that Brazil nut-sensitive individuals would be sensitive to the 2S albumin expressed in soybeans (Nordlee et al. 1996). Even today this abandoned product is used by both proponents and opponents of genetic modification of crops. For proponents that the product was abandoned in assessment is a demonstration that the evaluation system works; for opponents this is often hyped as a near-catastrophic error that only by luck prevented human victims. The concerns that emerged from this incident and from other experiments resulted in global regulatory authorities led by the World Health Organization introducing a risk-evaluation flow chart to determine the potential for allergens to be transferred by biotechnology engineering (Codex 2003; FAO/WHO 2001). The key features of the WHO decision tree are whether the prospective transgene is related to a known allergen by its sequence homology. This assessment is enabled by rapid growth of plant genome and allergen sequence archives. Additional assessments include the protein that may be resistant to the action of gastric proteases exposing the protein to the gut's immunological system and whether source organism of the gene possesses known allergens. Assessment of known allergens such as Brazil nut 2S albumin as potential transgenes would fail even a cursory evaluation and would not move forward.

There are other situations involving expressing a known immunogen, not an allergen, in soybeans such as the development of soybean-expressed vaccines (Moravec et al. 2007) or biopharma (He et al. 2016). This would require an efficient means of segregation from commodity beans, perhaps differentiating these beans by an intrinsic colour marker such as orange β-carotene (Schmidt et al. 2015a). Commodity soybean biotechnology products that have now been deregulated including Roundup resistance by the ESPS gene support the assessment that these transgene products are not allergenic (Chang et al. 2003; Hoff et al. 2007). Similarly, the silencing of intrinsic soybean proteins has been evaluated for alteration of intrinsic allergen content. Both Dupont-Pioneer and Monsanto have produced high-oleic oil soybeans by variations of gene silencing fatty acid desaturase (FAD2) resulting in impeding fatty acid desaturation from progressing beyond 18:1. High-oleic oil is an important trait to produce oils without *trans* fats needed to improve cardiac health, now mandated in industrial foods in the United States and other nations. The suppression of FAD2 does not alter soybean's intrinsic allergen content.

Other potential breeding or transgenic alterations of soybean do have the potential to alter the content of intrinsic soybean allergens. Suppression of one or more major storage proteins in soybean leads to proteome rebalancing, a process by which other intrinsic proteins replace the shortfall maintaining the standard protein content (Kinney et al. 2001; Schmidt and Herman 2008; Herman 2014). If allergenic proteins replace less allergenic proteins this could increase the altered soybeans allergenic potential. Such an outcome has been demonstrated in a transgenic soybean in which both glycinin and conglycinin storage proteins were suppressed that resulted in the soybean seed remodeling its proteome replacing the soybean storage proteins with P34, Kunitz trypsin inhibitor, and lectin allergens (Schmidt et al. 2011). Such a storage protein minus line with elevated allergens shows how manipulating the proteome of soybean without introducing any foreign gene products can significantly alter allergen content. This line was intended only for investigative purposes to study protein regulation but it is a caution that altering allergen content in soybean and other crops with or without expressing foreign transgene proteins is feasible. The current regulatory regime and other mandated assessments (Batista et al. 2007; Chang et al. 2003; Gizzarelli et al. 2006; Goodman and Leach 2004; Herman 2003; Herman and Burks 2011) should be sufficient to identify allergen issues before any consideration of deployment.

8 Summary

Soybean allergens are an important issue for health care and for the soybean industry. Soybean as one of the eight FALPA-regulated food allergens has a large impact on industrial food production. Today's markets have thousands of industry-sourced products that contain soybean-derived components. For most consumers soybean protein is an often hidden healthful and economic addition to the food supply. For a small percentage of the population that has soybean sensitivity, either classical allergy or FPIES intolerance, it is the broad penetration of soybean into the prepared food production that poses concern. FALPA-mandated labelling is useful for these individuals to avoid allergenic sources and risk but does not mitigate the risk in restaurant settings or for those unable or unwilling to read product labels. Much is now known about soybean allergens, but the growing incidence of soybean allergy and the emergence of clinical FPIES show that there is much more that needs to be understood. For soybean breeders and the biotechnology industry, food allergy is a critical issue for deployment of new products. Addressing the risk, prevention or mitigation of soybean hypersensitivity requires more research using animal models and in clinical investigations.

9 Where to look for further information

FDA: http://www.fda.gov/Food/ResourcesForYou/Consumers/ucm079311.htm

Food Allergy and Education Awareness, FARE: https://www.foodallergy.org

10 Acknowledgements and dedication

Soybean allergy-related research in the Herman laboratory has been supported by the USDA/NIFA, the Minnesota Soybean Association and the now-ended Illinois-Missouri Biotechnology Alliance programme.

This chapter is dedicated to the memory of Dr. Niels Nielsen who passed away in 2016. Niels was a prominent member of the soybean molecular biology community and it was his research programme that elucidated many of the details of the glycinin storage protein genes.

11 References

Ahn, K. M., Han, Y. S., Nam, S. Y., Park, H. Y., Shin, M. Y. and Lee, S. I. (2003), Prevalence of soy protein hypersensitivity in cow's milk protein-sensitive children in Korea. *J. Korean Med. Sci.* 18, 473–7.

Al-Muhsen, S., Clarke, A. E. and Kagan, R. S. (2003), Peanut allergy: An overview. *Can. Med. Assoc. J.* 168, 1279–85.

Baeverfjord, G. and Krogdahl, A. (1996), Development and regression of soybean meal induced enteritis in Atlantic salmon, Salmo salar L., distal intestine: A comparison with the intestines of fasted fish. *J. Fish Dis.* 19, 375–87.

Bakke-McKellep, A. M., Press, C., Baeverfjord, G., Krogdahl, A. and Landsverk, T. (2000), Changes in immune and enzyme histochemical phenotypes of cells in the intestinal mucosa of Atlantic salmon, Salmo salar L., with soybean meal-induced enteritis. *J. Fish Dis.* 23, 115–27.

Bando, N., Tsuji, H., Hiemori, M., Yoshizumi, K., Yamanishi, R., Kimoto, M. and Ogawa, T. (1998), Quantitative analysis of Gly m Bd 28K in soybean products by a sandwich enzyme-linked immunosorbent assay. *J. Nutr. Sci. Vitaminol.* 44, 655–64.

Barrows, F. T., Bellis, D., Krogdahl, Å., Silverstein, J. T., Herman, E. M., Sealey, W. M., Rust, M. B. and Gatlin, D. M. (2008), Report of the plant products in aquafeed strategic planning workshop: An integrated, interdisciplinary research roadmap for increasing utilization of plant feedstuffs in diets for carnivorous fish. *Rev. Fish. Sci.* 16, 449–55.

Batista, R., Martins, I., Jeno, P., Ricardo, C. P. and Oliveira, M. M. (2007), A proteomic study to identify soya allergens – the human response to transgenic versus non-trnasgenic soya samples. *Int. Arch. Allergy Immunol.* 144, 29–38.

Baur, X., Pau, M., Czuppon, A. and Fruhmann, G. (1996), Characterization of soybean allergens causing sensitization of occupationally exposed bakers. *Allergy* 51, 326–30.

Baur, X., Degens, P. O. and Sander, I. (1998), Baker's asthma: Still among the most frequent occupational respiratory disorders. *J. Allergy Clin. Immunol.* 102, 984–97

Beardslee, T. A., Zeece, M. G., Sarath, G. and Markwell, J. P. (2000), Soybean glycinin G1 acidic chain shares IgE epitopes with peanut allergen Ara h 3. *Int. Arch. Allergy Immunol.* 123, 299–307.

Bellou, A., Kanny, G., Fremont, S. and Moneret-Vautrin, D. A. (1997), Transfer of atopy following bone marrow transplantation. *Ann. Allergy Asthma Immunol.* 78, 513–6.

Bernstein, D. I., Gallagher, J. S., Grad, M. and Bernstein, I. L. (1984) Local ocular anaphylaxis to papain enzyme contained in a contact lens cleansing solution. *J. Allergy Clin. Immunol.* 74, 258–60.

Bilyeu, K., Ren, C., Nguyen, H. T., Herman, E. and Sleper, D. A. (2009), Association of a four basepair insertion in the P34 gene with the low allergen trait in soybean. *Plant Genome* 2, 141–8.

Branum, A. M. and Lukacs, S. L. (2008), Food allergy among US children: Trends in prevalence and hospitalizations. *NCHS Data Breif* 10, 1–8, Hyattsville, MD.

Burks, W. (2003), Peanut allergy: A growing phenomenon. *J. Clin Invest.* 111, 950–2.

Burrells, C., Williams, P. D., Southgate, P. J. and Crampton, V. O. (1999), Immunological, physiological and pathological responses of rainbow trout (Oncorhynchus mykiss) to increasing dietary concentration of soybean proteins. *Vet. Immunol. Immunopath.* 72, 277–88.

Calbrix, R., Schinckel, A., Stewart, T., Herman, E., Helm, R. and Nielsen, N. (2012), Development of tools to study immune-mediated allergenic responses to food and feed. In R. Wilson (Ed.) *Designing Soybeans for 21st Century Markets*. AOCS Press, Urbana, IL, pp. 239–52.

Candreva, A. M., Smaldini, P. L., Curciarello, R., Cauerhff, A., Fossati, C. A., Docena, G. H. and Petruccelli, S. (2015), Cross-reactivity between the soybean protein p34 and bovine caseins. *Allergy Asthma Immunol. Res.* 7, 60–8. doi: 10.4168/aair.2015.7.1.60.

Candreva, Á. M., Smaldini, P. L., Curciarello, R., Fossati, C. A., Docena, G. H. and Petruccelli, S. (2016), The major soybean allergen Glym Bd 28k induces hypersensitivity reactions in mice sensitized to cow's milk proteins. *J. Agric. Food Chem.* 64, 1590–9. doi: 10.1021/acs.jafc.5b05623.

Cantani, A. and Lucenti, P. (1997), Natural history of soy allergy and/or intolerance in children, and clinical use of soy-protein formulas. *Pediatr. Allergy Immunol.* 8, 59–74.

Chambers, L., Brown, A., Pritchard, D. I., Sreedharan, S., Brocklehurst, K. and Kalsheker, N. A. (1998), Enzymatically active papain preferentially induces an allergic response in mice. *Biochem. Biophys. Res. Commun.* 253, 837–40.

Chang, H. S., Kim, N. H., Park, M. J., Lim, S. K., Kim, S. C., Kim, J. Y., Kim, J. A., Oh, H. Y., Lee, C. H., Huh, K., Jeong, T. C. and Nam, D. H. (2003), The 5-enolpyrivylshikimate-3-phosphate synthase of glyphosate-tolerant soybean expressed in Escherichia coli shows no severe allergenicity. *Mol. Cells* 15(1), 20–6.

CODEX. (2003), Codex Alimentarius Commission, Alinorm 03/34: Joint FAO/WHO Food Standard Programme, Codex Alimentarius Commission, Twenty-Fifth Session, Rome, Italy, 30 June–5 July 2003. Appendix III, Guideline for the conduct of food safety assessment of foods derived from recombinant-DNA plants, and Appendix IV, Annex of the assessment of possible allergenicity, pp. 47–60.

Cordle, C. T. (2004), Soy protein allergy: Incidence and relative severity. *J. Nutr.* 134, 1213S–19S.

Curciarello, R., Smaldini, P. L., Candreva, A. M., González, V., Parisi, G., Cauerhff, A., Barrios, I., Blanch, L. B., Fossati, C. A., Petruccelli, S. and Docena, G. H. (2014), Targeting a cross-reactive Gly m 5 peptide as responsible for hypersensitivity reactions in a milk allergy mouse model. *PLoS One* 9(1), e82341. doi: 10.1371/journal.pone.0082341.

Dearman, R. J. and Kimber, I. (2009), Animal models of protein allergenicity: Potential benefits, pitfalls and challenges. *Clin. Exp. Allergy* 39, 458–68.

Deleuran, M., Ellingsen, A. R., Paludan, K., Schou, C. and Thestrup-Pedersen, K, (1998), Purified Der p1 and p2 patch tests in patients with atopic dermatitis: Evidence for both allergenicity and proteolytic irritancy. *Acta Derm. Venereol.* 78, 241–3.

Docena, G. H., Fernandez, R., Chirdo, F. G. and Fossati, C. A. (1996), Identification of casein as the major allergenic and antigenic protein of cow's milk. *Allergy* 51, 412–6.

Dodo, H. W., Konan, K. N., Chen, F. C., Egnin, M. and Viquez, O. M. (2008), Alleviating peanut allergy using genetic engineering: The silencing of the immunodominant allergen Ara h 2 leads to its significant reduction and a decrease in peanut allergenicity. *Plant Biotechnol. J.* 6, 135–45.

Dreau, D., Lalles, J. P., LeJan, C., Toullec, R. and Salmon, H. (1995), Hypersensitivity to soybean proteins in early weaned piglets: humoral and cellular components. *Adv. Exp. Med. Biol.* 371B, 865–9.

FAO/WHO. (2001), Evaluation of Allergenicity of Genetically Modified Foods; Report of a joint FAO/WHO expert consultation on allergenicity of foods derived from biotechnology, 22–25 January 2001. FAO/WHO: Rome, Italy, pp. 10–13. http://www.who.int/foodsafety/publications/biotech/en/ec_jan2001.pdf.

Fisher, A. A. (1985), Allergic reactions to contact lens solutions. *Cutis* 36, 209–1.

Friesien, K. G., Goodband, R. D., Nelssen, J. L., Belcha, F., Reddy, D. N., Reddy, P. G. and Kats, L. J. (1993), The effect of pre- and postweaning exposure to soybean meal on growth performance and on the immune response in the early-weaned pig. *J. Anim. Sci.* 71, 2089.

Fu, C. J., Jez, J. M., Kerley, M. S., Allee, G. L. and Krishnan, H. B. (2007), Identification, characterization, epitope mapping, and three-dimensional modeling of the alpha-subunit of beta-conglycinin of soybean, a potential allergen for young pigs. *J. Agric. Food Chem.* 55, 4014–20.

Gatlin, D. M., Barrows, F. T., Brown, P., Dabrowski, K., Gaylord, T. G. Hardy, R. W., Herman, E. M., Hu, G., Krogdahl, Å., Nelson, R., Overturf, K., Rust, M., Sealey, W., Skonberg, D., Souza, E. J., Stone, D., Wilson, R. and Wurtele, E. (2007), Expanding the utilization of sustainable plant products in aquafeeds: A review. *Aquacul. Res.* 38, 551–79.

Gilissen, L. J., Bolhaar, S. T., Matos, C. I., Rouwendal, G. J., Boone, M. J., Krens, F. A., Zuidmeer, L., Van Leewen, A., Akkerdaas, J. and Hoffmann-Sommergruber, K. (2005), Silencing the major apple allergen Mal d 1 by using the RNA interference approach. *J. Allergy Clin. Immunol.* 115, 364–9.

Gizzarelli, F., Corinti, S., Barletta, B., Iacovacci, P., Brunetto, B., Butteroni, C., Afferni, C., Onori, R., Miraglia, M., Panzini, G., Di Felice, G. and Tinghino, R. (2006), Evaluation of allergenicity of genetically modified soybean protein extract in a murine model of oral allergen-specific sensitization. *Clin. Exp. Allergy* 36, 238–48.

Goldberg R. B., Hoschek, G. and Vodkin, L. O. (1983), An insertion sequence blocks the expression of a soybean lectin gene. *Cell* 33, 465–75.

Gonzalez, R., Zapatero, L., Caravaca, F. and Carreira, J. (1991), Identification of soybean proteins responsible for respiratory allergies. *Int. Arch. Allergy Appl. Immunol.* 95, 53–7.

Goodman, R. E. and Leach, J. N. (2004), Assessing the allergenicity of proteins introduced into genetically modified crops using specific human IgE assays. *J. AOAC Int.* 87, 1423–32.

Gupta R. S., Springston, E. E., Warrier, M. R., Smith, B., Kumar, R., Pongracic, J., Holl, J. L. (2011), The prevalence, severity, and distribution of childhood food allergy in the United States. *Pediatrics* 128, e9–e17.

Han, Y., Lin, J., Bardina, L., Grishina, G. A., Lee, C., Seo, W. H. and Sampson, H. A. (2016), What characteristics confer proteins the ability to induce allergic responses? IgE epitope mapping and comparison of the structure of soybean 2S albumins and Ara h 2. *Molecules* 21(5), 622, E622. doi: 10.3390/molecules21050622.

He, L., Han, M., Qiao, S., He, P., Li, D., Li, N. and Ma, X. (2015) Soybean antigen proteins and their intestinal sensitization activities. *Curr. Protein Pept. Sci.* 16, 613–21.

He, Y., Schmidt, M. A., Erwin, C., Guo, J., Sun, R., Pendarvis, K., Warner, B. W. and Herman, E. M. (2016), Transgenic soybean production of bioactive human epidermal growth factor (EGF), *PLoS One*. doi:10.1371/journal.pone.0157034.

Helm, R. M., Ermel, R. W. and Frick, O. L. (2003) Nonmurine animal models of food allergy. *Environ. Health Perspect.* 111, 239–44.

Helm, R. M., Cockrell, G., Herman, E. M., Burks, L., Sampson, H. A. and Bannon, G. A. (1998), Cellular and molecular characterization of a major soybean allergen. *Int. Arch. Allergy Immunol.* 117, 29–37.

Helm, R. M., Cockrell, G., West, C. M., Herman, E. M. Sampson, H. A., Bannon, G. A. and Burks, A. W. (2000), Mutational analysis of the IgE-binding epitopes of P34/Gly m1. *J. Allergy Clin. Immunol.* 105, 378–84

Helm, R. M., Cockrell, G., Connaughton, C., Sampson, H. A., Bannon, G. A., Beilinson, V. D., Nielsen, N. C. and Burks, A. W. (2000), Soybean G2 glycinin allergen. 1. Identification and characterization. *Int. Arch. Allergy Immunol.* 123, 205–12.

Hepple, L. M., Sissons, J. W. and Pedersen, H. E. (1987), A comparison of the antigenicity of soya-bean-based infant formulas. *Br. J. Nutr.* 58, 393–403.

Hepple, L. M., Sissons, J. W. and Banks, S. M. (1989), Sensitization of preruminant calves and piglets to antigenic protein in early weaning diets: Control of the systemic antibody response. *Res. Vet. Sci.* 47, 257–62.

Herian, A. M., Taylor, S. L. and Bush, R. K. (1993), Allergenicity of various soybean products as determined by RAST inhibition. *J. Food Sci.* 5, 385.

Herman, E. M. (2003), Genetically modified soybeans and food allergies. *J. Exp. Bot.* 54, 1317–19

Herman, E. M. (2009), Biotechnology could combine positive traits in seed-based fish feed. *Global Aquaculture Advocate* Sept/Oct 2009 issue, pp. 88–90.

Herman, E. M. and Burks, A. W. (2011), The impact of plant biotechnology on food allergy. *Curr. Opin. Biotechnol.* 22, 224–30.

Herman, E. M. (2014), Soybean seed proteome rebalancing. *Front. Plant Sci.* 5, 437. doi: 10.3389/fpls.2014.00437.

Herman, E. M. and Schmidt, M. A. (2016), The potential for engineering enhanced functional-feed soybeans for sustainable aquaculture feed. *Front. Plant Sci.* 7, 440. doi: 10.3389/fpls.2016.00440.

Herman, E. M., Helm, R., Jung, R. and Kinney, A. C. (2003) Genetic modification removes an immunodominant allergen from soybean. *Plant Physiol.* 132, 36–43.

Hiemori, M., Bando, N., Ogawa, T., Shimada, H., Tsuji, H., Yamanishi, R. and Terao, J. (2000), Occurrence of IgE antibody-recognizing N-linked glycan moiety of a soybean allergen, Gly m Bd 28K. *Int. Arch. Allergy Immunol.* 122, 238–45.

Hoff, M., Son, D. Y., Gubesch, M., Ahn, K., Lee, S. I., Vieths, S., Goodman, R. E., Ballmer-Weber, B. K. and Bannon G. A. (2007), Serum testing of genetically modified soybeans with special emphasis on potential allergenicity of the heterologous protein CP4 EPSPS. *Mol. Nutr. Food Res.* 51, 946–55.

Hymowitz, T. (1986), Genetics and breeding of soybeans lacking the Kunitz trypsin inhibitor. *Adv. Exp. Med. Biol.* 199, 291–8.

Iliev, D. and Elsner, P. (1997), Generalized drug reaction due to papaya juice in throat lozenges. *Dermatology* 194, 364–6.

Jarvinen, K. M. and Nowak-Wegrzyn, A. (2013), Food protein-induced enterocolitis syndrome (FPIES): current management strategies and review of the literature. *J. Allergy Clin. Immunol.* 1, 317–22.

Jofuku, K. D., Schipper, R. D. and Goldberg, R. B. (1989), A frameshift mutation prevents Kunitz trypsin inhibitor mRNA accumulation in soybean embryos. *Plant Cell* 1, 427—35.

Josephs, L. M., Hymowitz, T., Schmidt, M. A. and Herman E. M. (2006), Evaluation of Glycine germplasm for nulls of the immunodominant allergen P34/Gly m Bd 30k. *Crop Sci.* 46, 1755–63.

Kalinski, A. J., Weisemann, J., Matthews, B. F. and Herman, E. M. (1990) Molecular cloning of a protein associated with soybean oil bodies which is homologous to thiol proteases of the papain family. *J. Biol. Chem.* 265, 13843–8.

Kalinski, A. J,. Melroy, D. L., Dwivedi, R. S. and Herman, E. M. (1992), A soybean vacuolar protein (P34) related to thiol proteases which is synthesized as a glycoprotein precursor during seed maturation. *J. Biol. Chem.* 267, 12068–76.

Katz, Y., Gutierrez-Castrellon, P., González, M. G., Rivas, R., Lee, B. W. and Alarcon, P. (2014), A comprehensive review of sensitization and allergy to soy-based products. *Clin. Rev. Allergy Immunol.* 46, 272–81. doi: 10.1007/s12016-013-8404-9. Review.

Kinney A. J., Jung, R. and Herman. E. M. (2001) Cosuppression of the 〈 subunits of α-conglycinin in transgenic soybean seeds induces the formation of endoplasmic reticulum-derived protein bodies. *Plant Cell* 13, 1165–78.

Knippels L. M. J. and Penninks, A. H. (2005), Recent advances using rodent models for predicting human allergenicity. *Toxicol. Appl. Pharmacol.* 207, S157–60.

Krogdahl, Å., Lea, T. B. and Olli, J. L. (1994), Soybean proteinase-inhibitors affect intestinal trypsin activities and amino acid digestibilities in rainbow trout (Oncorhynchus mykiss). *Comp. Biochem. Physiol.*107, 215–19

Krogdahl, Å., Penn, M. H., Thorsen, J., Refstie, S. and Bakke, A. M. (2010), Important antinutrients in plant feedstuffs for aquaculture: an update on recent findings regarding responses in salmonids. *Aquac. Res.* 41, 333–44

Le, L. Q., Lorenz, Y., Scheurer, S., Fotisch, K., Enrique, E., Bartra, J., Biemelt, S., Vieths, S. and Sonnewald, U. (2006a), Design of tomato fruits with reduced allergenicity by dsRNAi-mediated inhibition of nsLTP (Lyc e 3) expression. *Plant Biotechnol. J.* 4, 231–42.

Le, L. Q., Mahler, V., Lorenz, Y, Scheurer, S., Biemelt, S., Vieths, S. and Sonnewald, U, (2006b), Reduced allergenicity of tomato fruits harvested from Lyc e 1-silenced transgenic tomato plants. *J. Allergy Clin. Immunol.* 118, 1176–83.

Legendre, C., Caillat-Zucman, S., Samuel, D., Morelon, S., Bismuth, H., Bach, J. F. and Kreis, H. (1997), Transfer of symptomatic peanut allergy to the recipient of a combined liver-and-kidney transplant. *N. Engl. J. Med.* 337, 822–5.

Leonard, S. A. and Nowak-Wegrzyn, A. (2012), Clinical diagnosis and management of food protein-induced enterocolitis syndrome. *Curr. Opinion. Pediatr.* 24, 739–45.

Levy, Y. and Danon, Y. L. (2003) Food protein-induced enterocolitis syndrome: Not only due to cow's milk and soy. *Pediatr. Allergy Immunol.* 14, 325–9.

Levy, Y., Kornbrot, B. and Danon, Y. L. (1998) Soybean allergy in cow milk-tolerant infants. *Pediatr. Asthma Allergy Immunol.* 12, 253–9.

Li, D. F., Nelssen, J. L., Reddy, P. G., Blecha, F., Hancock, J. D., Allee. G. L., Goodband, R. D. and Klemm, R. D. (1990) Transient hypersensitivity to soybean meal in the early-weaned pig. *J. Anim. Sci.* 68, 1790–9.

Lilleeng, E., Froystad, M. K. and Krogdahl, A. (2007), Effects of diets containing soybean meal on trypsin mRNA expression and activity in Atlantic salmon (Salmo salar L). *Comp. Biochem Physiol. A Mol. Integr. Physiol.* 147, 25–36.

Lorenz, Y., Enique, E., LeQuynh, L., Fotisch, K., Retzek, M., Biemelt, S., Sonnewald, U., Vieths, S. and Scheurer, S, (2006), Skin prick tests reveal stable and heritable reduction of allergenic potency of gene-silenced tomato fruits. *J. Allergy Clin. Immunol.* 118, 711–18.

Mansfield, L. E. and Bowers, C. H. (1983), Systemic reaction to papain in a nonoccupational setting. *J. Allergy Clin. Immunol.* 71, 371–4.

Moravec, T., Schmidt, M. A., Herman, E. M. and Woodford-Thomas, T. (2007), Production of Escherichia coli heat labile toxin (Lt)B subunit in soybean seed and analysis of its immunogenicity. *Vaccine* 25, 1647–57.

Morita, H., Kaneko, H., Ohnishi, H., Kato, Z. and Kondo, N. (2011) Antigen-specific immune response to endotoxin-free recombinant P34. *Allergy* 66, 985–6.

Morita, H., Kaneko, H., Ohnishi, H., Kato, Z., Kubota, K., Yamamoto, T., Matsui, E., Teramoto, T., Fukao, T., Kasahara, K. and Kondo, N. (2012), Structural property of soybean P34 and specific IgE response to recombinant P34 in patients with soybean allergy. *Int. J. Mol. Med.* 29, 153–8. doi: 10.3892/ijmm.2011.841.

Naylor, R. L., Hardy, R. W., Bureau, D. P., Chiu, A., Elliott, M., Farrell, A. P., Forster, I., Gatlin, D. M., Goldburg, R. J., Hua, K. and Nichols, P. D. (2009), Feeding Aquaculture in an Era of Finite Resources. *Proc. Natl. Acad. Sci.* 106, 15103–10.

Nettis, E., Napoli, G., Ferrannini, A. and Tursi, A. (2001), IgE-mediated allergy to bromelain. *Allergy* 56, 257–8.

Niinimaki, A., Reijula, K., Pirila, T. and Koistinen, A. M. (1993), Papain-induced allergic rhinoconjunctivitis in a cosmetologist. *J. Allergy Clin. Immunol.* 92, 492–3.

Nordlee, J. A., Taylor, S. L., Townsend, J. A., Thomas, L. A. and Bush, R. K. (1996), Identification of a Brazil-nut allergen in transgenic soybeans. *N. Engl. J. Med.* 334, 726–8

Ogawa, T.,Bando, N., Tsuji, H., Okajima, H., Nishikawa, K. and Sasaoka, K. (1991), Investigation of the IgE- binding proteins in soybeans by immunoblotting with the sera of the soybean-sensitive patients with atopic dermatitis. *J. Nutr. Sci. Vitaminol.* 37, 555–65.

Ogawa, T., Tsuji, H., Bando, N., Kitamura, K., Zhu, Y. L., Hirano, H. and Nishikawa, K. (1993), Identification of the soybean allergenic protein, Gly m Bd 30K, with the soybean seed 34-kDa oil-body-associated protein. *Biosci. Biotechnol. Biochem.* 57, 1030–3.

Ogawa, T., Bando, N., Tsuji, H., Nishikawa, K. and Kitamura, K. (1995), Alpha-subunit of beta-conglycinin, an allergenic protein recognized by IgE antibodies of soybean-sensitive patients with atopic dermatitis. *Biosci. Biotechnol. Biochem.* 59, 831–3.

Ogawa, A., Samoto, M. and Takahashi, K. (2000,) Soybean allergens and hypoallergenic soybean products. *J. Nutr. Sci. Vitaminol.* 46, 271–9.

Orf, J. H., and Hymowitz, T. (1979), The inheritance of the absence of the Kunitz trypsin inhibitor in seed protein of soybeans. *Crop Sci.* 19, 107–9.

Orf, J. H., Hymowitz, T., Pull, S. P. and Pueppke, S. G. (1978), Inheritance of soybean seed lectin. *Crop Sci.* 18, 899–900.

Pastorello E. A., Conti, A., Pravettoni, V., Farioli, L., Rivolta, F., Ansaloni, R., Ispano, M., Incorvaia, C., Giuffrida, M. G. and Ortolani, C. (1998), Identification of actinidin as the major allergen of kiwi fruit. *J. Allergy Clin. Immunol.* 101, 531–7.

Petruccelli, S., Chirdo, F. G. and Añón, M. C. (2005), Immunochemical reactivity of soybean β-conglycinin subunits. *Food Agric. Immunol.* 16, 17–28.

Quarre, J. P., Lecomte, J., Lauwers, D., Gilbert, P. and Thiriaux, J. (1995), Allergy to latex and papain. *J. Allergy Clin. Immunol.* 95, 922.

Rozenfeld, P., Docena, G. H., Anon, M. C. and Fossati, C. A. (2002) Detection and identification of a soy protein component that cross-reacts with caseins from cow's milk. *Clin. Exp. Immunol.* 130, 49–58.

Rupa, P., Hamilton, K., Cirinna, M. and Wilkie, B. N. (2008), A neonatal swine model of allergy induced by the major food allergen chicken ovomucoid (Gal d 1). *Int. Arch. Allergy Immunol.* 146, 11–8.

Rust, M. M., Barrows, F. T., Hardy, R. W., Lazur, A., Naughten, K. and Silverstein, J. (2011), *NOAA-USDA Alternative Feed Initiative, The Future of Aquafeeds*, NOAA Technical Memorandum NMFS-F/SPO-124. NOAA Scientific Publications Office, Washington, DC.

Saeed, H., Gagnon, C., Cober, E. and Gleddie, S. (2016), Using patient serum to epitope map soybean glycinins reveals common epitopes shared with many legumes and tree nuts. *Mol. Immunol.* 70, 125–33. doi: 10.1016/j.molimm.2015.12.008.

Sahlmann, C., Sutherland, B. J., Kortner, T. M., Koop, B. F., Krogdahl, A. and Bakke, A. M. (2013), Early response of gene expression in the distal intestine of Atlantic salmon (Salmo salar L.) during the development of soybean meal induced enteritis. *Fish Shellfish Immunol.* 34, 599–609.

Santucci, B., Cristaudo, A. and Picardo, M. (1985), Contact urticaria from papain in a soft lens solution. *Contact Derm.* 12, 233

Schmidt, M. A. and Herman, E. M. (2008), Proteome rebalancing in soybean seeds can be exploited to enhance foreign protein accumulation. *Plant Biotech. J.* 6, 832–42

Schmidt, M. A., Barbazuk, W. B., Stanford, M., May, G., Song, Z., Hong, W., Nikolau, B. J. and Herman, E. M. (2011), Silencing of soybean seed storage proteins results in a rebalanced protein composition preserving seed protein content without major collateral changes in the metabolome and transcriptome. *Plant Physiol.* 156, 330–45.

Schmidt, M. A., Parrott, W. A., Hildebrand, D. F., Berg, R. H., Cooksey, A., Pendarvis, K., He, Y., McCarthy, F. and Herman, E. M. (2015b), Transgenic soya bean seeds accumulating β-carotene exhibit the collateral enhancements of oleate and protein content traits. *Plant Biotechnol. J.* 1, 590–600. doi: 10.1111/pbi.12286.

Schmidt, M. A., Hymowitz, T. and Herman, E. M. (2015), Breeding and characterization of soybean Triple Null; a stack of recessive alleles of Kunitz Trypsin Inhibitor, Soybean Agglutinin, and P34 allergen nulls. *Plant Breed.* 134, 310–15. doi: 10.1111/pbr.12265.

Schmied, J., Rupa, P., Garvie, S. and Wilkie, B. (2013), Immune response phenotype of allergic verus clinically tolerant pigs in a neonatal swine model of allergy. *Vet. Immun. Immunopath.* 154, 17–24.

Sewekow, E., Bimczok, D., Kähne, T., Faber-Zuschratter, H., Kessler, L. C., Seidel-Morgenstern, A. and Rothkötter, H. J. (2012), The major soyabean allergen P34 resists proteolysis in vitro and is transported through intestinal epithelial cells by a caveolae-mediated mechanism. *Br. J. Nutr.* 108, 1603–11. doi: 10.1017/S0007114511007045.

Shaffrali, F. C. and Gawkrodger, D. J. (2001), Contact dermatitis from soybean extract in a cosmetic cream. *Contact Derm.* 44, 51–2.

Sicherer, S. H., Eigenmann, P. A. and Sampson, H. A. (1998), Clinical features of food protein-induced enterocolitis syndrome. *J. Pediatr.* 133, 214–19.

Sicherer, S. H. and Sampson, H. A. (2007), Peanut allergy: emerging concepts and approaches for an apparent epidemic. *J. Allergy Clin. Immunol.* 120, 491–503.

Sicherer, S. H. and Sampson, H. A. (2010), Food Allergy. *J. Allergy Clin. Immunol.* 125, S116–25.

Smaldini, P., Curciarello, R., Candreva, A., Rey, M, A., Fossati, C. A., Petruccelli, S. and Docena, G. H. (2012), In vivo evidence of cross-reactivity between cow's milk and soybean proteins in a mouse model of food allergy. *Int. Arch. Allergy Immunol.* 158, 335–46.

Sun, P., Li, D., Li, Z., Dong, B. and Wang, F. (2008), Effects of glycinin on IgE-mediated increase of mast cell numbers and histamine release in the small intestine. *J. Nutr. Biochem.* 19, 627–33.

Tada, Y., Nakase, M., Adachi, T., Nakamura, R., Shimada, H., Takahashi, M., Fujimura, T. and Matsuda, T. (1996), Reduction of 14–16 kDa allergenic proteins in transgenic rice plants by antisense gene. *FEBS Lett.* 391, 341–5.

Tada, Y., Hiromori, A., Tatsuhito, F. and Matsuda, T. (2003), Effect of an antisense sequence on rice allergen genes comprising a multigene family. *Breed Sci.* 53, 61–7.

Takahashi, K, Banba, H, Kikuchi, A, Ito, M. and Nakamura, S. (1994), An induced mutant line lacking the α-subunit of β-conglycinin in soybean (Glycine max (L.) Merrill). *Breed Sci.* 46, 65–6.

Taliercio, E. and Kim, S. W. (2013), Epitopes from two soybean glycinin subunits are antigenic in pigs. *J. Sci. Food Agric.* 93, 2927–32. doi: 10.1002/jsfa.6113.

Taliercio, E., Loveless, T. M., Turano, M. J. and Kim, S. W. (2014), Identification of epitopes of the B subunit of soybean β-conglycinin that are antigenic in pigs, dogs, rabbits and fish. *J. Sci. Food Agric.* 94, 2289–94. doi: 10.1002/jsfa.6556.

Tsuji, H., Okada, N., Yamanishi, R., Bando, N., Kimoto, M. and Ogawa, T. (1995), Measurement of Gly m Bd 30K, a major soybean allergen, in soybean products by a sandwich enzyme-linked immunosorbent assay. *Biosci. Biotechnol. Biochem.* 59, 150–1.

Wang, X., Geng, F., Wu, J., Kou, Y., Xu, S., Sun, Z., Feng, S., Ma, L. and Luo, Y. (2014a), Effects of β-conglycinin on growth performance, immunoglobulins and intestinal mucosal morphology in piglets. *Arch. Anim. Nutr.* 68, 186–95. doi: 10.1080/1745039X.2014.919733.

Wang, T., Qin, G. X., Sun, Z. W. and Zhao, Y. (2014b), Advances of research on glycinin and β-conglycinin: a review of two major soybean allergenic proteins. *Crit. Rev. Food Sci. Nutr.* 54, 850–62. doi: 10.1080/10408398.2011.613534.

Wood, R. A. (2003), The natural history of food allergy. *Pediatrics* 111, 1631–7.

Xiang, P., Beardslee, T. A., Zeece, M. G., Markwell, J. and Sarath, G. (2002), Identification and analysis of a conserved immunoglobulin E-binding epitope in soybean G1a and G2a and peanut Ara h 3 glycinins. *Arch. Biochem. Biophys.* 408, 51–7.

Yaklich, R., Helm, R. and Herman, E. M. (1999), Analysis of the distribution of the major soybean allergen in a core collection of Glycine max accessions. *Crop Sci.* 39, 1444–7.

Yasuhara, T., Takai, T., Yuuki, T., Okudaira, H. and Okumura, Y. (2001), Biologically active recombinant forms of a major house dust mite group 1 allergen Der f 1 with full activities of both cysteine protease and IgE binding. *Clin. Exp. Allergy* 31, 116–24.

Zheng, S., Qin, G., Tian, H. and Sun, Z. (2014), Role of soybean β-conglycinin subunits as potential dietary allergens in piglets. *Vet. J.* 199, 434–8. doi: 10.1016/j.tvjl.2013.11.020.

Chapter 11

Nutritional considerations for soybean meal use in poultry diets

Justin Fowler, University of Georgia, USA

1 Introduction
2 Nutritional content of SBM
3 Anti-nutritive compounds
4 Genetically modified soybeans
5 Conclusion and future trends
6 Where to look for further information
7 References

1 Introduction

Soybeans are refined into three primary products: soybean oil, soy hulls and soybean meal (SBM). The poultry feed industry uses about 50% of the SBM produced in the United States each year. The meal is valuable to the animal feed industry because it remains an excellent source of amino acids that would otherwise be lost to the human food chain after we have removed the oil from the bean. Overall, it is estimated that (either through soybean oil or SBM) over 95% of U.S. soybean production finds its way into animal agriculture. The continued use of SBM in animal feeds depends on the ability of the meal to provide the nutritional components that make it attractive to animal producers, especially as those components are balanced with respect to the anti-nutritional factors that SBM contains. This balance has become increasingly important as price constraints have made SBM to compete with alternative ingredients such as dried distiller's grains (DDGs), as well as other oilseed meals such as canola, cottonseed or groundnut meals. In addition to these vegetable protein sources, there are a variety of protein sources such as animal by-product meals that can compete with SBM on a limited basis. Also, the amino acid requirements critical to optimal poultry performance can be met increasingly by synthetic sources in certain cases. Depending on the price considerations, it may sometimes become more advantageous to meet the essential amino acid balance of the birds with individual purified amino acids, instead of relying on SBM alone.

SBM's nutritional value is a function of both the composition of the meal and how it was processed. Despite the fact that SBM is used primarily as a source of essential amino acids in poultry feeds, there are some anti-nutritional factors that might limit its use in poultry diets. These are tied to (1) certain proteins called trypsin inhibitors that

http://dx.doi.org/10.19103/AS.2017.0034.27

are common among legumes; (2) the non-starch polysaccharides (NSP) such as mannans, galactomannans, raffinose and stachyose in the carbohydrate fraction of the meal; and (3) hormone-like substances, called phytoestrogens, that can mimic the action of the female hormone oestrogen in humans, especially with respect to egg-laying birds where the compounds can be deposited relatively efficiently in the egg (as compared to deposition in the muscle). Also, there is a growing consumer concern related to the use of genetically modified soybeans for both human and animal consumption that poses a threat to a sustainable production in the long term.

Despite the concerns that may limit the use of SBM as a feed ingredient, it still has a total digestibility by poultry of about 85% and remains the most effective single-ingredient amino acid source in poultry production. Poultry meat has been one of the most successful agricultural sectors in the last 50 years. During that time, the U.S. broiler industry has gone from small-scale, backyard businesses to a highly integrated, efficient industry which provides the most widely consumed meat in the world. Even over just the last decade, both the average market weight and feeding efficiency of broilers has continued to make significant improvements through advancements in genetic selection, rearing house management and technology, and also nutrition. The industry's demand for high-quality protein creates a competitive market for feedstuffs. Poultry companies being highly vertically integrated makes the industry sensitive to the costs of production and puts a high premium on feed ingredients of sufficient quality to keep feed conversion rates low, which is something SBM has helped achieve.

2 Nutritional content of SBM

Domestic animal agriculture uses approximately 97% of the SBM consumed in the United States each year (the poultry industry consumes about 50%). The average broiler diet contains about 25% SBM (higher levels are used in starter diets, and less is used as the birds age). The wide use of SBM as the predominant vegetable protein source in poultry diets is a function of its excellent amino acid profile and its consistency with respect to its nutrient content. The average coefficient of variation for SBM is as low as 1.7%. It is helpful for nutritionists to have an ingredient for which nutrient

Table 1 Comparative nutrient composition of soybean meal in comparison to other protein meals

	Moisture (%)	Crude protein (%)	Crude fat (%)	Dig lysine (%)	Dig methionine (%)	Dig TSAA (%)	Dig threonine (%)	AMEn (kcal/kg)
Soybean	12	47.8	1.0	3.02	0.70	1.41	2.00	2458
Cottonseed	10	41.0	2.1	1.70	0.51	1.13	1.34	2010
Canola	9	38.0	3.8	2.02	0.77	1.74	1.50	2110
Meat and bone	8	45.0	8.5	2.20	0.53	0.79	1.58	2375
Peanut	10	47.0	2.5	1.52	0.50	1.10	1.12	2677
Fish	8	62.0	9.2	4.70	1.70	2.20	2.75	2950

values do not change significantly between harvests or between seasons. Having a low coefficient of variation means that poultry producers can trust that there will not be any significant differences in the ingredient's nutrient profile. Included below is a portion of the nutrient profile of SBM from the annual Feedstuffs Ingredient Analysis Table, 2017 Edition (Table 1).

Close to half of the content of SBM is protein, and more specifically, that protein content consists of high levels of key amino acids that are important for poultry production (principally lysine). Further, not only does SBM have high amounts of these key amino acids required for poultry, the efficiency with which these amino acids are digested by the bird (the 'digestibility coefficient') are quite high. Table 2 shows the digestibility coefficients for five of the critical amino acids in poultry diets for five different oilseed meals, the ingredient with the highest value is highlighted in red.

So, not only are the total amounts of essential amino acids considered average to above average for SBM, the amino acids that it does contain are readily available for digestion when the SBM has been properly processed. The most competitive ingredient in Table 2 is canola meal, which has higher overall methionine content and, though still lower, a very similar digestibility coefficient for both methionine and lysine.

When we look at cereal grains (such as corn) alone, the bird's requirement for sulphur-containing amino acids are sufficiently met. However, the cereal grains fail to meet the bird's need for the basic amino acids, arginine and lysine. What SBM offers to poultry diets is an ingredient that has an abundance of the basic amino acids that are not provided sufficiently by staple grains such as corn. So when corn and SBM are blended at ~60:40 ratio, the only amino acids that are still deficient relative to the chicken's requirements are the sulphur-containing amino acids (methionine and cysteine) because even though the sulphur-containing amino acids in SBM are highly digestible when compared to the other oilseed meals (90% for methionine and 79% for cysteine), the total amount it contains are low relative to the requirement of the chicken, especially when compared to competitive oilseed meals such as cottonseed or canola meal.

Other plant protein sources can partially replace SBM in poultry feeds (such as cottonseed meal, peanut meal, canola meal or DDGs from corn), but these are not typically feasible unless they are used in combination with additional synthetic lysine supplementation to compensate for the reduced levels and reduced digestibility of the lysine in those alternative ingredients. However, there are anti-nutritional factors and other potential issues that may limit the use of any of these protein sources (Elkin, 2002), SBM included.

Table 2 Standardized ileal coefficients for the key amino acids in poultry. The highest coefficient in each row is highlighted in bold

	Soybean meal	Cottonseed meal	Canola meal	Meat and bone meal	Peanut meal	Fish meal
Lysine	**89**	65	80	68	76	87
Methionine	**90**	72	84	71	86	87
Cysteine	**79**	74	77	28	**79**	63
Arginine	**92**	88	87	76	91	85
Threonine	83	68	73	62	**85**	82

3 Anti-nutritive compounds

All feed ingredients contain some degree of anti-nutritive compounds that limit their use to some degree. All legumes (which include SBM) contain relatively high levels of protease inhibitors and carbohydrate-binding proteins. Fortunately, both of these are proteins and can be deactivated by heat treatment, which is why SBM will always be heat-treated before being used in poultry feeds.

The prevention of either under- or overheating SBM is critical to maintaining its desirable nutritional profile. Raw soybeans go through a process where the beans are dehulled and cracked, and the oil is extracted. The soybean oil is the primary product obtained from the beans, and this is done either through extrusion press or through solvent extraction. After the oil is extracted, the remaining meal is sent for drying and heat treatment (typically about 220°F for 90 minutes). This extra heating step is critical for keeping SBM as a desirable feed ingredient for monogastrics such as poultry, achieving a 'Goldilocks condition' of heating enough to inactive anti-nutritional compounds, but not too much so as to destroy nutrient availability.

3.1 Trypsin inhibitors

There are certain proteins called trypsin inhibitors that are common among legumes. Trypsin inhibitors can interfere with the metabolism and absorption of the protein that makes these feedstuffs useful as a source of amino acids (Liener and Kakade, 1980). Trypsin is an enzyme that the body uses to metabolize dietary proteins, thus a 'trypsin inhibitor' will prevent the proper utilization of the amino acids in the SBM by stopping the trypsin enzyme from metabolizing the dietary protein. Thus, a sufficient inactivation of trypsin inhibitors in the meal is necessary for poultry to get optimal utilization from SBM.

Luckily, trypsin inhibitors are themselves proteins, which means they can be denatured through the heat-processing step that SBM typically undergoes after the oil is extracted. If the processed meal is not heated sufficiently, then the trypsin inhibitors will not be denatured, which means they will make it into the GI tract of the bird and reduce the digestibility of the key amino acids (such as lysine) that SBM contains. The symptoms of this will be finding loose droppings that lack visible white uric acid crystals and contain visibly undigested feed.

To prevent the purchase of under-heated SBM, a urease activity index is frequently utilized by producers. The urease enzyme has no biological relationship with trypsin inhibitors, and the enzyme itself has no toxic effects to poultry. However, it is convenient that heat denatures the urease enzyme at approximately the same rate as the trypsin inhibitors, so a check for urease activity is able to provide a relatively quicker and cheaper assay that is able to indirectly determine the level of trypsin inhibitors still active in the SBM without going through the relatively more complicated assay of trypsin inhibitor activity. Thus, the lower the urease activity level (measured by a simple pH change in the sample), the less likely it is that the SBM was under-heated. A pH rise of 0.05–0.20 in the urease assay has been the accepted range for indicating sufficiently heated SBM. That lower boundary of 0.05 was initially accepted because there was a concern that if the value was lower than that, then there may be a risk of the SBM having been overheated. It was found out later on (Dale et al., 1987) that there really is no such thing as a 'lower limit' on the urease assay, as long as it is coupled with a protein solubility assay to check for the potential of the sample having been overheated (Batal et al., 2000).

The current question is if the upper boundary of 0.20 is still an acceptable indicator that birds will not be consuming sufficient trypsin inhibitors to impair their performance or if even that value may indicate too high levels of trypsin inhibitors for modern poultry diets. Based on the feed consumption of a 1950s-era broiler, and an estimation of the amount of trypsin inhibitors that would be in SBM with a 0.20 urease assay, birds at that time would have consumed about 1200 mg of trypsin inhibitors during 6 weeks. However, a modern broiler consuming SBM with a urease index at the lower limit of 0.05 will have consumed upwards of 2000 mg of trypsin inhibitors in the same amount of time. Simply put, with the current rate of feed consumption that modern broilers have, the historic range for urease activity (0.05–0.20) results in the birds consuming significantly more trypsin inhibitors than the birds that were being grown at the time the assay was first developed. Given the rate of feed consumption in modern broilers, minimizing the urease assay becomes more critical than it was 50 years ago. This makes coupling the urease assay with a measure of protein solubility all the more essential for evaluating SBM quality.

3.2 Browning reaction

Lysine is a deficient amino acid in cereal grains. It is through the use of SBM that poultry have a highly digestible source of lysine. However, the industry was cautious with respect to having an 'acceptable range' for the urease assay which evaluated trypsin inhibition of lysine digestibility because of the concern that if an accepted value was set too low, then there could be a risk of the SBM being overheated. So while all of the trypsin inhibitors will definitely have been denatured, overheating can cause its own set of issues. The ε amino group of lysine is a highly reactive part of the side chain of that amino acid and is something that is used for a variety of biological reactions unique to lysine. Crosslinks made with the ε amino group of lysine are important for forming structural proteins such as collagen or elastin, for example. However, that terminal amino group will also react with carbohydrates under excessive heat processing, which is known as the Maillard or 'browning' reaction. Once proteins undergo this reaction, they are more resistant to proteolytic digestion and, specifically, the bioavailability of lysine is significantly reduced (which was one of the amino acids that made SBM an attractive feed ingredient to begin with). The consequence is the same as what happens if there were high levels of trypsin inhibitors – all of the highly digestible lysine that made SBM a competitive feed ingredient in the first place is no longer available to the bird, and therefore makes it appear (biologically) as a poorly digestible feedstuff.

The bioavailability of lysine is therefore key to SBM's use in poultry feed. There is a possibility that some of basic amino acid analyses can be inaccurate in estimating the lysine content of heat-modified feedstuffs. Only lysine with a free ε amino group is considered biologically available for protein synthesis (Rutherford and Moughan, 2007; Kim et al., 2012). Rutherfurd (2015) evaluated the importance of determining both reactive lysine and bioavailable lysine and found that the guanidination reaction is the most suitable for determining the lysine content that is bioavailable. The guanidination reaction targets the unreacted ε amino group of lysine through a reaction of O-methylisourea with lysine to produce homoarginine. It has been found to be another basis for determining unreacted versus bioavailable lysine.

In total, the safest practice for a producer to evaluate SBM quality is to couple the urease assay (where a high value will indicate an under-heated SBM) with a measure of protein solubility or availability (where a low value indicates an overheated SBM).

3.3 Non-starch polysaccharides

'Fibre' is simply a shorthand way of referring to the carbohydrate fraction of a feed ingredient that is not readily digestible by the animal via endogenous means. Dietary fibre is made up of a diverse mixture of carbohydrate polymers that may or may not be associated with non-carbohydrate groups (such as proteins or fatty acids), which are typical in plant cell walls. As a group, this set of carbohydrates are referred to by the collective term of NSP. Though we think of SBM as a protein source, and think of NSP as a problem related to the more carbohydrate-rich ingredients such as corn or wheat, the carbohydrate fraction of SBM can affect its use as well. SBM contains relatively high levels of NSP, approximately 20% (3% soluble NSP and 16% insoluble NSP). These NSP consist mainly of mannans and galactomannans, as well as two indigestible galactosides: raffinose and stachyose (typically present at about 1% and 5% of the SBM, respectively).

Since poultry have a relatively high rate of passage for digestive material (approximately 3–4 hours) and lack a significant ability to utilize microbial assistance in the way that ruminants, horses or ratites can, the presence of NSP in the diet can result in reductions in their nutrient utilization and performance. The harm NSP can cause, biologically, is related to the increase in the intestinal viscosity that occurs when they are included at high levels in the diet. When the material in the digestive tract is more viscous: enzyme contact with food components is decreased, overall nutrient absorption is reduced and the turnover rate of intestinal cells is higher (Gee, 1996; Daenicke, 2000; Teirlynck, 2009), and the absorption of any of the other nutrients in the diet is reduced. When these effects of viscosity combine, it has been shown that a 4% increase in NSP content can result in a 10% reduction in the digestibility of the lipid, starch and protein components of a feed. Maximizing the nutrient utilization from ingredients in the diet is a necessity when it comes to feeding poultry. Doing so not only maximizes the productive efficiency of the birds (which is important given that feed cost can be as much as 70% of the cost of production) but also will reduce nutrient excretion into the litter (which can have negative environmental impacts once that litter is land-applied).

One option available to nutritionists to make the most of the NSP content in feed ingredients is to use exogenous NSP-degrading enzymes (NSP-ase). NSP-ase function by hydrolysing the carbohydrate polymers in the plant cell wall into smaller carbohydrate fragments, which creates an impact on the increase in viscosity that is characteristic of NSP. The criteria that determine a 'good' NSP-ase for poultry can be quite limiting. For one, they must be able to pass through the low pH of the upper GI tract and the high temperatures of feed formulation, and they also must work during the limited interaction time with the NSP substrates allowed for by the high rate of passage of digestive material in the bird's GI tract. So, among the thousands of enzymes that have been discovered, it is a much smaller number that are true candidates for inclusion in poultry diets (Leeson and Summers, 2001).

The enzymes that have potential for exogenous use in poultry diets have a set of key methods of action: hydrolysing chemical bonds that are not sufficiently degraded by endogenous mechanisms, eliminating the encapsulation of key nutrients by undigested compounds, breaking down insoluble NSP for later digestion in the hindgut and supporting the activity of endogenous enzymes that may have not yet reached full activity in younger animals. Latham et al. (2016) found that improvements in broiler growth performance with the inclusion of an NSP-ase (β-mannanase) were most pronounced in the earlier phases of production where the levels of SBM are at their highest. Further a review by Slominski

(2011) found that an NSP-ase 'cocktail' of enzymes can be more effective at improving the performance of birds fed even a highly digestible, corn-SBM diet.

Dietary NSP are non-digestible by poultry, and though the concentrations are relatively low in corn-SBM diets, it has been shown that a combination of enzymes improves broiler performance and energy digestibility (Kaczmarek et al., 2014). However, those authors noted that differences in particle size can contribute to differences in the effects reported in the literature. Though SBM is most often considered as a primary source of protein, its carbohydrate fraction contains close to 20% NSP (Irish and Balnave, 1993). Furthermore, De Keyser et al. (2016) found that using two-fold the recommended dosage of an NSP-ase resulted in an increased efficiency of total protein and amino acid digestibility, suggesting that liberating non-digestible carbohydrates can also improve protein utilization.

3.4 Isoflavones

Isoflavones are phytoestrogens (plant-derived molecules with oestrogen-mimicking activity) found primarily in legumes. Among the legumes, soybeans have the highest isoflavone concentration (ranging from 0.12% to 0.31%). There are predominantly twelve isoflavone compounds in soybeans, and those that have been particularly studied are glyciein, genistein, daidzein and a highly active daidzein-metabolite called equol. Their consumption has been linked to both beneficial effects in the treatment of chronic diseases such as cancer, osteoporosis and cardiovascular disease (Anderson and Garner, 1997; Adlercreutz, 1998; Potter et al., 1998), as well as negative effects on reproductive functions in animal studies (Davies and Hill, 1989; Urpi-Sarda et al., 2008) due to its interaction with oestrogen receptors.

While the health impacts of soy isoflavones are still open to investigation, there is undeniable pressure from consumers with regard to the use of soybeans. Consumers are becoming more selective in their diet choices, and there is an emerging concern among a segment of the market with respect to the presence of soy isoflavones in the diet. Studies that have examined the effects of these soy-derived phytoestrogens on human health have been inconclusive, and soybeans are not the only sources of phytoestrogen in the human diet (other sources include coffee, tea, nuts, flax, sesame, sunflower, garlic, onion, red wine, cabbage, broccoli, berries and alfalfa). What is unique to poultry is the fact that the laying hen is capable of efficiently packaging the contents of her diet into the eggs she produces. This is a fact that we have utilized to increase the content of beneficial nutrients in the human diet (such as omega-3 fatty acids and fat-soluble vitamins), as well as to alter the yolk colour to meet consumer preferences. This means that it is also possible for soy isoflavones remaining in SBM to be transferred from the diet and into the egg. Whether consumers consider these compounds beneficial or not, the egg may represent a source of isoflavones in a typical person's diet.

4 Genetically modified soybeans

A genetically modified organism (GMO) is defined as any organism that has had its genome altered by genetic engineering techniques. In its most basic form, genetic modification of plants and animals has been happening for thousands of years as a part of human agriculture, modifying the ancestral teosinte into the modern corn, for example. But, more specifically, 'GMO' tends to most often refer to plants that have had genes from other species inserted into their genome (rather than evolution through selective breeding). Genetic modification has significant implications for the sustainability of agriculture. It has been predicted for quite some time that food production will need to double by 2025 and triple by 2050 (Vasil, 1998). Both grain and oilseed production has shown increased yields in response to GM interventions, using less inputs to produce more output.

With respect to soybeans, Monsanto introduced the first GM soybean in the United States in 1994, and within 20 years, close to 90% of the total soybean cultivation utilized GM soybeans. 'Roundup Ready Soybeans' were produced to be resistant to the glyphosate herbicide, which interferes with the synthesis of aromatic amino acids. These amino acids are essential in animals (must be consumed in the diet), but can be produced naturally by plants, and glyphosate interferes with that production. Also, Edwards et al. (2000) evaluated a SBM from a GM line that was designed to result in higher crude protein.

Though there is consensus that food derived from GMOs pose no negative health-related impact to humans (Nicolia, et al., 2014), the average consumer is much less likely to view GMOs as 'safe', which has led to regulatory and labelling standards that differ greatly from country to country. During the process of digestion, proteins are broken down into tri- and dipeptides, plus free amino acids. So any proteins arising from the genetic modification should be completely destroyed in the upper GI tract. Ash et al. (2003) found no GM proteins from SBM in whole egg, egg albumin, liver samples or faeces of laying hens fed GM soybeans (even though the GM proteins could be found in the feed).

5 Conclusion and future trends

The poultry industry provides a significant portion of the global demand for soybeans, thus the sustainability of soybean production (as a whole) is linked to the value of SBM as a feed ingredient in animal agriculture. The excellent benefits that SBM brings as a protein source can only be capitalized on when it is balanced with the anti-nutritional factors that it contains, which is a function of how it was processed. These limiting factors include trypsin inhibitors, NSP and concerns over phytoestrogens and genetic modification.

The poultry industry's demand for high-quality protein creates a competitive market for feedstuffs. Despite the concerns that may challenge the use of SBM as a feed ingredient, it still has a total digestibility by poultry of about 85% and remains the most effective single-ingredient amino acid source in poultry production. The use of exogenous enzymes has also served to improve the competitiveness of SBM in poultry diets by shifting the carbohydrate fraction away from non-digestible NSP.

Future research topics should focus on what would maintain SBM's competitiveness within the animal feed market. Particularly, given the characteristics that make SBM beneficial for feeding poultry, attention should be given to those features as they relate to the potential for alternative ingredients to become more competitive in feed formulation.

Also, more research can be done with respect to the effects of trypsin inhibitors in fast-growing genetic lines with much higher rates of feed consumption than 20 or 30 years ago when initial estimations of acceptable values for the urease assay were first described.

6 Where to look for further information

There are organizations presently involved in the development of strategies to maintain a stable supply of soybeans for both human and animal agriculture needs: The United Soybean Board, USAID's Soybean Innovation Lab, U.S. Farmers and Ranchers Alliance, and the U.S. Soybean Export Council.

7 References

Adlercreutz, H. 1998. Epidemiology of phytoestrogens. *Bailliers Clin. Endocrinol. Mateb.* 12:605–23.

Anderson, J., and S. Garner. 1997. Phytoestrogens and human function. *Nutr. Today* 32:232–9.

Ash, J., C. Novak and S. E. Scheideler. 2003. The fate of genetically modified protein from roundup ready soybeans in laying hens. *J. Appl. Poult. Res.* 12:242–5.

Batal, A. B., M. W. Doublas, A. E. Engram and C. M. Parsons. 2000. Protein dispersibility index as an indicator of adequately processed soybean meal. *Poult. Sci.* 79:1592–6.

Daenicke, S., W. Bottcher and H. Jeroch. 2000. Replacement of soybean oil with tallow in rye-based diets without xylanase increases protein synthesis in small intestine of broilers. *J. Nutr.* 130:827–34.

Dale, N., M. Araba and E. Whittle. 1987. Protein solubility as an indicator of optimum processing of soybean meal. *Proceedings of the 1987 Georgia Nutrition Conference for the Feed Industry.* Atlanta, GA, pp. 88–95.

Davies, H. and J. Hill. 1989. The effect of diet on the metabolism in sheep of the tritiated isoflavones formononetin and biochanin A. *Aust. J. Agric. Res.* 40:157–63.

De Keyser, K., L. Kuterna, S. Kaczmarek, A. Rutkowski and E. Vanderbeke. 2016. High dosing NSP enzymes for total protein and digestible amino acid reformulation in a wheat/corn/soybean meal diet in broilers. *J. Appl. Poult. Res.* 25:239–46.

Edwards, H. M., M. W. Douglas, C. M. Parsons and D. H. Baker. 2000. Protein and energy evaluation of soybean meals processed from genetically modified high-protein soybeans. *Poult. Sci.* 79:525–7.

Elkin, R. G. 2002. Nutritional components of feedstuffs: a qualitative chemical appraisal of protein. In *Poultry Feedstuffs Supply, Composition and Nutritive Value*. Eds McNab, J. and Boorman, N. Poultry Science Symposium Series, volume 26.

Gee, J. M., W. Lee-Finglas and G. W. Wortley. 1996. Fermentable carbohydrates elevate plasma enteroglucagon but high viscosity is also necessary to stimulate small bowel mucosal cell proliferation in rats. *J. Nutr.* 126:373–9.

Irish, G. G., and D. Balnzve. 1993. Non-starch polysaccharies and broiler performance on diets containing soybean meal as the sole protein concentrate. *Aust. J. Agric. Res.* 44:1483–99.

Kaczmarek, S. A., A. Rogiewicz, M. Mogielnicka, A. Rutkowski, R. O. Jones and B. A. Slominski. 2014. The effect of protease, amylase, and nonstarch polysaccharides-degrading enzyme supplementation on nutrient utilization and growth performance of broiler chickens fed corn-soybean meal-based diets. *Poult. Sci.* 93:1745–53.

Kim, J. C., B. P. Mullan and J. R. Pluske. 2012. Prediction of apparent, standardized, and true ileal digestible total and reactive lysine contents in heat-damaged soybean meal samples. *J. Anim. Sci.* 90:137–9.

Latham, R. E., M. Williams, K. Smith, K. Stringfellow, S. Clemente and R. Brister, J. T. Lee. 2016. Effect of β-mannanase inclusion on growth performance, ileal digestible energy, and intestinal viscosity of male broilers fed a reduced-energy diet. *J. Appl. Poult. Res.* 25:40–7.

Leeson, S., and J. D. Summers. 2001. *Scott's Nutrition of the Chicken.* 4th edn. Guelph, ON: University of Guelph.

Liener, I. E., and Kakade M. L. 1980. Protease inhibitors. In *Toxic Constituents of Plant Foodstuffs*. Ed. Liener, I. E. New York, USA: Academic Press.

Nicolia, A., A. Manzo, F. Veronesi and D. Rosellini. 2014. An overview of the last 10 years of genetically engineered crop safety research. *Crit. Rev. Biotechnol.* 34:77–88.

Potter, S., J. Baum, H. Teng, R. Stillman, N. Shay and J. Erdman. 1998. Soy protein and isoflavones: their effects on blood lipids and bone density in postmenopausal women. *Am. J. Clin. Nutr.* 68:1375S–9S.

Rutherford, S., and P. J. Moughan. 2007. Development of a novel bioassay for determining the available lysine contents of foods and feedstuffs. *Nutr. Res. Rev.* 20:3–16.

Rutherford, S. 2015. Use of the guanidination reaction for determining reactive lysine, bioavailable lysine and gut endogenous lysine. *Amino Acids.* 47:1805–15.

Slominski, B. A. 2011. Recent advances in research on enzymes in poultry diets. *Poult. Sci.* 90:2013–23.

Teirlynck, E., L. Bjerrum and V. Eeckhaut. 2009. The cereal type in feed influences gut wall morphology and intestinalimmune cell infiltration in broiler chickens. *Br. J. Nutr.* 102:1453–61.

Urpi-Sarda, M.; C. Morand, C. Besson, G. Kraft, D. Viala, A. Scalbert, J. Besle and C. Manach. 2008. Tissue distribution of isoflavones in ewes after consumption of red clover silage. *Arch. Biochem. Biophys.* 476:205–10.

Vasil, I. K. 1998. Biotechnology and food security for the 21st century: A real-world perspective. *Nat. Biotechnol.* 16:399–400.

Index

Alfalfa mosaic virus 34–35
Allergens, in soybean
  and biotechnology 246–247
  clinical allergies 236–237
    infantile food intolerance 237
  IgE-binding proteins 238–241
  infant food allergies and
      intolerance 237–238
  intolerance in aquaculture 245–246
  mitigating 241–245
  overview 235–236
Anthracnose 113
Aphid-resistant varieties 156–161

Bacillus thuringiensis (Bt) 155
Bacterial diseases
  bacterial blight 28–29, 109
  bacterial pustule (BP) 29–31, 109–110
  bacterial tan spot 31–32
  bacterial wilts 32
  other 32–33
  overview 27–28
Bacterial pustule (BP) 29–31, 109–110
Bacterial tan spot 31–32
Bacterial wilts 32
Bean pod mottle virus 35–36
Biochemical resistance, and HPR 148
Biological control 182–183
Biotechnology, and soybean allergens 246–247
BP. *see* Bacterial pustule (BP)
Browning reaction 261
Brown stem rot (BSR) 113–114
BSR. *see* Brown stem rot (BSR)

Cercospora sojina 10–11
Charcoal rot 114
Coleopteran pests 148–149
Cover crops 180–181
Crop rotation 178–179

Disease-resistant varieties
  by bacteria
    bacterial blight 109
    bacterial pustule (BP) 109–110
  by fungal pathogens
    anthracnose 113
    brown stem rot (BSR) 113–114
    charcoal rot 114
    frogeye leaf spot (FLS) 114–115
    Phomopsis seed decay 115–116
    Rhizoctonia damping-off and root
      rot 116–117
    Sclerotinia stem rot 117–119
    soybean rust 119–121
    stem canker 121
    sudden death syndrome 122
  G. soja and perennial *Glycine*
      species 128–129
  genomics research 127–128
  nematodes
    root-knot nematodes 125–126
    soybean cyst nematode 123–125
  by oomycetes 110
  overview 105–108
  pathogen genomics 128
  Phytophthora root and stem rot 110–112
  Pythium rot and damping-off 112
  selection for 126–127
  transgene-mediated 129

FLS. *see* Frogeye leaf spot (FLS)
Frogeye leaf spot (FLS) 114–115
Fungal pathogens
  anthracnose 113
  brown stem rot (BSR) 113–114
  Cercospora sojina 10–11
  charcoal rot 114
  frogeye leaf spot (FLS) 114–115
  Fusarium species 14–15
  Macrophomina phaseolina 13
  overview 3–6
  Phakopsora pachyrhizi 12–13
  Phomopsis seed decay 115–116
  Rhizoctonia damping-off and root
      rot 116–117
  Sclerotinia stem rot 117–119
  soybean rust 119–121
  stem canker 121
  sudden death syndrome 122
Fusarium species 14–15

G. soja and perennial *Glycine* species 128–129
Genetically modified organism (GMO) 263–264
GMO. *see* Genetically modified organism (GMO)

Hemipteran pests 152–153
Herbicides 183
Heterodera glycines biology 51–52
Host-plant resistance (HPR)
  Bacillus thuringiensis (Bt) 155
  biochemical resistance 148
  breeding challenges 154–155
  and coleopteran pests 148–149
  developing aphid-resistant
      varieties 156–161
  and hemipteran pests 152–153
  and lepidopteran pests 149–152
  overview 145–147
  pubescence-related resistance 147–148
HPR. *see* Host-plant resistance (HPR)

IgE-binding proteins 238–241
Infant food allergies and intolerance 237–238
Infantile food intolerance 237
Insect resistance management (IRM) 95–97
Integrated pest management 88–89
Integrated weed management
  biological control 182–183
  cover crops 180–181
  crop rotation 178–179
  enhancing competiveness 179–180
  establishing goals 176–177
  factors limiting 185
  herbicides 183
    impacts of 176
  managing weed seed bank 177–178
  mechanical control 181–182
  overview 175–176
  pest complexes 184
  scouting/mapping 183–184
  soil management 182
Invasive species 90–94
IRM. *see* Insect resistance management (IRM)
Isoflavones 220–222, 263

Lance nematodes 68–69
Lepidopteran pests 149–152
Lesion nematodes
  diagnosis 63–64
  management 65

Macrophomina phaseolina 13
Mechanical control 181–182

Nematodes
  general biology 48–50
  lance nematodes 68–69
  lesion nematodes
    diagnosis 63–64
    management 65
  overview 47–48
  reniform nematodes
    life cycle and disease cycle 65–67
    management 67–68
  root-knot nematodes 125–126
    description 59–60
    diagnosis 62
    life cycle and disease cycle 61
    management 62–63
  soybean cyst nematode 123–125
    biotic disease interactions 53–54
    genetic diversity 54–56
    Heterodera glycines biology 51–52
    management 57–59
    origin and distribution 50
    population dynamics 53
    resistance to *Heterodera glycines* 56–57
Non-starch polysaccharides 262–263
Nutraceutical functions
  in cardiovascular disease 197–198
  effects of soybean oils 196
  effects on blood glucose and diabetes 198–199
  effects on blood pressure, kidney and arterial function 198
  effects on cancer risk 199–200
  effects on cognition 202
  effects on menopause 200–201
  effects on obesity 199
  other health or adverse effects 202–204
  overview 195–196
Nutritional values
  browning reaction 261
  description 258–259
  genetically modified organism (GMO) 263–264
  isoflavones 220–222, 263
  non-starch polysaccharides 262–263
  and nutraceutical functions
    in cardiovascular disease 197–198
    effects of soybean oils 196
    effects on blood glucose and diabetes 198–199
    effects on blood pressure, kidney and arterial function 198
    effects on cancer risk 199–200
    effects on cognition 202
    effects on menopause 200–201
    effects on obesity 199
    other health or adverse effects 202–204
    overview 195–196
  oil profile 214–217
  overview 209–210, 257–258
  protein content and profile 210–213
  saponins 222–223
  tocopherols 217–220
  trypsin inhibitors 260–261

Oomycetes
  disease-resistant by 110
  Phytophthora sojae 6–9
  Pythium spp. 9–10

Pathogen genomics 128
Phakopsora pachyrhizi 12–13
Phomopsis seed decay 115–116
Phytophthora root and stem rot 110–112
Pubescence-related resistance 147–148
Pythium rot and damping-off 112

Reniform nematodes
  life cycle and disease cycle 65–67
  management 67–68
Rhizoctonia damping-off and root rot 116–117
Root-knot nematodes 125–126
  description 59–60
  diagnosis 62

life cycle and disease cycle 61
management 62–63

Saponins 222–223
SBM. *see* Soybean meal (SBM)
Sclerotinia stem rot 117–119
Scouting/mapping 183–184
Soil management 182
Soybean allergens
and biotechnology 246–247
clinical allergies 236–237
infantile food intolerance 237
IgE-binding proteins 238–241
infant food allergies and intolerance 237–238
intolerance in aquaculture 245–246
mitigating 241–245
overview 235–236
Soybean cyst nematode 123–125
biotic disease interactions 53–54
genetic diversity 54–56
Heterodera glycines biology 51–52
management 57–59
origin and distribution 50
population dynamics 53
resistance to *Heterodera glycines* 56–57
Soybean meal (SBM) nutritional considerations for
browning reaction 261
description 258–259
genetically modified organism (GMO) 263–264
isoflavones 219–222, 263
non-starch polysaccharides 262–263
oil profile 214–217
overview 209–210, 257–258
protein content and profile 210–213
saponins 222–223
tocopherols 217–220
trypsin inhibitors 260–261
Soybean mosaic virus 36–37
Soybean rust 119–121
Soybean vein necrosis virus (SVNV) 37–38
Stem canker 121
Sudden death syndrome 122
Sustainable insect pest management
insect resistance management (IRM) 95–97
and integrated pest management 88–89
invasive species 90–94
overview 87–88
in the United States 89–90
SVNV. *see* Soybean vein necrosis virus (SVNV)

Tobacco ringspot virus 38–39
Tobacco streak virus 39
Tocopherols 217–220
Transgene-mediated disease-resistant 129
Trypsin inhibitors 260–261

Viral diseases
Alfalfa mosaic virus 34–35
Bean pod mottle virus 35–36
overview 34
Soybean mosaic virus 36–37
soybean vein necrosis virus (SVNV) 37–38
Tobacco ringspot virus 38–39
Tobacco streak virus 39